Body, Brain, Behavior
Three Views and a Conversation

TAMAS L. HORVÁTH
Departments of Comparative Medicine, Neuroscience and Ob/Gyn,
Yale University School of Medicine, New Haven, CT, United States;
Department of Anatomy and Histology, University of Veterinary
Medicine, Budapest, Hungary

JOY HIRSCH
Departments of Comparative Medicine, Psychiatry and Neuroscience,
Yale University School of Medicine, New haven, CT, United States;
Department Medical Physics and Biomedical Engineering, University
College London, United Kingdom

ZOLTÁN MOLNÁR
Department of Physiology, Anatomy and Genetics, Oxford Martin
School and St John's College, University of Oxford, Oxford,
United Kingdom
Einstein Visiting Fellow, Charité-Universitätsmedizin Berlin, Berlin,
Germany, Visiting Professor at Acibadem Mehmet Ali Aydinlar
Üniversitesi, Istanbul, Turkey

ELSEVIER

ACADEMIC PRESS
An imprint of Elsevier

Academic Press is an imprint of Elsevier
125 London Wall, London EC2Y 5AS, United Kingdom
525 B Street, Suite 1650, San Diego, CA 92101, United States
50 Hampshire Street, 5th Floor, Cambridge, MA 02139, United States
The Boulevard, Langford Lane, Kidlington, Oxford OX5 1GB, United Kingdom

Notices
Knowledge and best practice in this field are constantly changing. As new research and experience broaden our understanding, changes in research methods, professional practices, or medical treatment may become necessary.

Practitioners and researchers must always rely on their own experience and knowledge in evaluating and using any information, methods, compounds, or experiments described herein. In using such information or methods they should be mindful of their own safety and the safety of others, including parties for whom they have a professional responsibility.

To the fullest extent of the law, neither the Publisher nor the authors, contributors, or editors, assume any liability for any injury and/or damage to persons or property as a matter of products liability, negligence or otherwise, or from any use or operation of any methods, products, instructions, or ideas contained in the material herein.

British Library Cataloguing-in-Publication Data
A catalogue record for this book is available from the British Library

Library of Congress Cataloging-in-Publication Data
A catalog record for this book is available from the Library of Congress

ISBN: 978-0-12-818093-8

For Information on all Academic Press publications
visit our website at https://www.elsevier.com/books-and-journals

Publisher: Nikki P. Levy
Acquisitions Editor: Natalie Farra
Editorial Project Manager: Sara Pianavilla
Production Project Manager: Kiruthika Govindaraju
Cover Designer: Miles Hitchen

Typeset by MPS Limited, Chennai, India

Working together
to grow libraries in
developing countries

www.elsevier.com • www.bookaid.org

Contents

Preface

The idea of this book was sparked during a conversation between Natalie Farra from Elsevier and Tamas L. Horvath from Yale at the meeting of the International Society for Neurochemistry and the European Society for Neurochemistry in August of 2017, in Paris. Natalie pitched the idea whether to edit or write a book about the communication between peripheral tissues and the brain which has been the main research interest of Horvath over the last 30 years. Based on his experience of editing or providing chapters to previous books, Horvath thought that writing rather than editing a book on the topic will give more freedom to express provocative and timely ideas. However, rather than just concentrating on body brain interactions in the adult, extending the topic to developmental stages and to brain–brain interactions presented an even greater challenge. It was immediately obvious that an approach that involves multiple authors with nonoverlapping expertise would be the most effective way to formulate and discuss these complex issues. The three of us have three different, seemingly nonoverlapping interests and related expertise. It would be very unusual that we attend the same sessions in scientific meetings, yet the answers for the complex issues lie in the interphase of our areas. In the book we provide these different takes on brain research in three "classical" parts. In the first part, the development and the evolution of the brain is discussed by Zoltán Molnár. In the second part, the relationship between the brain and the rest of the body is approached from the perspective of metabolism by Tamas L. Horvath. Finally, in the third "classical" part of the book, the interaction between brains is put in the focus by Joy Hirsch. However, the fundamental goal of our collaboration is to emphasize that neuroscience, or brain research, is not a linearly evolved discipline. Rather, there are multiple ways of asking and answering questions in this field that is siloed and lacks coherence in the definite understanding of the relationship between structure and function. These thoughts were crystalized for us through a multiyear series of conversations we had while preparing our individual parts. These conversations not only helped us to identify topics for our individual chapters, but they also produced 19 discussion sessions where we freely associate various topics that are at the interphase of our areas.

Conversations between "prepared minds" are widely appreciated as a tool for creativity and are encouraged among scientists, particularly to facilitate cross–disciplinary interactions; however, these conversations are not often shared with a wider audience. Nonetheless, the origins of many scientific breakthroughs have been attributed to random encounters and spontaneous thoughts generated in the moment during a lively conversation. Like raw data, conversations that spark spontaneous ideas are conventionally embedded in new thoughts. This is partly because real-life conversations often ramble: they are not efficient and, perhaps more importantly, do not employ PowerPoint slides or any defined format. Here, we present examples of real conversations among three scientists representing a range of disciplines, including physiology, medicine, developmental neuroscience, social neuroscience, cognitive and behavioral sciences, and veterinary medicine. The questions in common were starting points. The goal was to expose old and new questions to innate creative processes ignited during casual conversations in order to explore opportunistic fusions.

This book is a kind of experiment that explores the use of conversation as a conduit for scientific thought and development. The creative process behind this book is based on weekly Zoom conversations with Tamas L. Horvath, Zoltán Molnár, and Joy Hirsch between the fall of 2020 and the summer of 2021, which represents the period of the midst of the COVID-19 pandemic. During this year, the global scientific community adapted virtual modes of communication that substituted face-to-face encounters with web-based interactions. Talking, just talking to each other, went online. As active scientists and academic professors from different life science disciplines, as well as friends who share a common interest in the future directions of our respective areas of research, we met once a week to talk—just talk! Taking advantage of the Zoom "record" feature and automatic transcription services, these conversations were captured and available to review. However, we did not initially have the intent to include them in this book. Nonetheless, when reviewed, it occurred to us that these conversations had led us to new ideas and perspectives, and we decided to integrate them as part of our discovery process. Our overall theme was organized (loosely) around interdisciplinary integration to expand our models and scientific approaches in ways that spark fresh ideas about established notions of medical science, research, and human health and well-being. It occurred to us that these conversations would not likely have happened in a conventional scientific setting such as a seminar or a large scientific meeting because of the specialty-based organizations of our institutions and disciplines. The unlikely circumstances that have brought us together were not related to our particular areas in science but are related to our common passions about the science that we do. Compiling these emergent conversations has provided a working model framework for creativity, and we invite others to engage in similar conversations intended to question traditional boundaries and illuminate novel approaches to the grand challenges of our time.

Acknowledgments

We thank Natalie Farra (Senior Acquisitions Editor, Neuroscience, Elsevier/Academic Press) for initiating this project, and we are grateful to Sara Pianavilla (ELS-SDG), Kiruthika Govindaraju (Senior Project Manager), and Ms. Indhumathi Mani (Copyrights Coordinator) for production.

Zoltán Molnár: I would like to thank Thomas Molnár, Nadia Pollini, Thomas Henning, Alexander Demby, Carlotta Barelli, and Jim McCormack for reading and commenting on his chapter and on some of the discussions. I greatly appreciated the dialogues I had with generations of medical and biomedical students at St John's College on our weekly tutorials over the last twenty years. I am grateful to Christine Reynet who drew my attention to the ARTE documentary on: *The Stomach: Our Second Brain* that we discussed on June 17, 2021 (Discussion 18), and for the numerous conversations about metabolism and the bigger picture of body—brain interactions. I am indebted to Sue Fontannaz who encouraged me with her insightful questions and discussions on nature and nurture at City of Oxford Rowing Club. I consider myself extremely lucky to have been constantly supported and encouraged by my wife, Nadia Pollini.

Tamas L. Horvath: I am indebted to Marya Shanabrough, the laboratory manager with whom I have worked since 1990 when I arrived at the United States and Yale. She has read and corrected my writings over the past three decades, including Chapter 2 of this book. Special thanks to Susan Andranovich, who assured permissions to publish all the figures in Chapter 2. I would also like to acknowledge the partnership of more than quarter of a century at Yale with my wife, Sabrina Diano (now at Columbia University) without whom much of our discoveries described in Chapter 2 could not have been made.

Joy Hirsch: I am very grateful to Ray Cappiello and Jen Cuzzocreo, Lab Manager and Research Assistant, who week by week assisted me with the archiving of the conversations and preparation of Chapter 3. I am also indebted to the members of my laboratory at Yale who have been my partners in the pioneering journey to develop the "neuroscience of two." The wonderful lessons that I have learned from my colleagues Tamas L. Horvath and Zoltán Molnár have expanded my horizons and confidence in the future of multidisciplinary approaches. We had so much fun! Finally, my husband, Jim Rothman, is the sunshine that has supported me throughout.

Joy Hirsch (New Haven), Tamas L. Horvath (West Yarmouth, Cape Cod), Zoltán Molnár (Oxford)

November 10, 2021

Introduction—Zoltán Molnár

How did we end up with the views we have today?
Brief description of scientific trajectories
Early years

I was born in Nagykőrös, a city about 1 hour South-East from Budapest in Hungary. From a young age I was attracted to understand how an organism is put together. When I was 6 the highlight of my week was to help my mother in the kitchen to get the chicken ready for cooking. Chickens would come with all internal organs in those days, so one had to open up the various cavities and remove them. I spent hours being fascinated by how the body of a chicken was put together. My father was an artist (ceramicist and sculptor) and I found some of his books such as Jenő Barcsay "Anatomy for the Artist" and tried to mimic some of these drawings of skeletons and muscles. I also had highly inspiring biology, physics, and chemistry teachers at primary and secondary schools. I have fond memories of László Kiss, Ethelka Rózsás, István Páhán, and József Jauch from the Arany János Gimnázium at Nagykőrös. Together with my older brother Elek and my younger brother Béla, we all ended up at medical school to the South of Hungary at the Albert Szent-Györgyi University, Szeged.

First exposure to science

I loved the preclinical subjects, anatomy, physiology, and pathophysiology. I started working and doing some teaching in the Physiology Laboratory in my free time and was exposed to sensory physiology (György Benedek, Gyula Sáry), sleep (Ferenc Obál), and pain (Miklós Jancsó). I loved the journal clubs on Monday evenings. I was delighted to participate in some of the scientific meetings on extra-geniculo-cortical visual system or sleep as an assistant. My task was to advance the slide projector when the speaker requested "next slide please." Of course, I soon became fascinated with the superb talks. I also loved the clinical subjects and had reasonable manual dexterity to become a surgeon. I decided to try pursuing neurological surgery and get a job in the University Clinic. As I or anyone from my family had never been member of the communist party, I also applied for a Hungarian Academy of Sciences—Soros Scholarship for a year to Oxford as a backup, in case my application for the residency job was turned down. Even in those years such factors played a role in the allocation of some of the jobs in addition to academic records. Nevertheless, I was successful in my application, and I started my residency in Mihály Bodosi's Neurosurgery Clinic. I very much enjoyed the experience, but it was very hard work. In spite of the long and irregular hours of work, I loved the medical teamwork. I even loved the trauma shifts because one could do quite a lot of actual operations, in contrast during the normal days, when the elective surgeries were performed by more senior colleagues. There were two aspects I did not like, however. One was to see some of the human tragedies that ended up in neurosurgery wards, such as advanced brain tumors or irreversible lesions. I found it very difficult to cope with the cases where there was very little hope for recovery. The other aspect I did not like was the fact that in those days doctors received "paraszolvencia", kind of gratuity directly from the patients. All doctors accepted this, and I hated the idea that patients pushed envelopes into their doctor's pocket. How can a health system allow this to happen? I hated even more that I started to accept these envelopes when I had difficulties paying the rent of my small apartment after a few months. I had difficulties looking myself into the mirror at the end of the day. When I heard the news about my Oxford scholarship Professor Mihály Bodosy (Neurosurgery) and György Benedek (Physiology), advised me to take up the scholarship and go to Oxford.

Starting science for real

Following advice of Péter Somogyi and János Szentágothay, I chose in the laboratory of Prof Colin Blakemore who was one of the most charismatic and brilliant neuroscientists in the UK with extremely broad interest in sensory perception, cognition, and brain development. Colin was a true pioneer in demonstrateing just how much the brain is shaped by the environment. There are of course some anatomical and genetic constraints, but the list of parameters that can be adjusted with experience is just endless. I also spent some time in the laboratory of Keisuke Toyama at Kyoto Prefectural Medical School to dissect developing cortical circuits and in the laboratory of Egbert Welker at the University of Lausanne, who was one of the pioneers in understanding how the primary somatosensory cortex of the mouse is patterned by the signals arriving from the whiskers. These topics influenced my own research when I returned to Oxford and set up my own laboratory in 2000.

Current scientific focus

What is the brain? A personal view

I look at the *brain as the product of a developmental process that evolved over millennia* and produced a structure that can generate highly complex functions. Our brain can design things, but our brain itself was not designed. It is the product of trial and error and selection. Adult structures can change and adopt to the needs of the organism within certain constraints, but the evolution of the radically different brain requires the change of its own development. Therefore the *evolution of the brain is the evolution of brain development*. The complex sensorimotor skills and cognitive capacities that can establish within those constraints will then determine whether that trait will be an advantage for selection and further evolution. This is why I decided to study both development and evolution of the brain. Our brain is the result of millions of years of experiments to produce a structure that can best serve our purposes. Our brain evolved to generate a structure that can be adopted to the individual's lifestyle. In spite of the overall similarities, *our brain reflects our previous life experiences.* The fine structure of the brain can reveal whether we dedicated time to some special hobbies, such as playing a musical instrument, whether we dance regularly or whether we are good in certain sports. Our brain becomes slightly different when we learn motor or language skills. The famous Spanish neuroanatomists Santiago Ramon y Cajal stated that "Every man if he so desires becomes sculptor of his own brain." However, there are biological limits to sensorimotor skills and to plasticity. Some of the sports require that we push the hand—eye coordination to the maximum limits when we return a very fast tennis serve or when we play a team sport and we have to make strategic decisions in a fraction of a second that can make all the difference. I am interested in how the brain assembles to generate a substrate on which these complex skills can develop, and what can go wrong with this assembly that can cause conditions that have an impact on the cognitive functions. The developing brain is not a smaller version of the adult brain! The cellular, molecular interactions follow different rules from the adult. Treating developmental conditions requires different strategies. I consider development the key to understand the brain.

Most of the neurons in your nervous system are born in utero, and subsequently, there is very little turnover for the rest of our lives. We are trapped with the same neurons for life, but the glial cells, epithelial cells making up our brain vasculature, and brain coverings are gradually replaced. Our brain is not only made up of neurons, but neurons represent half of the cells in the brain. The central nervous system contains astrocytes, oligodendrocytes, microglia, and ependymal cells that support the neurons in several ways. Our neuro-centric research to understand neurological conditions had to be broadened to make progress. Microglia is now a major target to prevent the development of conditions, such as epilepsy, autism, and schizophrenia, and also in neurodegenerative conditions, such as Parkinson's disease and Alzheimer's disease.

Charles Darwin stated: "community in embryonic structure reveals community of descent." Thus to understand how the human brain emerged during mammalian evolution, we need to understand the evolution of the development of the nervous system that produced our brain and in particular the enlarged cerebral cortex. Evolution act on the level of population, it builds on individual variability and modifications of development that will produce altered adult structures that have different limits for adaptation. These will be tested and selected on population level. I would like to modify Theodosius Dobzhansky's statement from "Nothing in biology makes sense except in the light of evolution" to "Nothing in biology makes sense except in the light of development."

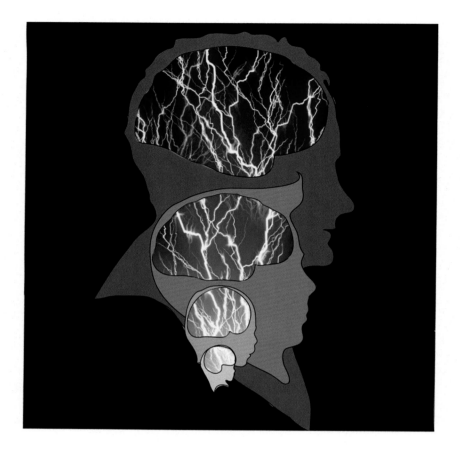

How should universities prepare to build neuroscience departments of the future?

My vision for neuroscience is *not to have a separate neuroscience department at all*, but to keep neuroscientists embedded into several departments, such as physiology, anatomy, pharmacology, zoology, pathology, neurology, and psychiatry, and only cultivate the association through virtual groupings. In fact, this is the arrangement we have at the University of Oxford. Neuroscience is a huge topic, and it draws views from various disciplines, such as anatomy and embryology, genetics, neurophysiology, molecular biology, neuropathology, neurology, and psychiatry, and the best is to benefit from this diversity and interactions. Neuroscience can be further subdivided into systems neuroscience, cellular neuroscience, and molecular neuroscience, but it is not justified to group research groups according to these categories, since most studies have to use all of these levels for comprehensive understanding of a particular scientific question or a clinical condition. The strict division of neuroscience to clinical, translational, or basic neuroscience is also problematic, since everyone is contributing to all these aspects, since only the relative proportions are different. Cellular neuroscience is not just cellular biology of the brain. Neuroscience can study different dimensions in time and space for its observations and can use a great variety of tools. Some neuroscientists use genetic models (reporter gene knock-in, bioindicators, electroporation, RNAi, viral vectors), and others use molecular biology (analysis of transcriptional profiles with bioinformatics), proteomics (self-assembling protein arrays, label free protein detections), classical anatomical analysis of circuitry, electrophysiology (patch clamping, voltammetry), imaging (time-lapse, spinning disc, confocal microscopy, multi photon microscopy, electron microscopy, magnetic resonance imaging, magnetoencephalography, positron emission tomography, single-photon emission tomography), behavioral analysis, and experimental psychology. Some neuroscientists have interest in development, evolution, comparative aspects, and others are interested in metabolism, endocrinology, and cardiovascular and respiratory medicine. While it might initially look practical to group research teams according to their model systems (drosophila, nematode, leech, zebrafish, chick, mouse, rat, ferret, nonhuman primate, and human), intellectually it might be better to interact with groups who address similar scientific questions even if they do so in different model organisms. Sometimes neuroscience groups are organized according to the topic they study, such as sensory, motor, memory, sleep, or cognitive neuroscience. While such groupings can be highly stimulating, it is important to have additional exposure to broader aspects. The nervous system cannot be understood in isolation; therefore the more links neuroscientists have to these other disciplines the better. About 30 years ago I was hoping that Oxford will establish some sort of neuroscience department, but now I can see the benefits that it did not happen. I visited some new neuroscience departments and institutes recently, and I was shocked to see just how similar the model organisms, techniques, and approaches were in the laboratories of the new principal investigators. Most groups used mice as a model system and the only difference was that they engineered the optogenetic stimulation to slightly different parts of the brain to detect the consequences of the manipulation of the different cell types. They all used similar multiphoton imaging setup, they all just investigated neurons and did not consider other cell types or the vasculature or the metabolism. What was even more alarming is to see just how similar their narrow views were of overall neuroscience. While such specialized institutes can have excellent output, I am not sure that such an approach will change the course of neuroscience for the longer run. Neuroscience requires integrative aspects and that requires interactions with other disciplines. Blaise Pascal claimed that "I hold it equally impossible to know the parts without knowing the whole and to know the whole without the parts in detail."

Introduction—Tamas L. Horvath

How did we end up with the views we have today?
Brief descriptions of scientific trajectories
Early years

I am from a small agricultural town of Hungary, Nagykőrös (the same place where a coauthor, Zoltán Molnár, of this book is from). I actually grew up in the very periphery of this town that itself is an appendix in the sandy Great Hungarian Plain between two better appreciated cities. I grew up in a family where my maternal side was medically educated, while my father and his father were veterinarians. This interface between human and animal medicine was in my breast milk and remains with me today leading a Comparative Medicine Department at Yale Medical School.

However, I disliked formal education from day care, through elementary, middle, and high schools in my hometown to the veterinary school later in Budapest. My dislike was not about the learning of new information, but the way it was taught and then asked to be recited and scored. I specifically did not like literature classes, where we were "taught" how to interpret one's writing, and if we differed, we were scored down. I recognize that this approach was not unique to my upbringing and remains to be a corner stone in formal education worldwide.

Emerging interest in science

It was at the beginning of high school when I started to be really interested in the intellectual endeavor of research. While still hoping that I can make it as a professional basketball player, I knew that I would have to go to professional school after high school. In Hungary, and Europe in general, there was no college system at that time; thus you have to make up your mind about adulthood early on. It has its pros and cons; nevertheless, I set out to get to vet school. During my high school years, the biggest impact on me was by my biology teacher, László (Laci) Kiss. He instilled an excitement about everything, including the Krebs cycle and genetics. The book that really made me decide that research is a way to go for me was the Double Helix by Watson.

I entered vet school with significant concern about my abilities to become a practicing veterinarian. Indeed, early on it became clear to me that I have neither the affinity nor the talent. My luck was that at the end of the second year, I got to know a young assistant professor at the Physiology Department in the Medical School of Szeged, Mihály (Misi) Hajós. He was the thesis advisor of my brother, who was a third-year medical student there. Misi was an enthusiastic and ambitious pharmacologist with interest and emerging works in crucial physiological and neurobiological issues. Mots impressively to me at that time, he was very much involved with shepherding my brother through his thesis, including taking him to the lab of Arvid Carlson in Gothenburg, Sweden, where Misi just finished his PhD work. The three of us spent a significant of time together during the summer of 1987, and that experience retriggered my interest in research. That coincided well with the fact that since my early years in veterinary school, I have been wondering what makes the difference between us, humans, and animals, specifically in association with our brains and behaviors. For the most of my life, I have been under the "impression" that we humans are "special," somehow an advanced product of evolution attested by all the "accomplishments" of human societies. In part because of my work but also due to the appreciation of other single and multicellular organisms, this belief of mine in the superiority of human behavior as the epitome of evolution has been challenged. In this book, I will go through an arbitrary set of arguments to illustrate how and why these changes occurred in my way of conceptualizing the brain and its role in the promotion of success in the environment. The writing of these paragraphs has occurred over the past 3 years, the last of which was under the "supervision" of the COVID epidemic. That alone has been an amazing experience with all its impact on our lives. Regarding the focus of this book, it was remarkable to see how a virus, an organism without a "brain" or anything close to it, could overtake human society around the globe at every level. Despite the potential to restrain this epidemic by man-made inventions (vaccines), it is clearer than ever that biological principles way outside the realm of any nervous system is capable of organizing and successfully self-propagating on the expense of any "higher" intelligence. Perhaps one day these principles will aid our efforts to better understand ourselves and our brain.

Starting research for real

I am a failed veterinarian whose lifeline was provided by coming to Yale to the Department of Obstetrics and Gynecology to study the neuroanatomy of the hypothalamus in relation to reproductive neuroendocrinology. Four days after arriving to Yale, on June 4, 1990, I started a project that was to explore the anatomical relationship between two subpopulations of neurons in the hypothalamic arcuate nucleus, a study of which implications stayed with me until today. I will discuss more of this later in the book. The reason I was brought to New Haven from Budapest was because I had some training in electron microscopy at the Vet School in Budapest by Ferenc Hajós, a human anatomist himself trained by János Szenthágothai, probably the most influential Hungarian neuroscientist in the world in the 20th century. Hajós has been friends with Csaba Léránth at Yale from their time together in the 1970s and early 1980s in the Department of Szenthágothai. Csaba needed a reliable and cheap labor from a trusted source, hence, my recruitment. Csaba Léránth was an amazing electron microscopist. There were several outstanding Hungarian electron microscopists in his generation, including Péter Somogyi, Miklós Palkovics, István Záborszky, and László Seres. Csaba's strength was his dedication to the quality of his work; he was a true craftsman. He was trained as a dentist, fixed up, and raced cars, so his approach to details was second to none. Being trained by him was one of the lucks in my life. It was a classical apprenticeship: he was uncompromising regarding quality and refused to offer me shortcuts or do things for me.

Current scientific focus

Despite the enormous efforts and resources devoted to neuroscience, we do not know much about fundamental brain-assigned functions, such as consciousness and schizophrenia; then great minds assumed about them hundred or hundreds of years ago. There is always the premise that the next "big thing" in neuroscience will solve a crucial piece of the puzzle and make breakthroughs that benefit the sick and society at large. For more than 30 years, the spending of governments, most notably the US Government, has been exponentially increasing on brain-related issues, such as how the brain works, what the underlying cause of Alzheimer's disease, dementias, Parkinson's disease, and schizophrenia; however, there are no breakthrough answers, no breakthrough solutions, and no breakthrough medications so far.

This lack of progress is accompanied by enormous advances to better understand specific processes and mechanisms related to neurons, neuronal connections, glia cells, and how they may be affected by genetics, epigenetics, the environment, and many other factors. For example, we understand how specific ion channels can affect ion flow in and out of various brain cells affecting many attributes of neurons and glial cells. Cellular biological discoveries advanced our understanding on how cells package, transport, release, and redistribute new and old molecules within and between cells. Through opto- and chemogenetics, we can see how signals can be selectively initiated at any part of the brain with impact on many brain functions, including the control of complex behaviors. Although these are remarkably elegant advances of fundamental biology of the brain, they did not solve the problem or offer feasible explanations for the aforementioned brain-associated complex functions and disorders. The promise has always been there, and, frequently a 5—10-year timeframe was offered for the solution to be delivered as long as the money flows. Despite all these hopes and promises delivered with all good intentions and sincere beliefs, there are no solutions yet.

I started my training as a neuroanatomist studying primitive brain circuits in relation to reproduction. These neuronal circuits reside in deep brain structures, for example, the hypothalamus. From early on, an intriguing aspect of these circuits for me was that they did not appear to specialize to support one function, rather they appeared to be involved in the multitude of regulatory mechanisms in support of organismal homeostasis. This very aspect of these brain regions made them historically unattractive for those who were pushing the frontiers of contemporary neuroscience. Understandably, those efforts could most rigorously be pursued by seemingly "orderly" structures with association with "higher" brain functions. In the recent past, this "orderly" approach on higher brain structures arrived to the hypothalamus and made it attractive to even hard core neuroscientists. My own efforts, on the other hand, have been to bring the "confusing" principles of hypothalamic structure and function to better understand functionality of the cortex, hippocampus, and other "higher" brain regions.

What is the brain?

I take the global point of view that the entire body is a brain and the contents of, for example, the cerebral cortex are a component of the mind, brain, and body system.

It is the brain that everybody believes makes us who we are. For thousands of years, from the earliest written texts to the societal and scientific views of today, the brain has been viewed as the center of our intelligence and the driver of the development and success of humankind. Over the last century, increasingly sophisticated methods have been brought to bear on the inquisition of the inner workings of the brain. The good, the bad, and the ugly of our existence and interactions among ourselves and with our environment have been attributed to the remarkable complexity and beauty of our brains. Undoubtedly, we all feel and believe that our experience in life is coded and decoded by our brains. We are convinced that the fact that we know who we are and that we can conceptualize the past, present, and future, and recall, at will, episodes from the past are because of our brain. This is a reasonable and logical assertion. There is little doubt that these attributes of us, as well as our ability to express ourselves via speech and writing, hinge entirely on us having a central nervous system, and the brain in particular. There is no doubt that the spinal cord as well as the peripheral nervous system are crucial for our existence, but other than writing, they are dispensable for all the other aforementioned attributes of us. It is not that they are not playing a role in the complexities of our behaviors but is that "higher" brain processing and consciousness do not hinge on these parts of the nervous system.

The brain is an organ that processes information from within the body and from the outside world, and in lieu of this information, it evokes adaptive movements within the body and the environment. This is the primary function of the central nervous system. What differentiates complex organisms with brains, including humans, from those without brains, such as trees, is the coordinated and adaptive locomotor responses that they can predictably manifest. These movements relate to the pursuit of oxygen, food and water, reproduction, and to escape from predators and dangers that jeopardize their survival. Although there has been a considerable amount of fascination about the brain and its function by scientists and the lay public, this tissue is not superior to any other tissues of the body originally classified by anatomy and then by function. In essence, the brain is not "more" in isolation and then the liver, the musculoskeletal system, the white fat, the pancreas, the kidney, or the heart. Removal of any one of these tissues terminates life. From this perspective, the brain is not more and is not less than these other organs. Where the brain, as an anatomical entity, differs from the other organs (by the superficial perspective of the state-of-the-art of our times) its crucial role is in coordinating the activity of the various organs with the environment and evoking adequate behavioral (locomotor) responses. This together with the so-called "higher brain functions," such as learning, memory, decision-making conceptualization, and consciousness, puts the brain into a "special" category. It may be, however, that this is a misconception, one that has been leading our understanding of the brain not much closer than the revolutionary concepts about the central nervous system emerged in the late 1800s and early 1900s. In my chapters of this book, I will aim to provide an alternative conceptual framework of how the brain integrates with the rest of the body in association with its physiological role and malfunctions.

How would you organize an ideal "neuroscience" department?

At the outset, I suggest to eliminate the silo of "Neuroscience." First, we know that the brain, which is the primary organ that we are talking about when we think of "neuroscience," consists of less neurons than other cell types, including astrocytes, microglia, oligodendrocytes, endothelial cells, and perivascular cells. Thus referring to studies of this organ as "neuroscience" is both ill-defined and arrogant by the current state-of-the-art. Unfortunately, semantics has enormous bias on science let it be in reference to molecular, cellular, tissue, or whole organismal processes. The brain, where neurons reside, does not exist and function in isolation from the rest of the body. Thus through a very primitive Socratic argument, it is unavoidable to conclude that understanding of the brain can only be achieved when its role in the entirety of the body is considered. This notion suggests then that "Neuroscience" should only be approached from the perspective of the organism and cannot, by definition, be argued for independently or from within the confinements of the brain. The most evolutionarily advanced cells of the cerebral cortex do not function or send out signals "ad hoc" to evoke or modify a behavior, thought or feeling, all of which revolves around a movement. Instead, they do it in lieu of inputs coming from within the body and the rest of the word (Fig. 1). Thus a future entity (department, center, institution) that aims to understand brain functions and malfunctions should, by definition, incorporate integrative physiology of the organs of the body in communication with the brain. In other words, "classical" neuroscience-brain research should be combined with integrative physiology.

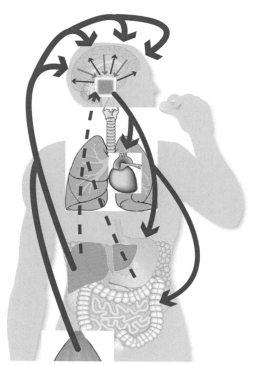

Figure 1 Schematic illustration of brain body communications in control of behavior.

Introduction—Joy Hirsch

How did we end up with the views we have today?
Brief description of scientific trajectories
Early biographical

I was born and raised on a single family multigenerational farm north of Salem, OR, USA. My parents and grandparents were descendants of the early pioneers to Oregon and the early Pilgrims to the continental United States arriving the same year as the Mayflower. I was the first of four children and the only girl in my family. A core value in the family was education, and college degrees were expected for my brothers, and thanks to the inspiration of my college-educated mother and grandmother, the priority for education was also extended beyond my brothers to include me. I wanted to be an astrophysicist long before I knew what that was. The first 8 years of my grade school were in a two-room school house (the same one that my father had attended) operated by an independent school district that was not incorporated into a larger network of schools. The local residents in the farming region where I grew up ran the school board and hired the teachers. There were two teachers for all eight grades. I graduated as valedictorian from my eighth grade class. The public high school was incorporated into the Salem school system and was institutionally minimal, mostly absent inspiration or opportunity, but within state guides for high school education. However, my subsequent undergraduate education in biology and basic sciences at the University of Oregon began to add color to my latent vision of making science my life. I worked to support myself through college, so academic activities were heavily balanced with work responsibilities. However, the opportunity door suddenly opened wide after graduation, I was offered a full scholarship to attend graduate school at Columbia University in the Department of Psychology. I had never been to the New York. My airline ticket was one way.

Entry-level science and first lessons

After graduation from Columbia, following a formative academic experience, came a faculty position at Yale, tenure, and a husband. I had also learned important life lessons about the management of a life in science from the point of view of a woman at Yale. And then there was the two-body problem. To be together, my husband and I both moved to MSKCC and Cornell in NYC where I started a functional imaging laboratory. This is when I began to understand that my theme in science was "pioneer," and I began to build the emerging neuroscience based on new imaging technology that fused brain function and brain structure, functional magnetic resonance imaging (fMRI). I started the research with the aim to map essential functions, such as language, motor, vision, and hearing, in the brain so accurately that the maps could be used for neurosurgical planning. With these maps, neurosurgeons were able to protect eloquent cortex that might have been displaced by a space occupying lesion that was targeted for resection. The science eventually branched to hypothesis-based investigations of cognition, perception, memory, as well as the biology underlying developmental disorders and neurological disease. The new noninvasive neuroimaging techniques and methods to investigate normal functioning human brain superimposed on high-resolution anatomical images were a scientific goldmine! However, this expanded mission and new direction of my work in cognitive neuroscience outgrew the mission of MSKCC/Cornell, and I joined the faculty at Columbia University Medical School as a director and founder of a new functional magnetic resonance center devoted to the investigation of the underlying neurobiology of brain structure and function. This center was aligned with national and international expansions in brain science, and the collective knowledge base of the underpinnings of human cognition and perception in the living human brain accelerated at "lightening-speed." It was a very exciting and productive time for neuroscience and my lab in particular. I loved the science and was grateful for the fast lane and opportunity to be a part of such an exciting discovery process. Then, suddenly and unexpectedly, there was a bump in the road at Columbia, and more lessons to learn about the fragile landscapes for science. Ultimately, I moved to Yale and took advantage of an opportunity to start a new generation of brain and behavior investigations that aimed to understand the neural underpinnings of live social interactions.

Starting science for real

Although human beings are irrepressibly social and interactive, our prior advances in understanding brain and behavior relationships have been limited to single participant functions due, primarily, to conventional neural imaging techniques, such as fMRI, that focus on single individuals in solo modes of behavior. As a result, the neural mechanisms for social and interpersonal interactions that underlie the sending and receiving of live cues that form the infrastructures for dynamic social connections between two people are not known. However, a new two-person neuroscience approach is enabled by an emerging neuroimaging technology, functional near-infrared spectroscopy. This technology enables the acquisition of hemodynamic signals simultaneously from two interacting individuals in addition to complementary information, including subjective behavioral responses, eye tracking, systemic responses, facial recognition, and voice recordings, the electroencephalography. My views about the brain have been shaped by this new exploration of brains in live and dynamic interactions and the grounding background from nearly two centuries of brain and behavior science. The profoundly fundamental nature of social functions is the motivating force that drives my new scientific exploration.

Current scientific focus

What is the brain? A personal view

My entry-level assumption based on the emerging "Neuroscience of Two" is that *a single brain is one half of a social unit*. A minimum social unit includes two brains (each owned by a functioning person) engaged in a social interaction where both are mutually influenced by the other in an active and reciprocal manner. The notion of a functional dyadic unit consisting of two single brains as a target of investigation has far-reaching consequences. These consequences influence how we conceptualize future directions in neuroscience. For example, this approach is intensely grounded in behavior and the many links to neurophysiology and social norms. The unit does not exist unless there is a dynamic exchange of information between two people. Conduits for exchanges of information include "micro-moments" of social interactions, such as eye-to-eye contact, components of gestures, facial expressions, a social touch, such as a hand shake, spoken words, and vocal tones. The conduits for dyadic social units (two brains) are created via sensory inputs and outputs and are universally influenced by social norms and purposes. This integrating overview lays a foundation for a pioneering focus on the neurophysiology of social interactions leading to questions, such as how should universities prepare to build leading neuroscience departments of the future?

How should universities prepare to build neuroscience departments of the future?

Unification of disciplines

As the result of approximately 400 million years of evolution, the brain has become a crowning achievement of biological complexity. This is due in part to the intimate connection between brain physiology, the conscious mind, and behavior. This grand union mandates that each be studied together as one integrated investigational target. A fundamental grasp of principles that lead to the understanding of the mind—brain interface remains a frontier in science, and also a generator of opportunities to design a new pathway forward. The interdisciplinary nature of brain, mind, and behavior leads to a paradigm shift that requires unification of knowledge based on the fusion of seemingly unrelated disciplines. The goal is to merge an enormously large range of disciplines including behavioral sciences, humanities, computational sciences, engineering, mathematics, and experimental, developmental, and theoretical biology under an organizational structure that addresses these challenges. This expansion of the domain of brain science encompasses science specialties focused over many scales starting at the cell and extending at the boundaries of society including disciplines outside of the traditional disciplines associated with understanding neurons and related physiology.

Harnessing human creativity

However, simple inclusion of these disciplines is clearly not a workable strategy for a leading neuroscience department of the future because of the siloed and inflexible organizational complexities of the traditional university system. Alternatively, however, a leading strategy might include the formation of multiple local "Discovery Pods" intended to generate cross-fertilization of disciplines. This would require bringing scientists together based on their accomplishments and their friendships (yes, I mean that) rather than their disciplines. I propose that an organization of "discovery pods" designed to support unplanned and natural connections between scientists from different disciplines could address the new challenge of large-scale integration using small-scale innovations. The goal is to enrich the academic environment to bias the probabilities toward innovation and discovery. I think of this method as a kind of creative combustion, where, when the right ingredients of a social moment come together, the spark of discovery ignites. The hope is that a few sparks might light a new pathway forward.

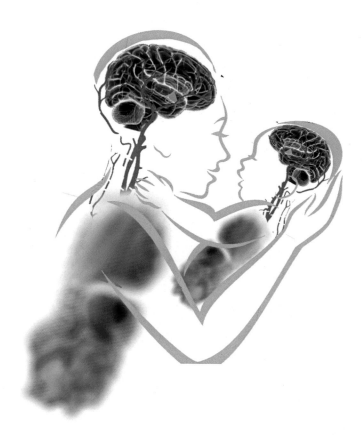

CHAPTER 1

Zoltán Molnár: the developing brain

A hemisphere of our brain is about the same weight as a basketball and a size of a mango and yet our brain is considered the most complex organ of the body with over 100,000,000,000 (100 billion) neurons along with many other cell types, such as astrocytes, oligodendrocytes, microglia, and tissues making up the vasculature and membranes that envelop the brain forming the meninges. Together with all these additional cells the count is around 200 billion. To give a comparison, the latest estimates for the number of stars in the Milky Way is between 200 and 400 billion. It is estimated that there are hundreds of different subtypes of neurons, and in total they make more than 100,000,000,000,000 (100 trillion) connections. These connections are modulated by a multiplicity of neural chemical factors that span spatial scales starting with molecules and progresses to cells, circuits, and systems. Establishing the trillions of connections in our brain, we rely on less than 30,000 genes. The establishment of this degree of specificity requires the interplay between the unfolding developmental program and the environment. The environment continuously acts on a well-choreographed and highly sophisticated and complex developmental program. The prevailing view is that this biological "hardware" leads to complex behavior including cognitive processes, emotions, perceptions, memories, and goal-directed actions.

There was a long-standing debate as to where the soul resides within the body. Hippocrates (460−370 BC) considered the brain as the seat of the mind. Galen (129−210 AD) localized imagination, reasoning, judgement, and memory in the cerebral ventricles. However, it was Thomas Willis (1621−1675) who suggested that it is the tissue of the cerebral cortex that is the seat of higher cognitive functions. The term "neurologia" or "neurology" was introduced to the World in Oxford by Thomas Willis in 1664 in his book *Cerebri Anatome*.[1] He studied the anatomy and the function of the nervous system using anatomical dissection and correlated his pathological findings to the detailed and accurate clinical observations he made of his own patients. He also compared the brains of various species establishing comparative neurology. The study of the nervous system evolved from his observations, and he is generally considered the founder of clinical anatomy and clinical neurology.

1.1 The Blind Men and the Elephant

Neuroscience is a huge topic, and it draws views from various disciplines. We use a great variety of methodologies for its study, as well as carrying out observations in multiple dimensions of space and time and emphasize very different aspects depending on our

Body, Brain, Behavior
DOI: https://doi.org/10.1016/B978-0-12-818093-8.00005-7
1

education, previous research areas and personal interests. Having discussions with Joy and Tamas was refreshing. Our points of views and research are very different, we study distinctive aspects of the functions of the nervous system and even have differences in defining what constitutes the brain or the nervous system. Tamas likes to include the entire body to this definition, while Joy advocates the study of more than one brain at the time. On the other hand, I consider the brain as the product of a developmental program that was altered by the life experience of that individual. That does not mean that either of us is wrong, but that we have different opinions based on the specialised area we each study. I would like to cite a poem by John Godfrey Saxe to illustrate our differences.

The Blind Men and the Elephant *by John Godfrey Saxe* (1872)
THE BLIND MEN AND THE ELEPHANT. A HINDOO FABLE.

I.
IT was six men of Indostan
To learning much inclined,
Who went to see the Elephant
(Though all of them were blind),
That each by observation
Might satisfy his mind.

II.
The *First* approached the Elephant,
And happening to fall
Against his broad and sturdy side,
At once began to bawl:
"God bless me!—but the Elephant
Is very like a wall!"

III.
The *Second*, feeling of the tusk,
Cried: "Ho!—what have we here
So very round and smooth and sharp?
To me 't is mighty clear
This wonder of an Elephant
Is very like a spear!"

IV.
The *Third* approached the animal,
And happening to take
The squiming trunk within his hands,
Thus boldly up and spake:
"I see," quoth he, "the Elephant
Is very like a snake!"

V.
The *Fourth* reached out his eager hand,
And felt about the knee.
"What most this wondrous beast is like
Is mighty plain," quoth he;
"T is clear enough the Elephant
Is very like a tree!"

VI.
The *Fifth*, who chanced to touch the ear,
Said: "E'en the blindest man
Can tell what this resembles most;
Deny the fact who can,
This marvel of an Elephant
Is very like a fan!"

VII.
The *Sixth* no sooner had begun
About the beast to grope,
Than, seizing on the swinging tail
That fell within his scope,
"I see," quoth he, "the Elephant
Is very like a rope!"

VIII.
And so these men of Indostan
Disputed loud and long,
Each in his own opinion
Exceeding stiff and strong,
Though each was partly in the right,
And all were in the wrong!

MORAL.
So, oft in theologic wars
The disputants, I ween,
Rail on in utter ignorance
Of what each other mean,
And prate about an
Elephant Not one of them has seen!

I look at the brain as the product of a developmental process that evolved for millennia.[2] Adult structures can change and adapt to the needs of the organism within certain constraints. This evolution of a radically different brain also requires the change and evolution of its own development.[3] The complex sensorimotor skills and cognitive capacities that can be established within those constraints will then determine whether that trait will be an advantage for selection and further evolution. Our brain is the result of millions of years of chance experiments to produce a structure that can best serve our purposes. It evolved to generate a structure that can be molded and adapted to the individual's lifestyle.[4] All humans have brains with very similar overall size, appearance, and microscopic structure that is different from other species. On average, we all have a brain that superficially has a similar shape with similar lobes, distinctive pattern of folds or bumps, known as gyri, and grooves, known as sulci. It weighs around 1350 g. Our developmental programs rely on the genes that we inherit from our parents. At conception two stacks of genetic cards get mixed up in a unique combination and we use these cards to build our brains. When and how we use these cards is regulated by the environment. The unfolding genetic signals are based on the cards, but it is the way we use them that can be adjusted according to our individual needs. A functional nervous system relies on precise spatial and temporal orchestration of gene expression, trillions of proper electrical and chemical connections between billions of cells, and an exact balance of cell types that navigate and integrate over great distances. As connections form between nerve cells and their electrical properties emerge, the brain begins to process information and mediate behaviors even during embryonic life. Some circuitry is built into the nervous system during embryogenesis; however, interactions with the world will continuously update the brain's functional architecture throughout life. The mechanisms by which these plastic changes occur appear to be a continuation of the process that sculpts the brain during development. To understand the brain and its devastating diseases, we need to reveal the mechanisms that produce it and the ways in which it can constantly change.

Despite the overall superficial similarities, our brain reflects our personal life experiences. The fine structure of the brain can reveal whether we dedicated time to some special hobbies, such as playing a musical instrument, dance regularly, or are good at certain sports. Our brains become slightly different when we learn motor or language skills. The famous Spanish neuroanatomist Santiago Ramon y Cajal stated that "Every man if he so desires becomes sculptor of his own brain." There are stunning examples of fascinating skills that one can develop if enough time and effort are dedicated to that practice. One can also recover some early injuries and transfer some of the functions to other brain regions. However, there are biological limits to our sensorimotor skills and to transfer of functions with plasticity. Some sports require that we push our hand-eye coordination to the absolute limit when we return a very fast tennis serve, perform a complex dance routine with our partner, play in an orchestra, or participate in a team sport, and we must make strategic decisions in a fraction of a second that can make all the difference. The two mango-sized hemispheres of our brain enable us to do all this and more!

1.2 Building the brain is like a house of cards

As a developmental neurobiologist, I would like to understand how the brain is formed, how it assembles from the hundreds of neuronal subtypes, with various glial cells, vasculature, and meninges to generate a substrate on which these complex skills can develop. What can go wrong with this assembly that can cause conditions that have an impact on the neuronal computations underlying complex phenomena, such as perception, thought, language, attention, episodic memory, and voluntary movement? Building the brain is like a house of cards. The early connections provide the foundation of the adult structure, and disruption of these may be the source of many developmental flaws. Cerebral cortical developmental disorders (including schizophrenia, autism, and dyslexia) and perinatal injuries involve cortical neurons with early connectivity. I wish to understand the early neural circuits during cortical development and disease and gain fundamental basic knowledge on neurogenesis, cell migration and formation of connections between various neuronal populations. There is an enormous flexibility built into this developmental program that we must get right to build our neurological hardware that will enable us to interact with our environment for the rest of our lives. There are conditions that cannot be avoided if you inherit a specific pack of cards (Huntington's disease, Down's syndrome), but you might be able to delay or alter these conditions. Understanding the neurobiology of development can help us elevate the platform from which we could develop treatments to eventually interfere with some conditions and avoid developing devastating syndromes (e.g., epilepsy, learning difficulties).

Neurons change their connections as the result of complex interaction with the environment. In fact, your brain reflects your entire life experience. It will help you recognize the people you meet, enable you to learn to read and to spell, ride a bike or a horse, dance, or play tennis. All these complex cognitive and motor functions require a machinery that is put together in an organized fashion. In fact, the interactions with the environment start from very early stages and our machinery is put together using these very early signals. Some of the components of this machinery are only present during development and subsequently no longer present in our mature brain. This makes repair much more difficult. How can you regenerate something so delicate and sophisticated if there is a damage in this structure? You cannot replace the knowledge of these complex behaviors if the substrate of this behavior is gone after a stroke or due to degenerative processes! For instance, think about the areas in your brain, which encode for language. If you speak Russian, this ability is maintained through enormous neuronal networks that are interconnected in a very specific manner to subserve these complex functions. If you then damage a piece of your brain, such as your cerebral cortex at the surface of your brain where some aspects of language are represented in the temporal—parietal lobes of your left hemisphere, the previously ingrained knowledge would be lost. In theory, you can replace the neurons with similar copies, but they must re-learn these abilities. Tamas and I learned Russian at primary school, like all Hungarian kids did at the time. Some of my rudimentary knowledge of Russian goes back to the time when I was at the Arany János Primary School in Nagykőrös, a small town in the middle of Hungary. If you also learned other languages at various stages of your life, they might be represented in slightly different locations. I started English at 12 and German at 19 and I started learning French after I met Nadia, now my wife, at the age of 30. I made a good effort to learn French, but never really became very fluent or mastered spelling. My Hungarian, Russian, and French might be localized in different parts of my cerebral cortex. There are some clinical examples that have demonstrated that damage to a particular location in a multilingual individual might affect one language more than the other. Such rare cases demonstrate the awesome sophistication of our cerebral cortical language functions. However, the cerebral cortex is just the tip of the iceberg. Highly automated motor programs are needed to deliver speech. This requires complex control of musculature through cranial nerves. The basal ganglia, brainstem, and cerebellum are all involved. Moreover, you also need to hear yourself to adjust your voice and accent. All these programs develop simultaneously while our brain is adapting to its environment. Replacing such neuronal circuits after a lesion in an adult is not trivial, yet neighboring neurons could adapt and take over some if these functions due to plasticity to some degree.

Establishment of these circuits and programs occurs during development of the brain. The developing brain is not a smaller version of the adult brain. Its cellular and molecular interactions follow different rules than the more mature stages. Even a few days can make a great difference in the activation patterns in the primary somatosensory cortex of a perinatal mouse brain.[5] Treating developmental conditions therefore requires different strategies and clinical conditions, such as perinatal stroke or childhood epilepsy, must be treated differently than in the adult. Study of the developing brain is also truly multidisciplinary; it includes developmental neuroscience, neuropsychiatry, neuroimmunology, neurotoxicology, and neuroepidemiology. The various developmental steps unfold as integrated events of neurogenesis, neuronal migration, differentiation, formation of connections, and establishment of representations based on spontaneous and sensory driven activity patterns.[6] These complex events are usually discussed separately to be more didactic, but they occur very much synchronously in a highly choreographed fashion. Dissecting causal relations is difficult. There are complex interactions during early brain damage, direct fetal brain infections, maternal immune response due to viral and bacterial infections, stress, over/malnutrition, and toxic chemicals.[7,8] These topics represent fundamental issues, and all have huge clinical importance because these conditions can have an impact on the entire life of the individual and their families. I deliberately want to touch "this part of the elephant" with my research and with my views, although I realize that there are many other body parts to explore (Fig. 1.1). I consider development the key to understanding the brain.

Figure 1.1 Different views of an elephant, different views of the brain. Developmental neurobiologists look at the evolution of developmental processes that generate the adult structure that is the substrate of cognitive functions. Neuroendocrinologists look at the entire organism for body and brain interactions. Cognitive psychologists extend all this to interbrain interactions to the level of a group or to the society.

1.3 Your brain is comprised of different cells with various birthdates, but most of your neurons are as old as you are

The generation of neurons of our nervous system is different from the rest of our body, they are born with you and will not get replaced. Our nervous system contains neurons and various glial cells, vasculature, and meningeal coverings. Most of our neurons are born in utero and we have them for the most part of our lives.[9] We are trapped with the same cerebral cortical neurons for life, but the glial cells, epithelial cells making up our brain vasculature and brain coverings are gradually replaced. How do we know all this? How can we visualize the turnover of our cells in our different tissues without administering some substances or label cells with radioactive pulses? How could we track the birthdates of cells in our body? When our cells divide, they must produce a new nucleus with a new copy of our genetic material DNA. The new DNA is synthesized during the last S-phase of division, and then, it will be present in that cell for the rest of its lives. There is some minor editing, but the components, nucleotides, remain largely the same in our DNA. We can tell the birthdate of a particular cell by giving the dividing cells nucleotides that have been labeled. Knowing the time of the administration of the labeled nucleotides that incorporated into the newly synthetized DNA, we can now tell the birthdate of the cells. These techniques are routinely used in animal experiments, but how can we do this in human? How could we feed labeled nucleotides to someone and examine the incorporation of this label into their brain at various stages?

There was an "experiment" conducted by mankind that resolved some of these questions. The background level of radioactivity has been constant for centuries, but it changed a lot over the last 50 years.[10] All living life forms had a constant C14/C12 ratio, that only varied by fractions of a percent over the preceding millennia, before humanity decided to test and explode hundreds of nuclear weapons above ground in the early 1950s. After more than 8 years of difficult negotiations, the rampant testing led the US Secretary Dean Rusk, the British Foreign Secretary Lord Home, and the Soviet Foreign Minister Andrei Gromyko to sign the Limited Nuclear Test Ban Treaty, in Moscow, on the 5th of August 1963. After this treaty, the C14/C12 isotopic ratio had reverted to the pre-1955 level (Fig. 1.2). Radioactive C14 levels peaked in 1963, before the signing and implementing of the nuclear treaty banning nuclear weapon tests which reduced the isotope levels in our environment (Fig. 1.2). I was born in August 1964; therefore I still have a testimony of the increased C14 levels imprinted in my brain, as a byproduct of the Cold War between the Soviet Union and the United States. My age group around the entire World incorporated far more radioactive carbon in our bodies than any generations that have ever lived before or

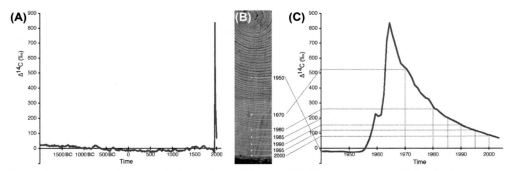

Figure 1.2 Changes of carbon-14 levels in our biotope. (A) Change in carbon-14 levels through the last 4000 years (the *y*-axis is change from normal, in parts per thousand). There is a sharp spike due to nuclear testing around 1963. (B) Cross-section through an old tree with the rings as time capsules with unique isotopic ratios. (C) Each ring is a different year and sampling revealed the carbon-14 curve with a peak in 1963. Subsequently, the carbon-14 levels have almost fully reverted to their pre-1955 levels. *From Spalding KL, Bhardwaj RD, Buchholz BA, Druid H, Frisen J. Retrospective birth dating of cells in humans. Cell. 2005;122:133−143.*[10]

after us (Fig. 1.2). However, some parts of our body constantly replace cells together with the nuclei that contain C14-labeled DNA, while others have very little turnover. Determining C14/C12 isotopic ratio was used to estimate the age of cells in various living tissues, including tree barks, our liver, skin, and brain. Such studies revealed that the isotopic ratios of my skin, gut, and blood having C14/C12 ratios corresponding to the atmospheric ratio of recent years and of Oxford where I have been living since 1988. However, my nervous system is a heterochronic hybrid of neuronal and glial constituents with the relative age of neurons and glia changing during my life. My neurons from 1964 are time-capsules of the high C14/C12 ratio in their DNA, which were synthetized when these neurons were generated and were born with me in Nagykőrös, Hungary. My other body parts were also born at the same time, of course, but have been completely replaced since. The lack of neuron replacement is very different from the rest of the cells in my body. My skin cells are replaced every 2—4 weeks, and my liver regenerates every 5—6 months. We do not regenerate our nervous system. In fact, we have the highest number of neurons after birth, and then, we lose neurons at a steady rate throughout life. Both your and my cerebral cortex was at the peak of its thickness when we were around 11 years old, and then, we all started to lose nerve cells and the thickness decreased steadily ever since.

As you now know, our brain is not only made up of neurons. In fact, neurons only represent less than half of the cells in the brain. Our central nervous system contains astrocytes, oligodendrocytes, microglia, and ependymal cells that support the neurons in several ways (Fig. 1.3). A better understanding of the contributions of these elements to our nervous system is immensely valuable. Our research to understand neurological conditions is largely centered around neurons, but it has to be broadened to make progress. We cannot just approach our "elephant" always from the very same direction (Fig. 1.1). The interactions between the neurons and other elements can have crucial importance during development and during our life.

The nervous system develops from a thin sheet of neural plate that folds into a neural tube. Astrocytes are derived from the neural tube and their C14 levels in average indicate that they are much younger than the neighboring neurons. Microglia account for 10—15% of all cells found within the brain. Microglia is a cell type that is not derived from the neural plate, it migrates to the developing brain from a different germ layer. They are from the embryonic mesoderm, which also gives rise to cells of the blood and immune system. Microglia are key cells in the overall maintenance of our brains. They are constantly monitoring our brains for lesions, viruses, bacteria, and regulating synaptic maintenance. Microglia are extremely responsive to small changes in local neuronal activity. During development their pattern changes dynamically depending on the stages of neurogenesis, synaptogenesis, synaptic pruning, or programed cell death. Microglia are considered the "gardeners" of the brain, pruning the neuronal connections according to their actual needs. Microglia can have several

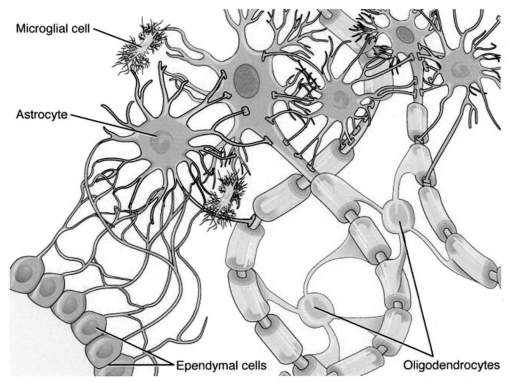

Figure 1.3 Cell types contributing to the nervous system includes astrocytes, oligodendrocytes, microglia, and ependymal cells. *From LumenLearning.com, Nervous Tissue, Anatomy and Physiology: https://courses.lumenlearning.com/ap1/chapter/nervous-tissue-2/.*

functional states and recent studies led to a greater appreciation of the different functional states and functional heterogeneity of microglia. Microglia are now a major target to prevent the development of conditions, such as epilepsy, autism, and schizophrenia, and in neurodegenerative conditions, such as Parkinson's and Alzheimer's diseases.[11]

1.4 Importance of connectivity in brain function

The second main difference between the brain and the rest of the body is its enormous variety of cell types. Our nervous system is built from hundreds of different types of neurons, which interconnect with highly intricate and specific connections. No other organ or tissues exhibit a comparable degree of complexity. Not only are all the brain cells and brain areas interconnected in very complex manners, but also they are positioned vis-a-vis one another in very particular and important ways. The cerebral cortex is over 80% of the volume and contains 20% of all neurons of our brain. It has a uniform six layered structure. All mammalian brains in any species have this uniform cortical arrangement with six layers, so we have a prototype circuit across the whole brain, with basic circuits following the same principles everywhere.[12] The brain receives sensory input from the sensory organs and this input is relayed through the thalamus to the cerebral cortex (Fig. 1.4). These layers differ in their cell composition, soma size, density, and shape. Each layer can have many different cell types (only

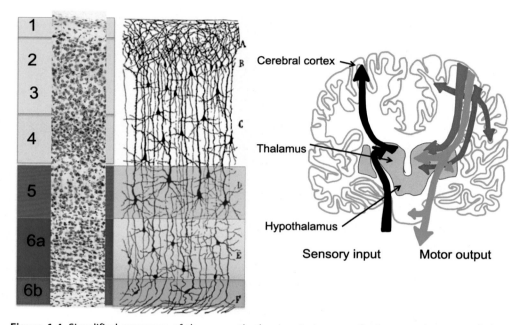

Figure 1.4 Simplified summary of the currently dominant views on the input and output relations to the cerebral cortex. Based on the composition, soma size, density, and shape of the cortical neurons, the cerebral cortex is traditionally divided up to six layers. The layering is based on the cell patterning after simple histological staining methods (left panel). Each layer contains different cell types. Only just a fraction is represented in Ramon y Cajal's drawing from 1909. The upper layers project within the cerebral cortex, whereas the lower layers (5—6) develop projections to thalamus, striatum, brainstem, and layer 5 even to spinal cord. Only layer 5(green) projects out to distant targets outside the skull. The thalamic input targets the middle of the cortex, layer 4. The cortico-thalamic projections originate from lower layers, layers 5, 6, and 6b (green, blue, and red). The cortex sends ten times more projections to the thalamus than the sensory periphery. The descending layer 5 projections send additional projections through side branches to numerous subcortical structures, including the striatum, thalamus, and basal pons. It is through these projections that motor commands send a copy to other brains structures before leaving the brain to the brainstem and spinal cord to elicit movement. Left panel contains a drawing from Cajal[14] and the right panel was drawn based on a section through the human brain from University of Wisconsin Comparative Mammalian Brain Collection (https://neurosciencelibrary.org/Specimens/primates/human/sections/coronal-cell/thumbnail.html).

just a fraction is represented in the drawing of Cajal from 1909 in Fig. 1.4 panel). The thalamic input targets the middle of the cortex, layer 4. In return the cortex sends projections to the thalamus. These projections originate from the lower layers, layer 5, 6 and 6b. Most of the input to the thalamus is not from the external environment, but from the brain itself. In fact, the projections from the cortex to the thalamus outnumber the input from the sensory organs 10 to 1. It is as if the brain is mostly listening to itself and only occasionally updates its internal information from the sensory signals from the environment.[13]

The generation of these complex cortical and subcortical interactions is the outcome of multiple genetic and environmental factors taking course during development. This development and the relationship of the connections between the thalamic projections that mediate the sensory input, and the cortex particularly attracts my interest.[5,6,15] It is through these projections that we receive information from the external world, and it is through these connections that we regulate our brain's state. The development of these pathways is key for the functional specialization and functioning of the cerebral cortex. Thus the thalamocortical projections are crucial for cortical development.

1.5 Generation of neuronal diversity

Our nervous system starts as a single layer of neuroepithelial cells that will produce neurons and other cells according to a preset choreography. The developmental steps presented evolved over millions of years, but the basic principles are preserved in all mammals.[2,16,17] This simple plate will form a tube with an inner and outer surface. Since the ventricles are inside the tube and later the brain, this surface is called ventricular. The outer surface is covered by the basal membrane and the pia mater; therefore it is called basal or pial surface (Fig. 1.5). Cell divisions occur in ventricular or, on cell biological terms, apical surface of the neuroepithelium. Neuronal progenitors divide to produce both more progenitors and neurons during our embryonic life. There are some progenitors that span the entire depth of the neuroepithelium with apical (ventricular) and basal (next to the basal membrane at the pial surface) processes. These elongated progenitors are also called radial glia (RG). This term RG was given to them before it was known that they are progenitors, but now we are stuck with this name, so sometimes we call them RG progenitors (Fig. 1.5). There are other progenitors without these long processes, and they are called intermediate progenitors. They are next to the lateral ventricles. The first-born cells congregate just below the pial surface in a zone that is called the primordial plexiform zone or preplate. It has been considered that the preplate stage in mammals resembles to the level of final adult organization in the reptilian dorsal cortex. In mammals, the preplate will split into the marginal zone and subplate-intermediate zones by the cortical plate (Fig. 1.5). The very first neurons of the human cerebral cortex arrive from outside the cortex via

Human 26 PCWs

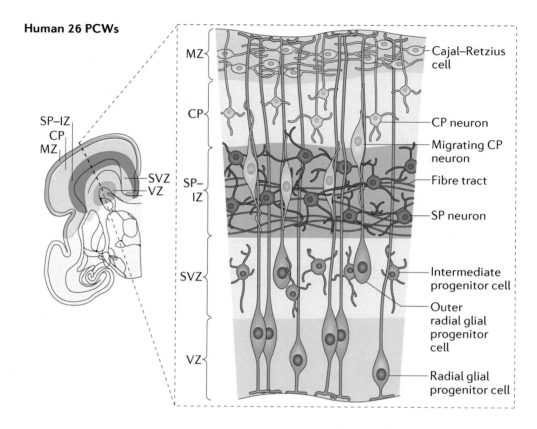

Nature Reviews | Neuroscience

Figure 1.5 Schematic representation of a coronal section through the left hemisphere of human brain gestational week 17. The neurons are generated from the progenitors situated at the ventricular surface of the neuroepithelium in the ventricular zone and subventricular zone. The earliest born cells form a layer at the pial surface, the so called primordial plexiform zone or preplate that will split into subplate-intermediate zone and marginal zone by the cortical plate. There are various progenitor cells in the ventricular zone (including radial glia progenitors, outer radial glia progenitors and intermediate progenitors). Layers of principal neurons are developed in the cortical plate and they are produced in an inside-first outside-last fashion, that is, first the deepest layer and then neurons for each consecutive layer migrate further above. *From Hoerder-Suabedissen A, Molnár Z. Development, evolution and pathology of cortical subplate neurons.* Nat Rev Neurosci. *2015;16(3):133–146.*[16]

tangential migration from still unidentified regions followed by the onset of local cortical neurogenesis to generate principal neurons.[18] The subsequently generated neurons bypass them, giving you an inside first outside last pattern development of the six layers in the brain.[17,19]

The various neuronal cell types are defined by genetic influences in the germinal zone, which is by factors that act on the transcription of the various progenitors.[20] Since different sectors of the neuroepithelium have differential expression of these factors in space and time, the produced neurons have different fate restrictions. By merely using the same transcription factors in slightly different order, length, or combination, different cell types can be generated. Segregation of neurogenesis in time can produce different neurons even if they derive from the same progenitors from the same sector of the neuroepithelium. Indeed, the details of clonal relations and clonal dispersion reveal highly heterogeneous progenitor populations. For example, knowledge of the properties of these progenitors can help us to understand clinical conditions where a particular progenitor type is affected due to its specific susceptibilities. Comparative studies revealed important species differences in the proportions of various progenitors and how neurons are born and specified.[3] These differences drove the evolution of the cerebral cortex.

1.6 Migration is key to newly born neurons to reach to the proper destination

Newly born neurons migrate to their proper location in the cortex. The migration can be radial, from ventricular towards pial surface, or tangential, parallel with the ventricular and pial surfaces. Since the fate of the neurons is largely determined during their last division, a precise neuronal migration is an important process for the formation of layers of cortical principal neurons. The newly born neurons have a destination address. They do not just pile up in the brain in a haphazard fashion. If they cannot make their journey because of molecular or cellular defects in their locomotor machinery, then their position is not ideal to form an optimally functioning neuronal network. The newly born principal neurons utilize the pial-directed process of radial glial progenitors as a migration scaffold (Fig. 1.5). In addition, the migration is controlled by cell-to-cell interaction between these neurons and their progenitors. If the molecular mechanism of migration is altered or is disrupted by environmental interventions (radiation, alcohol, drug abuse), then it can have a devastating impact on the functions of that altered cerebral cortex. The development of the brain depends on molecular and cellular interactions, and alterations in this choreography can have severe implications on further development. As neurons migrate towards their destination, they already start to extend axons and begin to form dendrites and therefore begin to establish neuronal circuits. Initially the clonally related neurons are coupled to each other through gap junctions that directly connect the cytoplasm of the cells, which allows various molecules, ions, and electrical impulses to directly pass through a regulated gate between cells. Therefore the chemical and electrical interactions are based on their common origin. The functional syncytium enables the earliest communication between these related neurons that will then start forming synapses. Thus those neurons that are already wired together will begin to fire together.[6]

We tend to associate the higher cognitive functions with the cerebral cortex. These functions rely on complex neuronal networks including thalamus, basal ganglia, hippocampus, claustrum, amygdala, and many other structures in the brain. We also identify the cortex as the site of abnormalities with no other structures in the CNS being considered as such. However, all the influences that I am discussing here could affect many other regions of the CNS. Current research is very heavily centered on the cerebral cortex. Large molecular and cellular screens identify the neuronal populations that are affected in an unbiased fashion, but some of the structures that should also be considered, such as the barrier functions of placenta, vasculature, or blood—brain barrier, are excluded. The contributors to the pathology can act in various combinations in various genetic backgrounds, making it extremely difficult to reveal associations. Moreover, some of the manifestations occur in 3—4 decades after the damage or perturbation, making it difficult to associate causal relations. These are the challenges that we currently face when trying to understand the neurobiology of cognitive conditions, such as autism, dyslexia, attention-deficit hyperactivity disease, and schizophrenia.

1.7 Brain evolution is the evolution of brain development

Not all brains look the same and the differences between mammalian, avian and reptilian brains can be spectacular. Some comparative neurobiologists can recognize the species just by looking at the brain or even a slice through the brain. Some brains are folded others are smooth, but all mammals have six layered cerebral cortices. Other vertebrates (turtle, iguana, crocodile) do not have six-layers in the cortex. Reptiles have an enormous ball of cells protruding into the lateral ventricle, called the dorsal ventricular ridge (Fig. 1.6). Birds and reptiles follow a different brain organization strategy. Birds and reptiles can outperform mammals in certain cognitive or sensory functions with differently organized cortical structures; just think about a crow, or an eagle. Crows can use tools, remember faces, communicate with each other in highly sophisticated manner. Eagles have 20/5 vision, whereas humans have 20/20. It means that what looks sharp to me from 5 m is just as sharp to an eagle from 20 m. The cognitive capacity of the crow and the superior vision with "eagle eyes" are accomplished with a brain without a cerebral cortex. Their brains operate without layered structure; they have a high-density cell conglomerate structure, the dorsal ventricular ridge (DVR), where neurons are arranged into nuclei instead (Fig. 1.7B). This considerable structural divergence of the pallium evolved over 315 M years. Each form perfected its own arrangement and found different solutions to build circuits that assist these functions. For instance, the visual representation in an iguana is just as sophisticated as in a ferret. We used electrophysiological methods to record from the brains of iguanas and

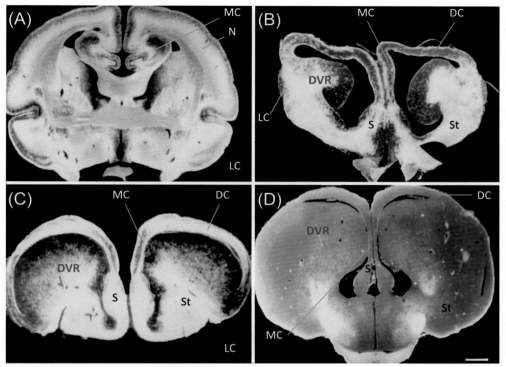

Figure 1.6 On the cross sections of four different brains. There are spectacular differences between forebrain organization in (A) marsupial, native cat, (B) turtle, (C) iguana, and (D) crocodile. Note the thicker dorsal cortex in marsupial (A) and the huge ball-like structure, the dorsal ventricular ridge (DVR) in (B)–(D) protruding into the lateral ventricle. Abbreviations: *St*, striatum; *MC*, medial cortex; *LC*, lateral cortex; *S*, septum. Scale bar 1 mm. *Modified from Molnár Z, Butler AB. Neuronal changes during forebrain evolution in amniotes: an evolutionary developmental perspective. Prog Brain Res. 2002;136: 21–38,[22] reproduced with the permission of Elsevier Science B.V. in Molnár Z, Métin C, Stoykova A, et al. Comparative aspects of cerebral cortical development. Eur J Neurosci. 2006;23:921–934.[23]*

we found neurons that respond similarly to visual stimulation as mammals. Iguanas have sophisticated subdivisions of the visual representation in the densely packed conglomerate of DRV cells.[21]

In the cross-sections of animal brains, one can see the differences between forebrain organization in marsupial, turtle, iguana, and crocodile (Fig. 1.6).[23] In these brains, different sectors of the forebrain reached prominence by changing early migratory patterns, changing local neurogenetic programs.[3] Marsupials have a similar dorsal cortex and turtle, iguana, and crocodile a huge ball-like structure protruding into the lateral ventricle that contains a densely packed conglomerate of cells. How do these differences in brain organization emerge? How did brain development evolve to produce the human brain?

Theodosius Dobzhansky stated, "Nothing in biology makes sense except in the light of evolution." I would like to make an argument that "nothing in biology (and especially in neuroscience) makes sense except in the light of development." Tom Lickiss, who did his graduate studies in my laboratory perfected my statement further to "Nothing in biology makes sense."

1.8 Nothing in biology makes sense except in the light of development

When I study development, I try to compare specific developmental mechanisms in various species. Changes in development drive evolution. The currency of evolution is variation, and this variation emerges from alterations in development. Evolution acts on the level of population, it builds on individual variability and modifications of development that will produce altered adult structures that have different limits for adaptation, and these will be tested and selected on a population level. It is 163 years since the first presentation of the idea of Natural Selection by Charles Darwin. It was since known that there is differential reproduction and heredity which leads some traits to become more common than others. Selection acts at the level of an entire population. Diversification within population can occur by tinkering with developmental regulatory networks. Since evolutionary change occurs not by the direct transformation of adult forms, it is the development that must change to make a difference. Therefore evolution is the change of development. It is the development that must evolve. Sometimes evolution is represented erroneously. Evolution does not work as it is portrayed in the "Right here right now" music video by English musician Norman Quentin Cook, also known by his stage name Fatboy Slim (https://www.youtube.com/watch?v=ub747pprmJ8), which demonstrates almost all misconceptions one can have on evolution. The "Right here right now" video portrays a single individual transforming from one adult structure into another during its life. Evolution never acts on an individual. As I mentioned, the currency of evolution is the variation within a population that selection can act on. Fatboy Slim was not the first to make such mistakes. In 1808 Johan Friedrich Meckel promoted a recapitulation theory.[24] He claimed that more complex organisms went through developmental stages of more primitive organisms. In 1892 Ernst Haeckel formulated his biogenetic law "Biogenetisches Grundgesetz," according to which the "ontogenesis is a brief recapitulation of phylogenesis" "Die Ontogenie ist die kurze Rekapitulation der Phylogeni."[25] However, development does not recapitulate the adult forms, it only shows some similarities at initial stages, then the diverging developmental processes produce the differences in the adult. Therefore comparative development is the best discipline to understand the workings of evolution. Charles Darwin stated, "community in embryonic structure reveals community of descent."[26] Darwin was inspired by Karl Ernst von Baer's ideas which were triggered to respond to Meckel's recapitulation theory. In 1828, von Baer wrote about the similarities of embryos: "I have two small embryos preserved in alcohol, that I forgot to label. At present I am unable to determine the genus to which they belong. They may be lizards, small birds, or even mammals."[27] Observing embryos from various species von Baer derived four generalizations known as "von Baer's laws." According to these laws animals begin as simple embryos that share common characteristics and become different as developmental trajectories diverge. They never pass through the adult stages of lower vertebrate species; they do not recapitulate evolution during development as portrayed in the "Right here right now" music video by Fatboy Slim. The von Baer-ian trajectories now have been confirmed with the modern tools of genetics and bioinformatics by comparing the gene expression patterns in various species. The similarities are spectacularly close at early development and later they diverge. We are particularly interested in the diverging mechanism of brain development.

Thus to understand how the human brain emerged during mammalian evolution we need to understand the evolution of the development of the nervous system that produced our brain and the enlarged cerebral cortex.[4] By comparing development, we can understand how more neurons and in much greater variety are produced in our brain. We can reveal mechanisms that enabled the neuronal production to evolve in different sectors of our brain. By examining the emerging differences between brains of mammals, reptiles, and birds we can get glimpses into the origins of the cerebral cortex.

1.9 Origins of the mammalian cerebral cortex

The production and assembly of neuronal circuits are different in the reptilian, avian and mammalian brains. In the avian pallium, different sectors contain different progenitors that produce different elements.[3] In mammals the diverse elements of the functional columns are produced within the same sector of cortical neuroepithelium. In avian brains, the neurons that assemble into functional units organize themselves perpendicular to boundaries of initial subdivisions. In the mammalian cortex, distinguishable radial columns are formed. Mammalian neurons form columns that extend across the layers of the cerebral cortex perpendicular to the pial surface (Fig. 1.7).

Large brains are populated with more diverse neuronal populations that could assemble in complex circuits and this enabled the behavioral complexity with advantages. This has only been achieved by diversification and amplification of output from progenitors (Fig. 1.7A and C). Comparisons of the various progenitor populations in amniotes, in a clade of tetrapod vertebrates comprising the reptiles, birds and mammals, give insight into the differences of how the cells of their nervous system are generated and where the differences emerge. Initially, germinative zones in the pallium of amniote brains display strong conservation prior to neurogenesis. However, as soon as neurogenesis starts, pallial neural stem cells in amniotes show regional variations and a general trend to specialize to produce different sets of neurons. As more elaborate hardware is required for complex information processing, a more diverse and densely

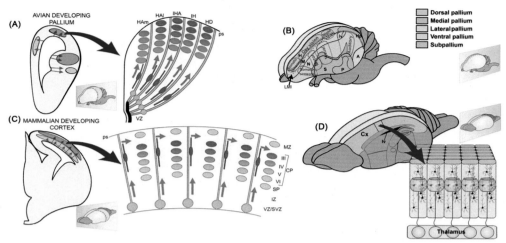

Figure 1.7 The production and assembly of neuronal circuits are different in the avian (A and B) and mammalian (C and D) brain. In the avian pallium (A), different sectors contain different progenitors that produce different neuronal types with different characteristics that assemble into nuclei rather than layers (pink and blue circles). In mammals, the diverse elements of the functional columns are produced within the same sector of cortical neuroepithelium (C) and the different neuronal cell types form layers (pink lower and blue upper layers). In avian brains, different segments of the neuroepithelium contain different progenitors that contribute with different neurons (pink and blue circles) that assemble into functional units organized perpendicular to boundaries of initial subdivisions (B). In mammalian cortex the distinguishable radial columns are formed (D). Mammalian neurons form columns that extend across the layers of the cerebral cortex perpendicular to the pial surface and contain various neuronal cell types. *From Montiel JF, Vasistha N, Garcia-Moreno F, Molnár Z. From Sauropsids to mammals and back: new approaches to comparative cortical development.* J Comp Neurol. *2016;524(3):630–645.*[28]

populated embryonic progenitor pool feeds the increased demand in neuronal number of the embryonic brain. In species with larger brains the neuronal epithelium with the progenitors adopts compartmentalization, so there is a much more extended and elaborate mitotic factory for the brain. We postulate that these divergences were likely initiated by disparities in the graded expression of morphogenic factors from the telencephalic signaling centers and evolved further over millions of years.[3]

Variations in the development of the neural tube enabled the production of the six-layered cerebral cortex, which further enlarged in primates. It is surprising that accumulation of subtle modifications from very early brain development accounted for the diversification of vertebrate brains and the origin of the neocortex. There is a strong correlation between the increase of upper layer complexity and the increase of progenitor populations between lizard, mouse, and monkey. The increase in the complexity of upper layers is accompanied with the increase of the progenitor pool during development (Fig. 1.8).[23]

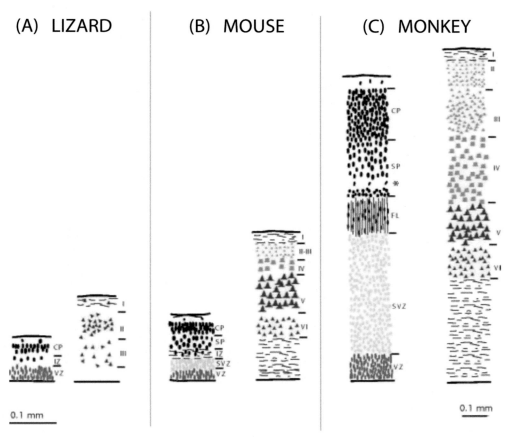

Figure 1.8 The increase of upper layer complexity and the increase of progenitor populations between lizard (A), mouse (B), and monkey (C). The left schematic panels depict the distribution of progenitors (pink and yellow) and the right schematic panels represent the layering in adults. Note the increase in the complexity of upper layers is accompanied with the increase of the progenitor pool during development. Abbreviations: *VZ*, ventricular zone; *SVZ*, subventricular zone; *FL*, fiber layer; *IZ*, intermediate zone; *CP*, cortical plate; *SP*, subplate. *From Molnár Z, Métin C, Stoykova A, et al. Comparative aspects of cerebral cortical development. Eur J Neurosci. 2006;23:921–934.*[23]

Morphogens are signaling molecules that are released from structures within or outside of the developing brain and can act over long distances and whose nonuniform distribution governs the pattern of development. In the forebrain, initially, faint differences in regional morphogen secretion promote further differences in telencephalic differentiation among amniotes. This starts to produce a differential developmental program that will lead to differential production of neurons that will lead to a wide variety of proportions and organization of sectors of the early brains in different vertebrates. It prompted different sectors to host varied progenitors and distinct germinative zones. These cells and germinative compartments generate diverse neuronal populations that migrate and mix with each other through the combination of radial and tangential migrations in a taxon-specific fashion. Together, these early variations had a profound influence on neurogenetic gradients, lamination, positioning, and connectivity.[3]

It is fascinating to explore the evolutionary path of the brain of reptiles, birds, and mammals, by comparing the developmental themes and variations that led to the building of the vertebrate brain. Variations from early neurogenic stages had a cumulative influence and were the main cause of vertebrate brain diversification. In 1828, von Baer got all this right and we now just use more and more sophisticated methods to study the details.[27]

We are beginning to identify the conserved and altered developmental mechanisms that led to the differential arrangements. Cell variety increased by segregating neurogenesis in space and time. The novel progenitor cells produced more numerous and more diverse neurons, which located and connected in different manners partly due to additional contributions of novel migrations of neuronal populations. It is remarkable how evolution could produce such a highly complex and efficient yet still plastic structure, our brain (Fig. 1.9).

The developmental sector that produces the mammalian neocortex is well identified in mammals and its homologs are also known across vertebrates.[3] This sector evolved its divergent relevance due to the varied power of signaling centers during development. The differences in size and the action of morphogens promoted the appearance of new precursor cell types. Consequently, divergent, and more populated germinative zones appeared across embryonic brains. Accompanied by novel cell populations, which can also migrate from external sources, the brain evolved to diversify the neuronal production. All these evolutionary variations, together, changed development that allowed the production and existence of the neocortex. We are beginning to discover these cellular and molecular factors in a range of animal models. However, much more detailed lineage studies are required to map out the homology of brain structures to unravel the evolutionary history of our own brains. This entire evolutionary process that altered the development and thus changed the adult form was not directed. It was not aiming for any particular organization and took millennia to take place. It may sound surprising, but given enough time and variation, incredibly efficient and complex structures emerged by chance, such as our brain.

1.10 Conservation and divergence during development and evolution

The similarities between cell groups have been a central issue in comparative developmental biology. Initially these comparisons were based on the gross morphology and shape of the region (Fig. 1.6). With the introduction of a few histochemical markers, some of these early assumptions had to be revised.[28] We can now survey the complete set of genetic messages, encoded in ribonucleic acids (RNAs) that produce proteins,

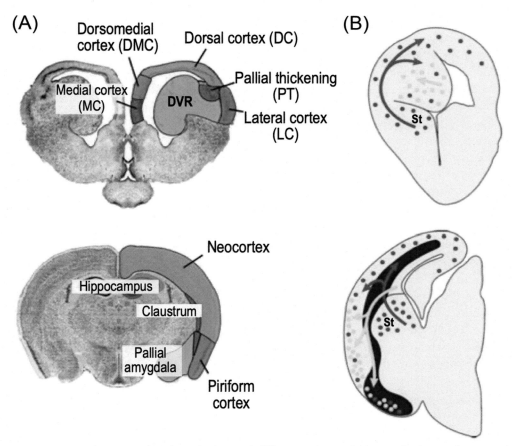

Figure 1.9 Reptile/mammalian homologies and differences in cortical organization and neuronal migration. Pallial regions in turtle (top) and mouse (bottom). (A) These regions are defined by neuroanatomy and transcriptomics. Colors represent proposed homologies, based on current anatomical, developmental, and transcriptomic data. (B) In mammals, the territory that expresses the transcription factor Pax6 is indicated in *dark blue*. Inhibitory, GABAergic neuronal precursors (*red dots*) originate from subpallial sources and migrate tangentially into the pallium in both mammals and sauropsids according to highly conserved mechanisms. Excitatory, pyramidal-type neuronal precursors (*yellow dots*) of the lateral migratory stream traverse the Pax6 territory to reach lateral pallial regions in mammals but remain in situ within the dorsal ventricular ridge in sauropsids. Despite its extensive target area, the lateral migratory stream is a subset of the radially migrating pallial neurons. Interestingly, the tangentially migrating GABAergic cells have similar origin from Dlx gene expression territories from the medial ganglionic eminence (origin of *red arrows*) and they migrate dorsal to the cortex in mammal and dorsal ventricular ridge and dorsal cortex in reptiles. *From Silver DL, Rakic P, Grove EA, et al. Evolution and ontogenetic development of cortical structures In: Singer W, Sejnowski TJ, Rakic P, eds. The Neocortex. Strüngmann Forum Reports. Vol. 27; 2019. © 2019 Massachusetts Institute of Technology and The Frankfurt Institute for Advanced Studies, by permission of The MIT Press.[30]*

found in a group of cells or even a single cell. Combined with bioinformatics one can now compare cell types according to the specific set of ribonucleic acids that carry the messenger instructions from DNA, which itself contains the genetic instructions required to produce the specific set of proteins for that particular cell type, ensuring the development and maintenance of lifeforms. The complete set of messages that a cell contains is the transcriptome. This contains a huge family of RNA molecules, including messenger RNA. We can compare expression levels, sequence differences (due to genomic variation, RNA editing, and alternative splicing), and reveal differences in spatial organization (nuclear, cytoplasmic, synaptic, macromolecular complexes) or in temporal regulation, since only subsets of the transcriptome are measured at any one time. Such comparative studies confirmed that during certain stages of development (e.g., gastrulation, see below) there is extremely similar gene expression that is very highly conserved across species. However, after this crucial stage of development different mechanisms can evolve. This supports "von Baer's laws" according to which animals begin as simple embryos that share common characteristics and become different as developmental trajectories diverge.

During the early development of the embryo the formation of a three-layered structure, the gastrula is a vital step. This early process is termed gastrulation. It seems to be a step that every embryo must go through, and evolution did not tinker with these very fundamental mechanisms of the formation of the trilaminar germ disk, probably because it did not result with a viable organism. Lewis Wolpert, evolutionary developmental biologist claimed that "It is not birth, marriage, or death, but gastrulation which is truly the most important time in your life"[29] (Wolpert, 2008). During gastrulation, cell movements result in a massive reorganization of the embryo from a simple spherical ball of cells, the blastula, into a multilayered organism. Interestingly, the developmental mechanisms can also diverge before gastrulation. Cellular and molecular interactions are vital, but before and after the interactions the developmental steps can diverge. Examining the gene expression patterns before, during and after gastrulation revealed that there is a developmental transcriptomic hourglass where gastrulation remained relatively unchanged over long periods of time, the expressed genes and their patterns are very highly conserved. When I explain this to my class of biomedical and medical students, I remind them that during the exams they all must show a certain amount of knowledge to pass. Students show huge variation in their level of knowledge before and after their exams. Some of them knew the material months before others just crammed it in a few days before. Some of them will remember the principles we taught them and shall use these in their professions for decades, others will select a different discipline and shall forget most of it. However, during the exams they all must show a certain level of knowledge. This is the same with the single layered hollow sphere of cells which is reorganized into a trilaminar structure of gastrula. They all must have a high level of conservation to produce a trilaminar germ disk to produce a viable embryo, but they can change what is happening before or after that crucial event.

1.11 Evolution of neuronal types

We can now profile neurons not only based on their shape, their projection site, and physiological characteristics, but also based on their gene expression pattern. While the DNA is very similar in all neurons, the specific gene expression is not. When we did some of the first sequencing experiments of the adult mouse cerebral cortex, we found that 2% of genes account for 50% of poly(A) RNA. In an adult mouse brain, 75% of genes account for 99.9% of poly(A) RNA.[31] The complexity of the molecular differences is fascinating. There can be the same gene, but in different isoforms expressed in adjacent layers of the cerebral cortex. Some RNAs that never get translated (long intergenic noncoding RNAs) can have highly specific expression patterns.[32] Moreover, genes enriched in expression for specific layers are significantly associated with specific diseases and functions. Superficially, there is little conservation of gene coexpression between avian and mammalian brains. Projection neuron cell type identity can be determined by retrograde labeling from selected targets combined with individual transcriptomic analysis. These studies have now identified mammalian cell types with specific gene expression patterns.[30] Such studies revealed that in mammalian projection neurons some key genes expressed early in development led to their differentiation into a specific direction. Upper layer neurons have a different set of genes (e.g., SATB2, RORB, and RFX3) that are switched on the expense of the lower layer associated genes (e.g., CTIP2, TBR1, and SOX5). These important master genes can have cross-suppressing effects at early stages that lead to the differentiation into specific overall directions. Initially we assumed that similar molecular switches might be present in reptilian and avian brains and that revealing the markers used in mammalian brains could help us reveal homologous populations in reptilian and avian brains (Fig. 1.9). However, in the reptilian brain these switches did not evolve or have evolved in a very different fashion. For example, in mouse neocortex, some of these genes are known to repress each other's expression in postmitotic cells (e.g., Satb2 and Bcl11b; Tbr1 and Bcl11b). In reptiles, these two sets of genes can both be expressed and some glutamatergic neurons coexpress genes that are enriched in mammalian L2/3, L4, and L5a intratelencephalic neurons (SATB2, RORB, and RFX3), as well as genes specifying L5b and L6 corticofugal projection neurons (CTIP2, TBR1, and SOX5).[33] Interestingly, the GABAergic neurons show spectacular similarities in their gene expression in reptiles and mammals, and they group very similarly if we use their gene expression as the basis for clustering. The diversity of reptilian cortical GABAergic neurons indicates that the interneuron classes known in mammals already existed in the common ancestor of all amniotes, whereas the glutamatergic neurons had different and much more divergent evolutionary paths.

In collaboration with Chris Ponting and Grant Belgards, my laboratory performed a comparative analysis between chick and mouse to relate the expression patterns of their 5130 most highly expressed one-to-one orthologous genes.[34] Our idea was to isolate regions from avian and mammalian brain regions (Fig. 1.10) and to explore the regions that had the most similar gene expression pattern.

Our exploration in the differences and similarities of the region-specific transcriptome of the avian and mammalian brains included "Specificity comparison," which aimed at understanding whether there was a particular alignment of expression specificities of all genes in one region in one species with any region in the other species.

We also performed "marker comparison" when the top marker genes of each region were objectively defined in one species (each in turn) and assessed if they showed a tendency to be expressed more likely than expected in any region of the other species. Surprisingly, the only significant overlaps of markers and coexpressed genes were in the striatum, hippocampus, oligodendrocytes, and nidopallium. Cross-species overlap of gene coexpression modules using Weighted Gene Coexpression Network Analysis and "the cross-species preservation of module comparisons" revealed conserved striatal and hippocampal gene expression networks. There was a convergent gene coexpression network across chicken nidopallium and mouse layer IV (Fig. 1.11). However, lineage tracing clearly revealed their different origin from different sectors of the forebrain neuroepithelium during development (Fig. 1.9). It is fascinating that they still converged to use the same gene expression networks (Fig. 1.11). The two cell groups both receive thalamic input and serve similar roles in the sensory circuitry. It would be interesting to explore the possibility that thalamic input might trigger this convergent gene expression network in these nonhomologous neuronal populations. Apart from these conserved striatal and hippocampal gene expression networks and the convergent gene coexpression network across chicken nidopallium and mouse layer IV, we found no further significant similarities at adult stages. Our regional comparisons suggested divergent evolutionary pathways between regions in avian and mammalian brains.

Recent comparisons of single cellular transcriptomics of songbird neuronal circuits that are involved in vocalization revealed some specific transcriptional similarities between glutamatergic neurons in the song motor pathway and the mouse neocortex although some of these structures were not generated from the same sector of telencephalic neuroepithelium. Therefore they originate from different parts of the developing brain. The fact that there are some converging molecular pathways in converging nonhomologous neuronal circuits is highly interesting. Why would nature select to operate these circuits with similar molecular logic despite diverse origin? Mammalian neocortical layers are made of new cell types generated by diversification of ancestral

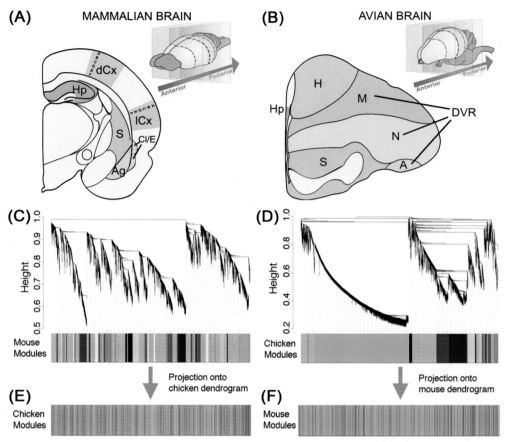

Figure 1.10 Schematic depiction of the regions studied for similarity in gene expression patterns in the mouse and chick brains. We observed little conservation of gene coexpression between adult avian and adult mammalian brains. Brain regions involved in the evolutionary debate of neocortex homology studied in mouse (A) and chicken (B) were dissected from serial sections in the anteroposterior axis; the claustrum (*dashed lines*) is located in more anterior sections of mammalian brains, and the arcopallium (*A*) is located in more posterior sections of avian brains. (C) Topological dissimilarity dendrogram and coexpressed mouse modules. Gene pairs that branch from one another lower on this dissimilarity dendrogram have more similar expression patterns across the dissected samples than other pairs. A color band corresponds to a group of genes sharing a similar expression pattern across samples; thus larger bands reflect more genes that share a pattern. (D) Topological dissimilarity dendrogram and coexpressed chicken modules. (E) Relative positions in the chicken dendrogram (represented in B) of one-to-one orthologs of genes in the mouse clustering (represented in A). (F) Relative positions in the mouse dendrogram (represented in A) of one-to-one orthologs of genes in the chicken clustering (represented in B). A to F in mouse, cortical layers; *Ag*, basolateral amygdala; *dCx*, dorsal cerebral cortex; *E*, endopiriform complex; *H*, hyperpallium; *Hp*, hippocampus; *lCx*, lateral cerebral cortex; *M*, mesopallium; *N*, nidopallium; *S*, striatum. *From Belgard TG, Montiel JF, Wang WZ, et al. Adult pallium transcriptomes surprise in not reflecting predicted homologies across diverse chicken and mouse pallial sectors.* Proc Natl Acad Sci U S A. *2013;110(32):13150−13155.*[34]

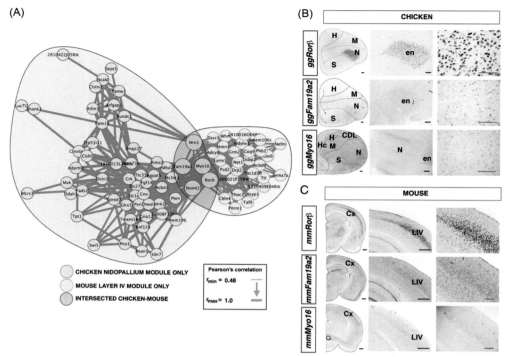

Figure 1.11 Convergent gene coexpression modules marking chicken nidopallium and mouse layer IV. (A) Expression correlations between genes in significantly overlapping gene coexpression modules expressed in chicken nidopallium and mouse layer IV. (B) RNA-in situ hybridization of convergent nidopallium markers. (C) RNA-in situ hybridization of convergent layer IV markers. (Scale bars: black, 500 μm; gray, 200 μm.) *Cx*, neocortex; *en*, entopallium; *LIV*, neocortical layer IV. (Scale bars: black, 500 μm; red, 200 μm). *From Belgard, TG, Montiel, JF, Wang, WZ, et al. Adult pallium transcriptomes surprise in not reflecting predicted homologies across diverse chicken and mouse pallial sectors.* Proc Natl Acad Sci U S A. *2013;110(32):13150−13155.*[34]

gene regulatory programs. In reptilian brains, some of these pathways coexist and evolved into a different direction.

Transcriptomic analysis of regions and individual cells from adult and developing brains can be very informative, but to really understand the homologies cell lineage is essential.[3] Therefore evolutionary biologists and developmental biologists must work together. These areas drifted from one another because of several reasons according to Raff (1996). Evolutionary biologists look at causality in selection, whereas developmental biologists search for proximate mechanisms (Fig. 1.12). It is like they approach our elephant mentioned in Fig. 1.1 from very different directions. Evolutionary biologists look at genes as sources of variation and as key elements creating diversity and change; whereas developmental biologists look at them as directors of function and they prefer universality and constancy. "Evolutionary biologists cite selection as the 'cause' of the presence of some

Over the Ice for Ontogeny and Phylogeny

TABLE 1.1. Differences between Evolutionary Biologists and Developmental Biologists in Their Views of Major Biological Qualities

Quality	Evolutionary biologists	Developmental biologists
Causality	Selection	Proximate mechanisms
Genes	Source of variation	Directors of function
Variation	Central role of diversity and change	Importance of universality and constancy
History	Phylogeny	Cell lineage
Time scale	10^1–10^9 years	10^{-1}–10^{-7} years

Figure 1.12 Differences between evolutionary biologists and developmental biologists in their views of major biological qualities from Rudolf A. Raff (1996). *From* The Shape of Life, Genes, Development, and the Evolution of Animal Form. *The University of Chicago Press. ISBN: 0-226-70266-9.*

structure, whereas developmental biologists cite genetic and developmental mechanism as the 'cause.' Hypothesis making in evolutionary biology is quite different from that in developmental biology. Hypotheses in developmental biology focus on mechanistic details, such as the presumptive functions of domains in genes, mechanisms for generating patterns in cell sheets, or cell behaviors. Hypotheses in evolutionary biology often use historical scenarios. They can include hypotheses on the effects of long-term selection, on the effects of environmental change, or on the causes and effects of extinction" (Raff, 1996). The history and timescales are also very different, evolutionary biologists study phylogeny in millions or billions of years (10^1–10^9 years); developmental biologists study cell lineage in days and milliseconds (10^{-1}–10^{-7} years). But as I previously argued development is key for evolutionary biologists. Evolution includes the evolution of developmental mechanisms.

Over the last two decades enormous progress has been made in understanding developmental mechanisms and developmental disorders in a handful of selected organisms. Most research is done on mice, but it would be important to use various other organisms. Genetics tools and modern anatomical and imaging methods have enabled great strides in understanding brain development, but it would be highly beneficial to extend these investigations to additional species from these selected models. Comparative developmental approaches can reveal different specializations that subserve similar functions in different brains, highlighting the contribution of specific relevant biological mechanisms to brain organization and function, while comparative developmental studies explore how brains achieve the features that distinguish them from other species. Combining both approaches and approaching our "elephant" from more directions can illuminate the evolutionary processes that made us human.

1.12 **The brain is a computer that is switched on while it is constructed**

The functional nervous system relies on precise spatial and temporal orchestration of gene expression, billions of proper electrical and chemical connections between millions of cells, and an exact balance of cell types that navigate and integrate over great distances. As connections form between nerve cells and their electrical properties emerge, the brain begins to process information and mediates behaviors even during embryonic life. There is a great deal of construction going on. Some circuitry is built into the nervous system during embryogenesis; however, interactions with the world continuously update and adapt the brain's functional architecture throughout life. The mechanisms by which these plastic changes occur appear to be a continuation of the process that sculpts the brain during development. To understand the brain and its devastating diseases, we need to reveal the mechanisms that produce it and the ways in which it can constantly change.[6]

We usually compare the brain to a computer. There are indeed some similarities, they both have complex circuits that process some form of electricity, and they compute according to their algorithms to resolve or respond to some form of external input. However, this comparison cannot be extended to the developing brain, and I would like to draw attention to numerous differences. Computers are put together by someone else; their boards are assembled in factories and then switched on to run programs that we select to install on them. The computer can run different programs without substantially changing the printed boards of the electric circuits. The construction of our brain is very different. Our brain is switched on and running programs while it is being assembled and constructed. The initial programs are self-generated, they use early and largely transient networks that are only present in the developing brain and only have little remnants in the adult structure, but they are vital for the construction of the neuronal machinery that we call brain.[6] Various aspects of the construction depend on the early programs that are run through the developing nervous system. This is very different from the construction of a computer, and therefore I always avoid this analogy in describing brain development to biomedical and medical students whom I teach.

The key takeaway is that our brain at birth is not a blank sheet of neuronal tissue a *tabula rasa*. When a newborn child opens its eyes for the first time, the brain is already prepared to process this information, although it takes months to reach the full capacity of vision. In the cerebral cortex, the same neuronal modules and activity patterns are transiently present to generate "a ready to go" network. So, neuronal activity is important much earlier than previously thought and we have to understand which factors control and modify this activity. However, we do not yet know whether all spontaneous activity patterns fulfill a particular functional role in development, what activity patterns are normal and abnormal or how different spontaneous activity

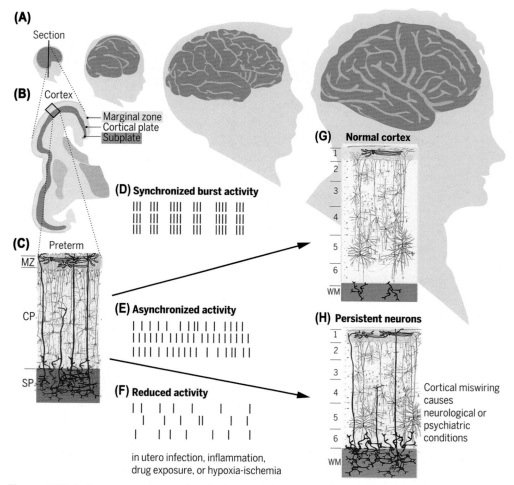

Figure 1.13 Early spontaneous synchronized neuronal activity sculpts cerebral cortical architecture. (A) Outlines of developing brains from the embryonic stage to adult. (B and C) The earliest generated cells reside in the marginal zone and subplate after the split of the preplate by the cortical plate (see also Fig. 1.5). The subplate and marginal zone neurons are much more mature than the CP and they contain the earliest synapses. (D to H) During subsequent development the early subplate- and marginal zone-driven circuits get gradually transformed to the adult-like six-layered cortex. All this is the natural consequence of normal development and requires spontaneous synchronized burst activity (D) that also controls other developmental steps, including myelination and programmed cell death (apoptosis), re-arrangement of somatodendritic morphology of neurons and their connections, and formation and awakening of synapses. The majority of subplate neurons are only present during early stages of development, then they disappear as the natural consequence of normal cortical development. Only a fraction survives in rodents as layer 6b neurons or in primates as interstitial white matter cells (G). Altering normal development can change their distribution and fate. Conditions, such as hypoxia—ischemia, drugs, infection, or inflammation, may alter spontaneous activity (E and F) during prenatal and early postnatal stages. Altering early cortical circuits can alter the activity in the circuits and these altered activity patterns may disturb subsequent developmental programs, including apoptosis (H). The altered development can be compensated to some extent, but some can lead to significant changes in the circuits that can manifest at later life. Surviving subplate neurons that persist in white matter or layer 6b may support altered circuits that could cause neurological or psychiatric disorders. Their altered distributions and increased numbers have been reported in frontal cortex in schizophrenia in human. *From Molnár Z, Luhmann HJ, Kanold PO. Transient local and global circuits match spontaneous and sensory driven activity patterns during cortical development.* Science. *2020;370(6514).*[6]

patterns interact. These activity patterns need examining together with detailed inquiry into key neural circuits.

During the last two decades, a major focus has been placed on genetics, whereas the fundamentally important role of electrical activity from the earliest stages of brain development has only recently become more emphasized again. During early development of the mammalian brain, transient neuronal populations integrate spontaneous and externally generated activity patterns to form mature cortical networks.[6]

Imagine that one has to build a complicated building. This requires elaborate transient scaffolds. If there are issues with the building work, one cannot understand what went wrong unless we also look at the scaffolds. Sometimes you cannot remove the scaffold because the building work went wrong and without the scaffold the building would collapse. Residual scaffolds could indicate issues with the permanent building itself. We need to understand the elaborate mechanisms behind how these transient scaffolds are established and used during brain development and how they change when there are alterations in the building work or there is altered sensory input.[16] Crucially, the early transient circuits both process and produce activity patterns that are vital for further maturation of the central pathways. There is an awful lot of building work that has been done before birth. Even then most sensory inputs are wired up to the thalamus and cortex and the motor outputs are ready to increasingly take control. Consequently, the sensory nervous system feeds into circuits that are preformed to some extent, although it is far from fully mature. In fact, the early stages of development rely on matching internally generated patterns to the various influences of the external world. It is vitally important to better understand the transition from early transient to permanent neuronal circuits, as impairments at the earliest stage of brain development can lead to the development of neurological and psychiatric conditions several years or even decades after altered development. These early impairments, namely genetic abnormalities, or early pathological conditions, such as intrauterine infection, drug exposure, or mild hypoxia, may potentially influence and disturb early brain activity and cause cortical miswiring at later stages of development (Fig. 1.13).[7,8] The difficulty is that examining brains at the time of diagnosing a neurological or psychiatric disorder does not inform clinicians what happened during the developmental years, sometimes decades prior. Given the brain's amazing ability to change, what we see is the brain's long-term response to early events. Moreover, given that the early transient circuits are long gone, it is difficult to speculate what happened to them in early life unless we can dissect the causal relations during these early stages. During development, sensory evoked activity slowly takes over from spontaneous activity. Thus controlled sensory stimulation might be able to alter the developmental trajectory.

1.13 How to link cognitive conditions to their developmental origins?

The developing brain has exceptional reserves. In human, it evolved to be able to adapt to its individual environment during our lifetime, but the cost was that there is a relatively protracted and vulnerable developmental period. An improved understanding can help to identify ways to prevent abnormalities. The challenge is to link anomalies of early circuits to the altered cognitive conditions. Some alterations in the circuit do not necessarily manifest in noticeable changes or abnormalities since there are spectacular reserves in the central nervous systems to perform these functions. Others lead to devastating conditions, such as intractable epilepsy and severe learning difficulties. The challenge for researchers is to identify the key parameters to monitor in the clinic. We still have a long way to go to establish the best practice to manage neonatal hypoxia—ischemia, fetal alcohol syndrome, maternal drug abuse, and intrauterine infection. These insults act on highly complex and integrated events of brain development at stages when the circuits contain transient elements (Fig. 1.13). The abnormalities require very different therapeutic interventions that must consider the specific steps of early development. These interventions are key to improve the chances of recovery. We made some progress over the last two or three decades, but we are still very far from elevating the platform from which we can tackle numerous brain developmental conditions that can have an impact on the life of individuals and their families for the rest of their lives.

Imagine you are playing a piano concerto. If you removed a couple of keys, you might get through the piece, but sometimes the sound would be excruciating. That is what can happen with gene activation during brain development. We all have our piano keys given to us at our conception and often we get through our "concerto" of life without major difficulties, but sometimes life can be devastating if one key is missing. Altered expression of a single gene can have a devastating consequence on brain development and subsequent brain function. We get the set of keys, from our parents. Their set of keys get mixed up in the fertilized egg which we then inherit to get through life.

Most cognitive conditions have developmental origins and they are prevalent in the general population (schizophrenia 1:100; autism 1:68; dyslexia 1:10; and attention-deficit hyperactivity disorder 1:30), but we are just at the very beginning of understanding how the complex interactions between an unfolding genetic program and the environment produce subtle alterations in neuronal cell numbers, proportions, differentiation, and circuit assembly that may manifest in the form of altered conditions or diseases. Building the human brain relies on complex cellular interactions, and the human brain has a very prolonged developmental period in comparison to our closest living relatives, which probably contributes to the huge cortical expansion of the human brain.[4] On the other hand, the most recently evolved traits have specific vulnerabilities that we are just beginning to appreciate. The methods of developmental biology research have become more refined, including single-cell RNA sequencing, allowing the discovery of much more subtle cellular alterations in healthy development and disease.[8,36] We are also beginning to appreciate both the power and the limitation of animal models. We can now study cell lineage, migration, and connectivity directly in human with normal or pathological conditions. We are beginning to understand why similar disruptions can cause different cognitive conditions, and why dysfunction in different pathways can lead to similar disorders.[18]

It is important that we approach brain developmental disorders from many different angles, just as we have to approach our "elephant" from many different directions (Fig. 1.1). Some of the causes of brain development might not even lie within the nervous system. There are complex interactions between nutrition, immune system, vasculature, and barriers in the brain or even in the placenta.[7,8] At early stages, the baby's brain development heavily relies on the mother. During development, mom and baby form a unit—metabolic, endocrine, social etc. Our current approach to brain development is very biased to the study of neurons whereas the placenta, vasculature, blood—brain barrier, and glia are all key at these early stages and their abnormalities can lead to abnormal development and subsequent abnormal brain function. We cannot just consider neurons when we look at the origins of brain developmental conditions. A much broader approach is needed. Therefore when developmental neurobiologists are looking into how the components of our nervous system are born, migrate, get assembled, and fine-tuned to perform their functions—how their interconnections remain plastic and how they are removed when they are no longer needed; they have to keep an open mind and consider these factors that are actually not even located in the brain.

Developmental neurobiologists want to understand to what extent the different computational functions of the brain are determined by the unfolding of the genetic program on the one hand and modified by the environment on the other hand. As our brain develops, our whole life experience gets embedded in its structure. Our experiences shape our brain—how many languages we speak, whether we play tennis, use our right or left hand. Unlike a computer, the brain is being constructed while it is already switched on and it is running programs that are essential for its own assembly. As we discussed during the creation of the cerebral cortical neural circuits, an early born cell type, the "subplate," is present that helps build the more permanent circuits (Figs. 1.5 and 1.13). In a way, these cells function as transient dynamic scaffolds. They are also the first recipients of input to the cerebral cortex and start to process the information when the rest of the brain cells is not even born. When their work is finished, they largely disappear. But in people with a cognitive problem, for instance schizophrenia, autism, or epilepsy, many more survive. These cells remain below the cerebral cortex among the subcortical fibers as interstitial white matter cells; these are the "shadows of the subplate" (Fig. 1.13).

Neurodevelopmental disorders are extremely important to study because they can have a huge impact on the entire life of the individual and their families. If we rank diseases according to the years lost due to ill-health, disability, or early death (disability-adjusted life year), then the brain developmental conditions feature very high on this list. The overall disease burden from perinatal brain injury and childhood epilepsy continue to have a very high disease burden. Cognitive conditions can have an impact on one's entire life. To change this, we need basic science to understand the mechanisms. My group is trying to understand what the consequences are if the "subplate cells" persist in larger numbers, how can we link problems with the persistent cells to cognitive dysfunctions (Fig. 1.13). Some of the keys on the piano are only needed at the earliest stages of the "concerto of life," but they are essential to establish melodic, harmonic, or rhythmic material key to the rest of the piece. Over millions and millions of years of selection, we have evolved this extremely powerful, awesome computational device that is so well suited to generate creativity in art and science—provided we have all the keys and the right input from our earliest stages of development. The rest depends on what we use it for. Development of the brain does not occur in isolation. Development of the brain relies on all the interactions with our environment through our sensory organs, the initiation and execution of our movements and subsequent feedback. Many of our complex skills develop via social interactions with others. Huge areas of our brain are dedicated to recognizing features of faces, language, and complex situations within a social group. The developing brain needs these interactions for its development, and this is why understanding the early alterations is so important to understand. The environment, such as interactions with others, also has a huge influence for the rest of our lives. Babies exposed to daycare have different upbringing from babies that were largely brought up in a family, not to speak of some of the conditions that one could observe in orphanages.

1.14 Mother and baby form a unit during and after pregnancy

These early developmental processes in the fetal brain are regulated to a large extent by a great variety of factors derived from the mother—such as simple nutrients or hormones or sensory stimuli, such as talking or singing. A normal maternal environment can be disturbed by various factors. From clinical observations and animal experiments it is known that maternal infection, stress, malnutrition, or toxic substances, can have a profound impact on cortical development and the offspring can develop a variety of neurodevelopmental disorders leading to cognitive conditions. Each developmental program can be vulnerable at various stages. Neurogenesis in various progenitors, neuronal migration, differentiation, early activity patterns can all be susceptible to specific environmental interactions at specific stages. There is selective vulnerability of cortical neuronal subtypes to maternal-derived insults, which depend on the time, place, and unique features of a cell type. These insults can affect numerous components or region of the developing brain.[7,8]

During development the nutrition of the fetus is entirely dependent on the mother, and it is provided through the placenta. The placenta is an organ that develops according to the instructions of the baby's genome within the mother's uterus during pregnancy. The placenta facilitates nutrient, gas, and waste exchange between the maternal and fetal circulations, but it also provides a barrier between the maternal and fetal circulations. These are separated from each other by the materno-fetal barrier (Fig. 1.14). The placenta is also an important endocrine organ producing hormones that regulate both maternal and fetal physiology during pregnancy, where mother and baby form a metabolic unit for the generation of certain hormones. The placenta

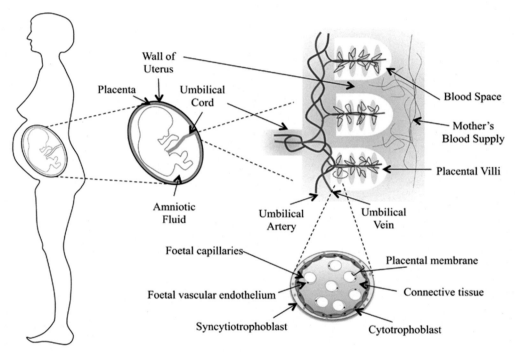

Figure 1.14 Mother and baby has two separate circulatory systems, but they form a metabolic and endocrine unit during pregnancy. The figure provides a schematic illustration of the placental structure and materno-fetal barrier in the placenta. The placenta is formed according to the genetic instructions of the baby. The fetus develops in the uterus isolated from the maternal circulation. The maternal blood enters the intervillous spaces from the spiral arteries. There is transfer from the intervillous space across the placental barrier to the fetal capillary network that joins the umbilical vessels. The syncytiotrophoblasts, cytotrophoblasts, and connective tissue that make up the placental membrane are the key structures of the placental barrier. Developmental conditions and abnormalities must consider the bigger picture, including the placenta. *From Stolp H, Neuhaus A, Sundramoorthi R, Molnár Z. The long and the short of it: gene and environment interactions during early cortical development and consequences for long-term neurological disease.* Front Psychiatry. *2012;3:06.*[7]

attaches to the wall of your uterus, and the baby's umbilical cord arises from it. The role of the placenta in regulating the maternal environment is somewhat overlooked when discussing brain development and brain developmental abnormalities. Recent evidence suggests the placenta is actively involved in responses to altered nutrition, inflammation, genetic conditions, and stress. Understanding the consequences of Zika virus infection requires the consideration of the impact on placental vasculature[37] and the nutritional status of the mother.[38] I believe that the placenta is that part of our "elephant" that we seldom touch when we would like to understand the brain and its development and therefore pieces of our puzzle remain missing.

1.15 Influence of maternal environment on the baby's prenatal development

It is known that when the normal maternal environment is disturbed due to special diet, maternal infection, stress, malnutrition, or toxic substances, it might have a profound impact on cortical development and the offspring can develop a variety of neurodevelopmental disorders that may manifest at a much later stage. Alterations in maternal gut microbiome during pregnancy, in response to infection, diet, and stress has been associated with abnormalities in brain function and behavior of the offspring.[39] Maternal depression, autoimmune disorders, and hypoxia have also been proposed to increase risk for neurodevelopmental abnormalities in the offspring.[7] Maternal-derived insults have to pass several major barriers before they reach the fetal brain, the placenta, and blood—brain barrier being the major ones (Fig. 1.15).

Association between maternal infections by viruses and bacteria and neurodevelopmental disorders in the offspring was established in epidemiological studies. There are various infectious agents, such as Zika virus, cytomegalovirus (CMV), influenza, rubella, and others that can alter the development of the fetus and its central nervous system. These pathogens can elicit their effects directly acting on the fetus or indirectly through the maternal immune response or both of these in combination. The congenital infections that spread from mother to fetus can have very specific effects on neuronal progenitors, various barrier functions, and blood vessels of the developing fetus or placenta or migrating neurons in the brain.[7,8] The maternal immune response to infections is mediated by inflammatory cytokines that can cross to the fetal circulation and alter developmental processes. It is a paradox that the pathogens might be restricted to the mother, yet the immune response might influence the neuronal development of the baby, but it may be the main factor in the neurodevelopmental disturbances in the baby. Infections trigger the innate immune response in the mother via activation of Toll-like receptors on macrophages, which leads to an increased release of proinflammatory cytokines in the maternal blood. Maternal cytokines cross the placenta and spread into the developing fetus, including the fetal brain. Receptors for these

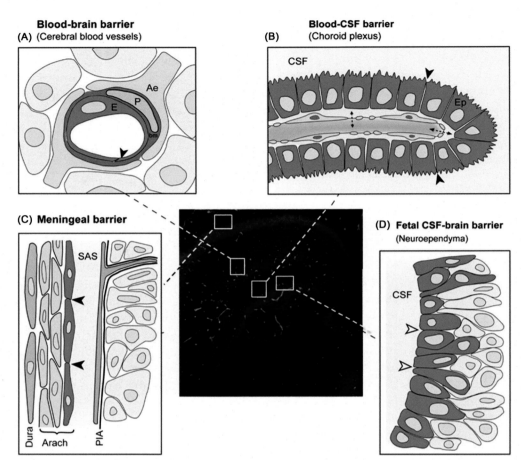

Figure 1.15 Barriers in the brain that separate the nervous system from the blood and cerebrospinal fluid. The central image represents an embryonic brain (*blue*) with dividing cells (*red*) and vasculature (*green*). The origins of the various barriers are depicted with small white boxes from the ventricular zone (A), choroid plexus (B), meninges (C), and the inner lining of the ventricles (D). Barrier interfaces in the brain are indicated in blue in the schematic diagrams of the blood—brain barrier (A), blood—cerebrospinal fluid barrier (B), meningeal barrier (C), and fetal CSF—brain barrier (D). The primary site of the blood—brain barrier (A) is the endothelial cells of the cerebral blood vessels (E), supported by pericytes and astrocytic endfeet. In the choroid plexus (B), the epithelial cells are the barrier interface of the blood—cerebrospinal fluid barrier, as the endothelial cells in the stroma are fenestrated. The meningeal barrier (C) is at the boarder of the arachnoid membrane and the subarachnoid space. The fetal cerebrospinal fluid—brain barrier (D) exists during early development, at the endfeet of the radial glial progenitors. This forms a barrier between the cerebrospinal fluid and the developing parenchyma. *From Stolp H, Neuhaus A, Sundramoorthi R, Molnár Z. The long and the short of it: gene and environment interactions during early cortical development and consequences for long-term neurological disease. Front Psychiatry. 2012;3:06.*[7]

cytokines are expressed in neuronal progenitors of the developing brain. In animal experiments, IL-6 and IL-17a have been demonstrated to mediate the major effects of maternal inflammation that acts on cortical development that can lead to alterations in the behavior at adult stages.[7,8]

1.15.1 Maternal stress and nutrition

Various stressful events during pregnancy have been associated with an elevated risk of developing neuropsychiatric disorders in the offspring. Immune response modulated by glucocorticoids or other hormones in pregnant women during stress is linked to the neurodevelopmental disturbances in the fetus in conditions, such as schizophrenia. Stress has a profound effect not only on glucocorticoids but also on neuromediators, growth hormones, insulin, and prolactin. Maternal stress hormones could further impair endocrine function on the placenta itself, impacting nutrient transport and placental growth.

Maternal stress can be mediated to the developing embryo via a great variety of immune and neuroendocrine factors. These effects can be selective to a particular neuronal subtype at a particular period of cortical development.

Maternal over- or under-nutrition both have influence on fetal brain development. The obesity resulting from the former condition correlates with a higher incidence of psychiatric disorders in the offspring. Protein deprivation in the maternal diet has a major effect on fetal brain development. The effects vary depending on the stages when the deprivation occurs, and it appears that an earlier gestational period, such as the first half, is the most vulnerable. There are historic cases where severe maternal undernutrition during famine increased the risk of schizophrenia, psychoses, depression, and other psychiatric disorders. Deficiency in vitamin A contributes to psychiatric disorders, such as schizophrenia while too high maternal levels of vitamin A are associated with microcephaly and other brain malformations in the offspring. Taking large amounts of vitamin A can harm the unborn baby. It is advised, if you're pregnant or thinking about having a baby, not to eat liver or liver products, such as pâté, because these are very high in vitamin A. It is also advised not to take supplements that contain vitamin A.

The developing neural tube is formed by alterations of the structure of the neural plate at early stages of development. These morphological changes will initiate the neuronal plate to fold and close into a tube through a process that depends on the kinetics of neurogenesis, neuronal migration, and differentiation. The prevalence of neuronal tube closure defects varies between 0.5 and 2 per 1000 births in countries without folic acid supplementation. Failure of closure at the cranial end of the neural tube results in anencephaly. Failure to close at the caudal end results in spina bifida, which has varying severities based on the level of neural tube closure. Occulta is the least severe, characterized by a patch of hair or change in pigmentation of the lower back, and is seen in 1 in 10 people. Meningocele is more severe, resulting in a cyst, formed by the meninges herniating between the vertebrae. However, the nervous system in this case is unaffected. The most severe form of spina bifida is myelomeningocele, where the spinal cord is completely exposed—this will render the individual paralyzed to varying degrees, and can be fatal. There is now a decreased prevalence of spina bifida due to the increased use of folic acid by women before conception and in their first trimester, with folic acid being important for reducing the incidence of this condition, through mechanisms which are unknown, but it is thought to be associated with the fact that neurulation is highly sensitive to metabolic changes. Fortification of flour with folic acid was made mandatory in numerous countries decades ago, but some have still not introduced it up to the present day. The reluctance to do so was due to studies by the Institute of Medicine that implied that patients who received folic acid 5 or more mg/day to treat vitamin B12 deficiency developed neuropathy.[40] This led to the suggestion that there is a tolerable upper intake level of folate (which includes natural food folate as well as synthetic folic acid) which has been arbitrarily set for 1 mg/day based on flowed arguments. It has been demonstrated that the upper intake level is unnecessary and should be removed and allow folic acid fortification.[40] Maternal deficiency of folic acid not only cause to spina bifida but can also lead to impaired cortical architecture in the offspring and abnormal behavior in animal models. Banning a harmful substance should not be more important than adding a beneficial substance. Both are important for the outcome of a pregnancy.

It is well known that the gastrointestinal system of the mother acts on the mother's brain through neuronal connections and through a variety of hormones produced in the mother (see Chapter 2: Tamas Horvath. The Hunger View on Body, Brain, and Behavior). However, it is not sufficiently appreciated that the same hormones can also act on the developing brain of the fetus. Animal models demonstrated that maternal high-fat diet and obesity impact fetal brain development via several routes, the most prominent being chronic inflammation and hormonal regulation by obesity-related hormones, such as leptin. Leptin can have direct effects on the mother's central nervous system in widespread areas, including hypothalamus, cerebral cortex, and hippocampus. Leptin produced in the mother is also transported through the maternofetal barrier to the fetal circulation and it enters the fetal brain. In the fetal brain, it binds to its receptors that show very widespread distribution including the germinal zones that are producing the neurons for the brain. Through these interactions maternal leptin can have impact on brain development.[41] In the developing cerebral cortex, leptin modulates the proliferation of progenitors. It has been demonstrated that caloric restriction lowers concentrations of leptin in maternal blood, and this can induce programmed cell death in the cortical precursor cells in the fetus in the developing cortex. Whether leptin acts directly or indirectly on neuronal progenitors is not yet known.

High levels of leptin stimulate neuronal differentiation, whereas low levels support an undifferentiated precursor state. In addition, leptin may also alter cell lineage decisions; it promotes progenitor differentiation towards oligodendrocyte cell fate. All these effects can have long-term consequences on the proportions of neurons produced and can change the circuitry which can manifest in altered behavior later in life.

1.15.2 Maternal alcohol consumption, drug abuse, medications, and smoking

Drug abuse, smoking, alcohol, medications, and other toxic substances alter brain development. The outcome of these alterations depends on the timing. It seems that the period during the peak of neurogenesis is the most vulnerable time for maternal-derived insults. Although the placenta and fetal blood—brain barrier (Figs. 1.14 and 1.15, respectively) do provide two levels of defense against toxic chemical penetration, it has been shown that many substances are able to penetrate the placenta and large molecules could reach the fetal compartment, albeit at a slower rate. Air pollution by particulate matter is a common problem in urban areas and has been associated with a risk of abnormal cognitive development, autism, and learning difficulties. Maternal smoking has a strong correlation with a risk of cognitive impairment in the offspring.

Exposure of the fetus to an excess of alcohol results in multiple, often severe abnormalities in brain development, collectively referred to as fetal alcohol syndrome (FAS) or fetal alcohol spectrum disorders. One in 100 pregnancies have some degree of alcohol involvement and can have mild FAS. Alcohol causes reduced brain size and neuronal migration abnormalities (Fig. 1.16). Animal experiments demonstrate the disruption of cortical progenitor proliferation and disruption of neuronal migration after maternal alcohol administration. Drugs, such as marijuana, cocaine, and opiates, cross the placental barrier and could similarly reach the fetal brain. Cannabinoid receptors are expressed widely in the developing nervous system, including the neuronal progenitors.

Valproates are medications used to treat epilepsy and bipolar disorder and prevent migraine headaches. However, valproate administration during pregnancy can be associated with neurodevelopmental disorders in the offspring. Prenatal exposure due to the maternal use of even common medicines, such as paracetamol, could affect fetal cortical development due to the increase in cytokine release to the fetal blood, which derives from the inflammatory response of the placenta to the medicine.

The child with fetal alcohol syndrome (right) has a smaller, less folded brain, with no interhemispheric connections (callosum) and reduced cerebellum. The MRI brain images of FAS children show agenesis (absence) of the set of fibers (corpus callosum) that connect the two halves of the brain. It also demonstrates enlarged lateral ventricles in the occipital cortex.

The lower panels show that the difference is size and the FAS brain is much smaller. Smaller-than-normal head circumference at or below the 10th percentile is

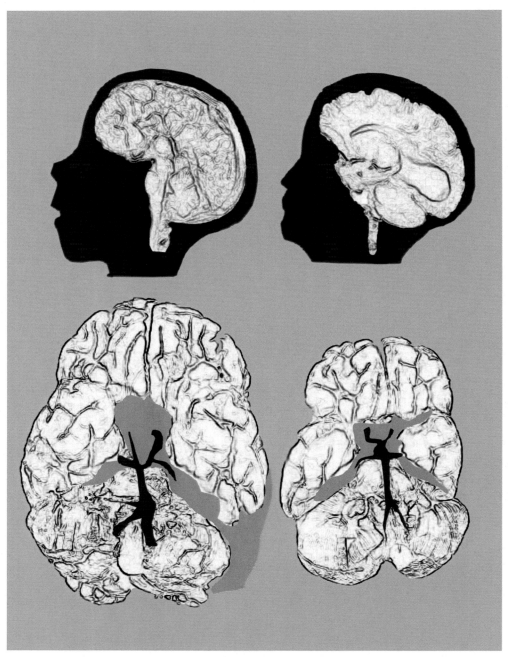

Figure 1.16 Comparisons of healthy (left) and fetal alcohol syndrome-affected brain (right) imaged with magnetic resonance imaging (sagittal views) upper panels and inferior (ventral) view of dissected brains.

one diagnostic criterion for alcohol-related neurodevelopmental disorder (ARND). ARND is associated with intellectual disabilities and problems with behavior and learning, tend to do poorly in school, has difficulties with math, memory, attention, and judgement, and displays poor impulse control. Upper MRI images are schematically drawn based on: https://medicalxpress.com/news/2013-09-affirm-dangers-pregnancy. html. Lower brain schematic drawings are based on: https://www.clinicaladvisor.com/slideshow/slides/fetal-alcohol-spectrum-disorders/.

1.15.3 Multiple hit etiology of neurodevelopmental conditions

Insults rarely occur in isolation and the neurodevelopmental conditions arise due to combined effects of genetic predisposition and multiple environmental insults. The symptoms can manifest in different manners, often years after the insults. Maternal inflammation and stress, toxic substances, nutrition, and ZIKA infection can have more devastating and more immediate effects dependent on the combinations. For instance, ZIKA congenital syndrome is much more prevalent in regions where there is low protein diet due to maternal malnutrition.[38] In most of the subtle cognitive conditions or developmental abnormalities the onset of the condition cannot clearly be linked to a specific genetic or environmental insult. In diseases, such as schizophrenia and autism, there is evidence for both genetic contributions and environmental insults, but neither one accurately predicts disease. In such conditions, it is postulated that there is a multiple hit etiology. In these conditions, the neuropathology is not strongly linked to a particular region of the brain and these subtle abnormalities cannot be linked to one specific genetic or environmental insult. Instead, a wide variety of broadly linked functional systems are affected. This can include genes involved in synapse formation and maintenance, cell–cell signaling, and trophic pathways, and injury widely affecting GABA, dopaminergic, and glutamatergic transmission. It has been hypothesized that the first hit in schizophrenia may affect neurogenic and cell specification pathways, such as Notch, while the second hit may have a greater effect on functional integration.[42]

Additive cytotoxic effects of low dose alcohol and hypoxia exposure in primary neuron cultures have been demonstrated. There have also been many studies showing that inflammation combined with hypoxia can cause increased neurological injury that will have long term impact. There is also some evidence that insults during early life may contribute to neurodegenerative disease later. For example, inflammation during embryogenesis (E10.5 in the mouse) results in loss of dopaminergic neurons in the basal ganglia, and increased sensitivity to the 6-hydroxydopamine neurotoxin later in life.[45] This substantial decrease in dopaminergic neurons may also increase the susceptibility of affected individuals to "normal" age-induced cell loss and may account for cases of extremely early onset Parkinson's disease.

It is also important to consider the timing of insult in relation to the stages of brain development. The term "window of vulnerability" is commonly used when considering developmental neurological injury. However, it is not yet clear if this vulnerability is intrinsic to specific periods of development (e.g., proliferation or synapse formation) and therefore may move from region to region along with the normal developmental gradients of these processes. Alternatively, it has been suggested that different cells may become selectively vulnerable to damage because of the different intrinsic maturation process. An example of this is found in subplate neurons, which are thought to have increased vulnerability to injury in the prenatal human due to an early birth date and therefore earlier maturation of glutamate receptor populations (Figs. 1.5 and 1.13). This leads to increased sensitivity to excitotoxicity at a time of development where there is an increased risk of hypoxic–ischemic injury.

Cerebral palsy provides an example of the consequence of the stage of brain maturity at the time of insult on subsequent neuropathology. Injury at term, for example, umbilical cord asphyxia during delivery, causes cerebral palsy with a typical pattern of gray matter loss, as well as damage to the hippocampus and cerebellum. Whereas preterm injury, leading to cerebral palsy, is generally associated with white matter and subcortical gray matter damage. There are a few confounding maturation steps that underlie these differences in regional damage. White matter is beginning to be myelinated in the preterm period in humans, and specific populations of precursor oligodendrocytes have been shown to be particularly susceptible to damage, which corresponds to this period. In contrast, areas of term damage are typically described as "watershed" damage, indicating that these brain regions are at the edge of the vascular zones, therefore increasing susceptibility to hypoxic damage. In preterm white matter, similar watershed regions have been suggested due to the pattern of vascular development. However, it was determined that differences in vascular blood flow could not account for the different magnitude of damage in the region. Therefore the different patterns of damage have been associated with selective sensitivities of developing cells at the time of insult. This is an example whereby different insults at different stages of brain development can cause distinct but related neuropathologies that contribute to the one clinical diagnosis.

1.16 Functional localization in the brain

Localization of function in the brain is a very old concept. Initially it was believed that faculties of sensation, memory, reasoning, and movement occupied each ventricle. The cerebral cortex was initially considered merely for cooling the blood by Aristotle. Lesion and stimulation studies in adult brains in animals provided gradual proof that there were topographical "primary" sensory and motor maps.

Vesalius (1543) realized that the brain controls the spinal cord which controls the muscles.[43] In 1649 Robert Boyle described in one of his letters that damage to the left

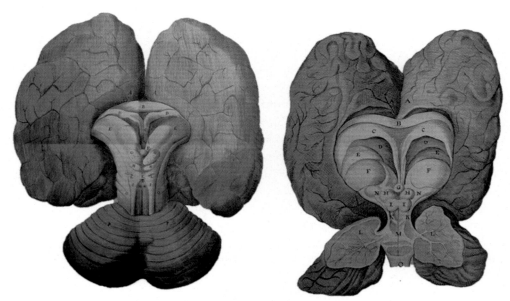

Figure 1.17 Thomas Willis' comparisons of a normal adult brain (left) with a brain from a boy with signs of learning difficulties and intractable epilepsy (right) from Cerebri Anatome of Willis (1664). Willis associated the functional alterations with the altered structure of the cerebral cortex that had smaller, but more frequent folding in the boy with intractable epilepsy. It is very likely that Willis described a congenital brain developmental condition, polymicrogyria syndrome. *From Molnár Z. Thomas Willis (1621—1675), the founder of clinical neuroscience. Nat Rev Neurosci. 2004;5(4):329—335.*[46]

side of brain causes right sided paralysis and loss of speech. Based on his comparisons of brains of human and animals and comparisons of brains of patients with learning difficulties and epilepsy (Fig. 1.17), Thomas Willis (1664) already stated that "The cerebrum is the primary seat of the rational soul in man, and of the sensitive soul in animals. It is the source of movements and ideas."[1,44]

Are different aspects of behavior controlled from different parts of the brain? While the cerebral cortex was considered the seat of higher cognitive functions for hundreds of years, the acceptance of the concept of cortical localization was not a straightforward process. I shall review the history of this process to illustrate that some of our current approaches might be still very outdated because we still allocate brain regions to various functions based on our interests, outdated concepts of mental or sensory categories. We might use increasingly sophisticated methods for our investigations, but we might be still using categories for mental functions that are obsolete and might not reflect how our brain actually functions. We wish to allocate regions with defined boundaries to functions, such as decision-making, learning, memory, perception, arousal, attention, initiation, and execution of movement, that rely on large distributed overlapping networks.

To some extent phrenology, the discipline that involves observing and/or feeling the skull to determine an individual's psychological attributes caused a great deal of damage to the concept of cortical localization. In the early 1800s Franz Joseph Gall (1758–1828) and Johann Gaspar Spurzheim (1776–1832) introduced the theory of cranioscopy, later phrenology—that mental characteristics and behavioral functions are localized in "organs" distributed across the surface of the brain and which can be analyzed by palpating bumps on the skull. Gall stated that the "cranium is only faithful cast of the external surface of the brain, and it consequently, but a minor part of the principal object." Validity of Gall's approach depended on assumptions: that the size and shape of the cranium reflected the size and shape of the underlying portions of the cerebrum; those mental abilities were innate and fixed; and that the relative level of development of an innate ability reflected the inherited size of its cerebral organ. Unfortunately, none of these assumptions proved to be valid. Therefore phrenology was destined to fail. In 1843 the French physiologist François Magendie "father of experimental physiology" referred to phrenology as "a pseudo-science of the present day."

After all the damage caused by phrenology, the concept of functional localization in the cerebral cortex had to be gradually reestablished. This period is reflected in the debate between the German physiologist Friedrich Leopold Golz (1834–1902) and the Scottish neurologist and psychologist Sir David Ferrier (1843–1928) who both performed lesion studies in the frontal cortex and studied behavior. Both attended the same conference in London in 1881[47] and presented their hypotheses. Goltz had a more unitary view of brain function. He performed bilateral frontal cortical ablations in dogs and concluded that after extirpation of large parts of both hemispheres of the brain, "intelligence is diminished." However, he also claimed that it is impossible to produce a complete paresis of any muscle. He did not observe sensory or motor symptoms after smaller cortical lesions, but after lesioning large areas he noticed some loss of perception and clumsy movements in his dogs. He only noticed some crude regional variations depending on the location of parietal or occipital cortical lesions in his dogs. Ferrier, who lectured on the same conference in London produced evidence that a circumscribed lesion produced clear and reproducible functional deficits in monkeys. Lesioning the leg representation in the precentral gyrus elicited contralateral spastic paralysis of the leg, with rigidity in the affected limb, an exaggerated patellar reflex very similar to the upper motorneuron syndrome observed in human after stroke. Subsequent examination of the brains by a committee of conference participants demonstrated that in fact both lesions caused degeneration of the descending pyramidal tracts in the exhibited dog of Golz and the hemiplegic macaque of Ferrier.[48] Langley who was attending the conference and was a member of the committee involved Charles Sherrington in the study of the dog specimen. This was Sherrington's first ever publication on documenting the degenerated descending pyramidal projections in the brainstem and in the spinal cord.[49] He later spent some time in the laboratory of Golz in Strasbourg.[50]

1.16.1 Methods to study localization in the nervous system

The basic concepts for the study localization in the central nervous system did not change over the last century. We use three basic methods: lesions, stimulation, and imaging. Initially lesions were mechanical (stroke or gunshot wounds in human; cooling, injection of excitatory amino acids, that keep the fibers relatively intact in animal studies). We can elicit lesions with various genetic methods in very selected cell populations at very specific timepoints. These lesions can be permanent or transient. Stimulations were initially performed with electrodes, but now we can use optogenetics or pharmacogenetics to transiently and selectively activate or inactivate a particular cell group through activating or inactivating engineered channels on subgroup of cells in the freely moving behaving animals. Imaging also came a long way just from light microscopy to the other methodologies, electron microscopy, computer tomography, functional magnetic resonance imaging (fMRI), positron emission tomography, magnetoencephalography, early receptor potential, 2-deoxyglucose, expression of immediate early genes, and voltage-sensitive and Ca^{2+}-sensitive dyes that can be imaged in freely moving and behaving subjects. These studies confirmed that the primary sensory and motor areas are topographic. In the motor cortex, it was the work of Fritsch & Hitzig, Ferrier, Hughlings-Jackson, Penfield; in the visual cortex—Munck, Ferrier, Innoue, and Holmes; in the somatosensory cortex—Head and Holmes, Auditory Cortex—Shaefer, Ferrier, Luciani, Munk, and Henschen; and in the cerebellum—Rolando, Flourens, Gowers, Bolk, and Holmes reestablished the localization principles.

I shall only give examples for the lesion and subsequent imaging studies in the visual representation in the occipital cortex. The maps in the primary visual cortex (Brodmann Area 17) of the occipital cortex were mapped out by examining patients who suffered gunshot wounds causing localized lesions. The same visual maps were proposed by three exceptional neurologists: Henschen, Inouye, and Holmes who arrived at very similar conclusions independently. All their observations were based on the examination of the visual deficits observed in patients with gunshot wounds. At the turn of the previous century the bullets were just a piece of metal that went through the tissue on impact. It was after the first World War that spin stabilization was introduced, and the bullets were made more damaging because they changed their shape or exploded on impact. This brought the development of fully jacketed, Dumdum, partially jacketed "soft point," hollow-point and the devastator bullet where the canister is filled with explosive within the hollow tip of the bullet. These bullets would no longer cause simple penetrating wounds, but much larger and extensive damage on impact. For cortical localization studies one would need very specific circumscribed lesions. However, even during the first World War the gunshot wounds were not simple, and they were difficult to interpret. The lesions can generate bone fragments, it can cause penetrating or perforating lesions. The bullet can have tangential migration along the inner lining of the skull or dura; it can ricochet off the skull or the dura, or career across several areas, or can have spontaneous migration after impact which can make the interpretation of the path of the bullet difficult. Correlation of an entrance wound, and plain X-ray data alone may be deceiving, since the bullet can migrate at a later stage. The gunshot wounds can generate subarachnoid, subdural, confluent intracerebral, and intraventricular bleeding, as well as contusion and fiber damage. The bullet is damaging not only the cortical tissue but also the adjacent white matter. It is a miracle that Henschen, Inouye, and Holmes could produce such precise maps from examining their cohort of patients.

1.16.2 Representations in the cerebral cortex

The cerebral cortex represents over 80% of the volume of the human brain. The cortex has a universally 6 layered structure. However, the proportions of the different layers that contain different types of neurons change across the cortex. These changes in cell composition are called cytoarchitecture. The cytoarchitectonic differences in the cortex reflect the differences in neuronal density, different soma size, somato–dendritic morphology, and connectivity and circuits of those given areas. For instance, the primary visual cortex has a very thick layer 4 that contains the neurons that receive the visual thalamic input, whereas the primary motor cortex has a very prominent layer 5 that is the source of the major motor output to spinal cord, basal pons, superior colliculus, thalamus, and striatum. The cytoarchitectonic differences match the functional specializations because they reflect the different circuitry that is required for the different sensory or motor computational functions of those given cortical areas. Therefore the cytoarchitectonic maps that were produced over hundred years ago are still being used today (Fig. 1.18).

Study of brain's functions relies on three basic groups of methods: lesion, stimulation, and recording. For recording of brain activity, we explore the reactions of single cells or many cells during a particular condition, often after some sensory stimulus or motor performance. We can then record the activity of these cells by recording changes in electroencephalogram (EEG) with small sensors that are attached to the scalp to pick up the electrical signals produced by the surface of the brain. However, EEG field potentials can only provide a measurement of general activity. Thus to get a more direct signal at the level of single-cell resolution, we need recordings from next to the cells (via extracellular recording) or inside the cell (via intracellular recordings or patch-clamp).

We can also explore the indirect changes associated with neuronal activity, such as changes in blood flow. fMRI is an excellent method that is based on this principle and

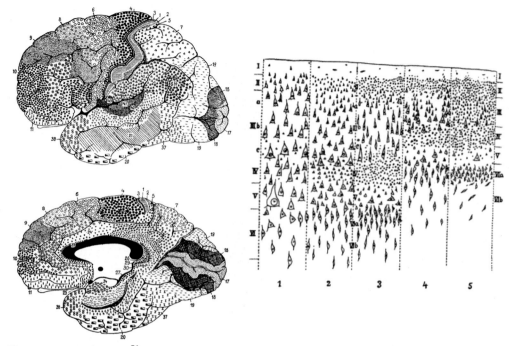

Figure 1.18 Brodmann's[51] showing the patterning of the cells in various cortical areas and illustrations of the left cerebral hemisphere of the human brain, viewed from the side (top) and the right hemisphere from the medial aspect. Brodmann's areas, defined by simple observation of stained sections of the cortex, are shown with different stippling and shading. They bear Brodmann's numbers, that were given by Brodmann according to the order he examined the cortical samples, which are still widely employed to describe cortical areas today. The right panel demonstrates the different cytoarchitectonic types of the cerebral neocortex as defined by von Economo and Koskinas.[52] *Left panel is from Brodmann K.* Vergleichende Lokalisationslehre der Grosshirnrinde *(in German).* Johann Ambrosius Barth; 1909,[51] *right panel is modified from von Economo C, Koskinas GN.* The Cytoarchitectonics of the Human Cerebral Cortex. *Oxford University Press; 1929.*[52]

used very widely to explore localization of blood in the brain. The concept of fMRI emerged from the original work of Charles Roy and Charles Sherrington (1890).[53] Their collaboration produced evidence of intrinsic vasomotor activity in cerebral blood vessels. fMRI is based on this hemodynamic response to a sensory stimulus or a motor function. It is an indirect measure of neuronal activity with spatial resolution of millimeters and this method comes with several advantages including its noninvasiveness, lack of radioactivity, and its great potential for longitudinal studies (ideal for assessing drug-related effects or monitoring patients).

The visual representations that were deducted from the collection of patients with gunshot lesions by Henschen, Inouye, and Holmes can be now measured with fMRI imaging much more efficiently and accurately. fMRI studies have modeled brain activity elicited by visual patterns and have reconstructed the representation of the patterns from brain activity. Dougherty and colleagues used expanding ring stimuli map eccentricity and rotating wedge stimuli to map polar angle to explore the sizes of visual areas, including V1, V2, and V3 (Fig. 1.19).[54] Using fMRI, they were able to correlate the phase of the best-fitting sinusoid for each voxel to the position in the visual field which elicits the maximal response for that voxel. The pseudo-color phase maps allow us to relate the visual fields and the cortical retinotopic maps. Fig. 1.18 demonstrates examples for the left hemisphere of one subject. When Dougherty and colleagues examined multiple subjects, they revealed that the sizes of visual areas varied by about a factor of 2.5 across individuals; interestingly, the sizes of V1 and V2, but not V1 and V3 correlated within individuals.

fMRI allows for the comparisons of representations between individuals and can follow the development or plasticity of these maps within the same individual. It can also reveal the re-mapping of cortical areas after trauma, and it is not restricted to a

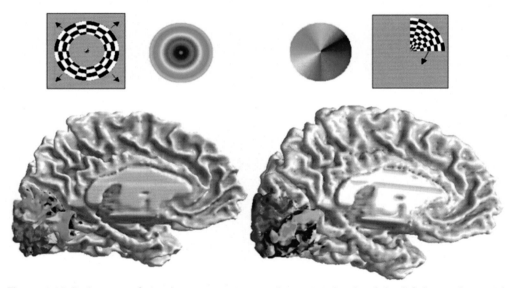

Figure 1.19 Retinotopy of visual representations in the occipital pole of the left hemisphere with functional magnetic resonance imaging using expanding ring stimuli map eccentricity (left) and rotating wedge stimuli map polar angle (right). The phase of the best-fitting sinusoid for each voxel indicates the position in the visual field that produces the maximal activation for that voxel. Thus these pseudo-color phase maps indicate the retinotopic maps, with the fovea at the back of the occipital pole and periphery more anterior (left lower panel) and the upper visual field (monitored by the upper retina) at the upper bank of the calcarine sulcus and lower visual field (monitored by the lower retina) in the upper bank of the calcarine sulcus. Data are shown for the left hemisphere of one subject. *From Dougherty RF, Koch VM; Brewer AA, Fischer B, Modersitzki J, Wandell BA. Visual field representations and locations of visual areas V1/2/3 in human visual cortex. J Vis 2003;3:586–598.*[54]

particular region that is exposed but includes the entire brain. Human brains can thus be compared with animal models and fMRI can help identify homology with them. Such studies revealed that the proportions of sensory representations are related to the density of peripheral receptors. Primary sensory and motor areas are topographic, and the exact mapping depends on various factors. In the visual system, the ventral and dorsal pathways analyze the "what" and "where" simultaneously[55] which can be a huge advantage to recognize whether a friend or a predator is approaching you and whether they are coming to socialize or view their dinner in you. Ventral stream is critical for visual perception whereas the dorsal stream mediates the visual control of skilled actions.[56] The different visual areas have slightly different circuitry that can extract different parameters from visual stimuli. Elston et al.[57] drew the attention to the correlation between the tangential dendritic field area of basal dendrites of layer III pyramidal neurons and modalities of visual processing. The increase in the basal dendritic field area of layer III pyramidal cells may allow more extensive sampling of inputs as required by higher-order processing of visual information.

fMRI proved to be a very useful method to reveal arealization in the brain. It revealed more complex representations, such as representation of language in multilinguals. The method has been used to study higher cognitive functions, such as episodic, procedural, and recognition memory, planning, emotion, judgement, generation and understanding of language, reading, maths, and even influence of religion on pain perception. Quantitative modeling of human brain activity with fMRI revealed crucial insights about complex cortical representations, such as language, initiation of movements, action, or intention. One ambition is to use the signals revealed with fMRI to read someone's mind, to explore the inner thoughts of individuals and generate some brain decoding devices. In theory, it is possible to detect dynamic brain activity under naturalistic conditions with noninvasive methods and decode these activity patterns to draw inferences of the cognitive functions. While in theory it is possible, the time and space resolution of EEG or fMRI hinder these efforts. When we record EEG, we are picking up very distant signals with little spatial accuracy. This can be improved by placing electrode arrays directly onto the brain surface, but this is only done in patients undergoing epilepsy surgery to localize the ectopic activities. fMRI is noninvasive, but the blood oxygen level-dependent signals measured via fMRI are indirect, and very slow, so it has been difficult to model brain activity elicited by dynamic real time stimuli. Nishimoto et al.[58] reported an encoding model that describes fast visual information and slow hemodynamics by separate components. They recorded blood oxygen level-dependent signals in occipitotemporal visual cortex of human subjects who watched a selection of natural movies. To generate a decoder that produced reconstructions of the viewed movies from the pattern of activation in individual voxels Nishimoto and colleagues combined encoding models with sampled natural movies that the subjects watched repeatedly. They then fit the model separately to individual voxels and revealed how early visual areas represent the information in these selected natural movies. From the pattern of the voxel activation in primary visual areas they could then reconstruct the major patterns of these movies. It is a highly exciting, but also somewhat frightening prospect that new methodologies can emerge that will be able to read our inner thoughts and feelings.

Examining the structure, connectivity, and activity of someone's brain can reveal a lot about that person. We can tell age, handedness, motor abilities, and language representations. No wonder that we now use imaging studies to reveal finer and finer details of representations. The fact that fMRI can produce some quantitative signals that reflect different brain activity in defined conditions can also be misleading. We can fool ourselves that we correlate cognitive processes with patterns of brain activity that can be measured by fMRI, whereas the real difference is due to something completely different, such as eye movement, muscle tone changes, and metabolic differences. The quest by imaging neuroscientists to localize all the functions in the brain is often criticized and ironically termed "neophrenology." The critics state that it is almost like we are reliving the time of phrenology again but with more modern and more direct approaches. Yes, the lowest level of cognitive functions, such as representations in the primary sensory and motor areas, motor execution, and sensory representations, are indeed topographic and well localized (see Fig. 1.18). There is also some form of middle-level topography in motor planning and object recognition, mediated by areas close to appropriate primary cortex. However, some of the highest-level cognitive functions, such as language, maths, logic, music, love, or hate, might not be strictly localized. These functions are mediated by many interconnected areas, "distributed networks." Moreover, we still allocate functions according to our own preconceived ideas and concepts if the brain has to be organized according to these historic and anthropocentric concepts. Textbooks describe various functions to be localized to specific regions in the brain: face representation to the right fusiform gyrus, visual word forms—left fusiform, landscapes, man-made objects, other shapes, color—inferior temporal gyri. Languages (including deaf sign languages) are localized to—left angular gyrus, space, visual guidance of movement, body image to—right posterior parietal cortex, spatial decisions—left dorsolateral prefrontal cortex, conscience, sexual desire, motivation, initiative—right orbitofrontal cortex. However, one could make an argument that the physical processing of the sensory stimulus and the conscious experience of it and whether it is generating fear, anxiety, interest, or pleasure depend on how large, distributed networks are working all together.

For instance, some form of visual perception was assigned to inferior temporal cortex, but in certain conditions it is also crucial for memory. Visual perception might be much more strongly linked to other modalities, movement, and memory than we previously thought. Also, high level of cognitive functions to detect salient features might be much more strongly linked to autonomic functions that we currently largely ignore. Our brain mostly talks to itself and only occasionally we update the internal image based on the environmental signal (Fig. 1.4). We select the information before it gets to us depending on our brain state that is largely regulated by our autonomic nervous system that is reporting from our body. These inputs for the brain state control are very seldom considered for the localization studies. The very same neuronal network can produce very different patterns of activity depending on the influences from the hypothalamus, liver, or gut. This makes it extremely difficult to sustain our current definitions and categories for further research on localization of functions.

Whether we can go any further with allocating brain areas to cognitive functions than this is highly questionable. There are some very famous experiments by Marco Iacoboni's group who scanned the brains of people watching Super Bowl commercials and concluded that: ...A Doritos spot stimulated "mirror neurons" associated with empathy and connection, while an Emerald Nuts advert prompted little brain response....[59]

Networks of linked areas mediate higher functions in our brain. Representations of these are present across the entire organ, not just in cerebral cortex. For example, sensorimotor links require subcortical cerebellar and basal ganglia "switch boards" to link all parts of the cortex to motor areas. Some parts of the thalamus, the higher-order thalamic nuclei, are integral parts of cortico-cortical communication.

Our brain processes information through many parallel systems. For example, within the visual system, there are many parallel maps. This arrangement could give huge selectional advantage to the individual because it reduces the time needed for analysis. Grouping cells together that respond and process selective aspects of a stimulus in the sensory system might reduce the wiring needed to connect these cells. Perhaps the short connections can link them more efficiently and having less neurites for connections is metabolically less costly. Moreover, having topographical maps can allow easy lateral inhibition with minimal wiring.

There are not only regional differences within a cerebral hemisphere but also differences between the two hemispheres. If we observe a large number of brains, it is obvious that the left and right hemispheres have different dimensions and the two hemispheres have different folding patterns with slightly different sulci and gyri. It was Geschwind and Levitsky[60] who described that there is an asymmetry of the superior temporal lobes in human brain: the left planum temporale is several times larger on the left in 65% of brains; on the right it is larger in only 11%.[60] Geschwind also established that there is a right-left asymmetry of the sylvian fissure in both human and ape. It was Juhn Wada[61] who established methods to anesthetize hemispheres separately with intra carotid injection of sodium amytal, and this allowed him to test the independent functions of the left and right cerebral hemispheres. His studies revealed that language, for example, was mostly localized within the left hemispheres. To study differences in activation and connectivity of the two hemispheres, a combination of fMRI and diffusion tensor imaging has been used. These further established the left-right differences. Moreover, some of these studies suggested that there might be differences in lateralization in male and female brains. Some studies suggest that male brains are typically much more lateralized than female ones, although this is challenged by other studies. It is interesting to contemplate how the right and left hemispheres started to specialize in different functions during evolution. The left hemisphere allocated a large part to language at the expense of space recognition which is now more represented in the right hemisphere. The two hemispheres work in synchrony and these specializations can only be exposed when their connections are severed, and specific experimental tests are formulated to access the two hemispheres separately.

These differences in areal representations and hemispheric differences in the adult brain are spectacular and the question is how do they emerge? Do we have a developmental program to generate these differences or is it something that the environment imposes on the developing brain? Could it be the combination of both?

1.16.3 Comparative aspects of cortical areas

Human cerebral cortex is big and convoluted, but not the biggest and not the most convoluted brain in the animal kingdom. An elephant brain is much bigger, and a dolphin brain has more convolutions.[4] So, what is so special about the human brain? Isotropic fractionator methodology established by Herculano-Houzel and Lent[62] revealed that in the human brain we have around 20 billion neurons. That is the largest number of any species studied so far. In comparison, the elephant's much bigger brain has 10 billion loosely packed, chimpanzee 6.5 billion densely packed neurons.[63] If we compare a primate and a rodent brain with similar size it is becoming apparent that the primate brain has a much larger number of neurons, they are smaller and more densely packed. For instance, the Owl Monkey has a brain that weighs 16 g and it contains 1.5 billion neurons. The South American rodent, the Agouti has a brain of 18 g with 0.9 billion neurons.[64] This suggests that the primate brains have different scaling rules compared to rodents. The human brain follows the primate rules and fits into the general primate scheme. If we compare the number of cortical areas in a rodent, such as the rat with the brain of a primate, such as the owl monkey, then it is becoming apparent that the owl monkey has many more subdivisions within the cortex than the rat. The size of the brain is very similar in rat and owl monkey, yet the differences in the number of cortical areas is spectacular. Therefore the size of the brain is just one factor, but the number and packing density of the neurons might have a more profound impact on the type of cortical representations that can be accomplished in a brain.

There are relative differences in cortical organization in selected mammalian species.[65] Some species have larger cortical somatosensory representation, such as the nocturnal mouse. Opossum, a diurnal marsupial has a relatively large visual cortex, a ghost bat, that is using echolocation, a large auditory cortical representation. Cortical areas are allocated to modalities that are the most useful for those animals in that habitat (Fig. 1.20).

Representation in the brain is not constant, it can change even in the adult brain. There is a great deal of plasticity in representations, including in the cerebral cortex. If we repeatedly use 2—3 fingers for touch for half an hour a day for a few weeks, then the training expands the existing representation of the fingers in the cortex. When human subjects are trained to do a rapid sequence of finger movements for 3 weeks (10—20 min/day) they will show a larger region of activation with MRI during task; the change persisted for several months. In similar experiments a monkey was trained for 1 h/day to perform a task that required repeated use of the tips of fingers 2 and 3 and occasionally 4. Three months after training, the representation in area 3b was substantially larger as revealed with direct electrode recordings. There are extreme examples for altered cortical representations in professional musicians or sport personalities. Indeed, their brain activation in specific tasks can be very different from that of an individual who did not dedicate several hours a day from the age of 6 to play a musical instrument or master a particular sport. Also, there are great differences between individuals who have some sensory deficit from early stages.

Use of sensory receptor arrays determines the amount of cortex devoted to a particular sensory system (sensory domains). There are examples for unusual morphological specializations and the cortical magnification of the regions devoted to processing inputs. The large bill of the platypus is an important and extremely sensitive electro sensory organ to detect pray. The bill representation is enormous in the platypus cortex and spans several cortical fields (Fig. 1.20). Another extreme example is the star-nosed mole that has a structure arranged around the nostrils that consists of an array of 22 appendages (rays), 11 on each side. These rays are sensitive tactile sensors used to explore food items and the immediate surround. In the cortex, these rays form representations that appear band-like on tangential sections in both S1 and S2, similarly to the patterns we are more used to in the barrel cortex in the mouse. These examples emphasize the impact of peripheral sensory morphology on the functional organization of the neocortex. Both mammals have evolved specializations in peripheral morphology and use of specialized body parts, which is accompanied by changes in cortical organization.[65]

1.16.4 Development of cortical fields

One of the greatest questions in developmental neurobiology is to understand how cortical areas specialize in various functions. How are the cytoarchitectonic areas manipulated to produce the areal differences we discussed in Fig. 1.18? Are these cortical areas preprogrammed or are they the consequence of our interactions with our environment during early development?

There were two basic polarized ideas for cortical arealization. One emphasized the role of inbuilt genetic programs that produce these differences and another that

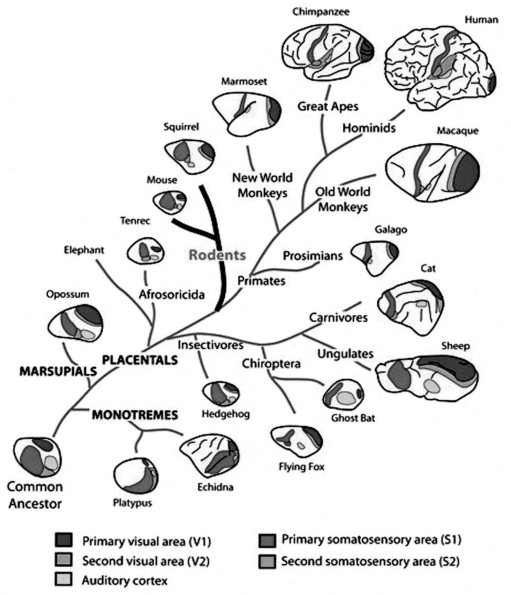

Figure 1.20 Relative proportions of cortical surface areas allocated to vision (*blue*) auditory functions (*yellow*), and somatic sensation (*red* and *pink*) in the brains of major orders of mammals revealed by the combination of electrophysiological, anatomical, and histochemical techniques. The evolutionary tree depicting the phylogenetic relationship of species studied with the organization of selected sensory fields on the lateral view of their left hemisphere (anterior left, medial is up, posterior right) and including the organization in the postulated ancestor. Certain cortical regions, such as S1, S2, A1, V1, and V2, are common to all mammals and most likely are homologous areas that arose from a common ancestor. *From Krubitzer L, Campi KL, Cooke DF. All rodents are not the same: a modern synthesis of cortical organization. Brain Behav Evol. 2011;78(1):51—93.*[65]

emphasized the role of environment. Rakic[17] observed that there are regional differences in the developing cortical neuroepithelium even before the completion of the neurogenesis. Since the neurons are produced in the ventricular and subventricular zone and then migrate radially to the cortex, Rakic proposed that the early differences in gene expression patterns in the germinal zone might provide a "protomap" of the future cortical areas. According to these ideas, some of the cortical areas are preprogrammed within the neuroepithelium before the generation and migration of cortical neurons, and before the connectivity of that area is established. Due to these programs, the areal differences would develop with or without external influences. Indeed, Pasko Rakic demonstrated that primary visual cortex developed in the absence of retinal input in monkeys where the retina was removed during early in utero life.[17]

According to the proponents of the environmental signals to determine arealization, Creutzfeldt, van der Loos and O'Leary, the cortical neuroepithelium itself does not contain the information to produce these regional differences. All cortical areas would have a general developmental pattern, "protocotex" and the regional differences must be imposed on the cortex by the external connectivity. According to this hypothesis, the cortex is a clean slate that has no predetermined goals; it is a "tabula rasa" that will acquire the regional differences as the result of the external environment. After decades of research, we now know that the truth is somewhere in between.[66] While there is strong evidence for the presence of molecular gradients and regional differences in the early cortical neuroepithelium and these differences emerge with or without external thalamic input, is also clear that they are not sufficient to form sharp boundaries of cortical areas or specialized cytoarchitectonic features in the absence of sensory input mediated through the thalamus. However, heterotopic transplantation and gene expression studies also demonstrated that the cortex in not a "tabula rasa," there are substantial areal differences from the earliest stages of development. The challenge is to dissect the intrinsic and extrinsic mechanisms that are involved in generating these differences.

I dedicated a large part of my scientific career to the understanding of thalamocortical development and plasticity.[5,6,15,67,68] The thalamus mediates almost all sensory input to the cerebral cortex from very early stages. Thalamocortical projections arrive at the telencephalon in a topographically organized fashion during embryonic development. At this stage, the earliest generated group of cells within the preplate is already split into subplate and marginal zone by the migrating cortical plate neurons (Fig. 1.5). Thalamic axons arrive at the cortex very early and start to accumulate below the cortex prior to the birth of most cortical neurons, and before their migration is complete. Therefore they can start to impose regional differences onto the cortex. It has been demonstrated that cortical neurogenesis, neuronal migration, neuronal differentiation, synapse formation, and pruning can be all altered by changing thalamocortical interactions. The subplate below the cortical plate is generated earliest and provides

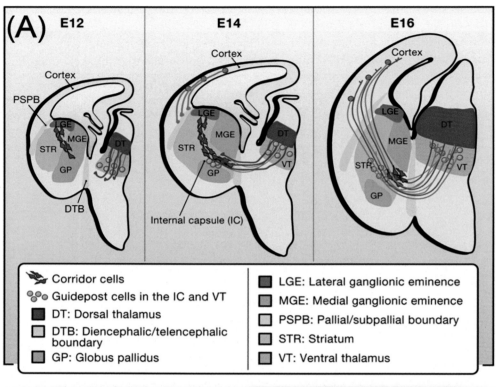

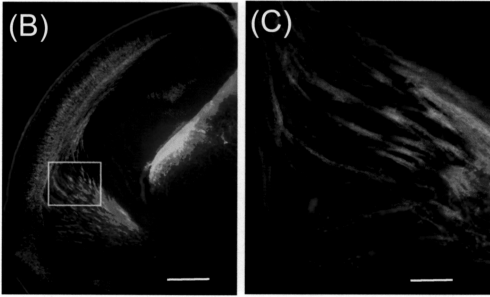

(Continued)

a relatively stable platform in the developing cortex during this period where the thalamocortical projections already form functional synapses to mediate environmental influences. Meanwhile the compartment of still migrating cortical cells, the cortical plate, is increasing in thickness, and new cells are added to it in an inside-first and outside-last fashion (Fig. 1.5). The time spent by thalamocortical axons within the subplate (so-called "waiting period") varies considerably among species and underwent a great protraction during evolution within the primate lineage. The bigger the brain, the longer the thalamocortical projections accumulate below the cortex, and the bigger the subplate is and the longer these influences can alter the cortical circuit assembly. The thalamic projections have areal specificity from the earliest stages.[5] The initial matching between the cells in the 3-dimensional volume of the thalamus to the 2-dimensional sheet of the cerebral cortex is established before the thalamic projections enter the cortical plate. These early connections can be rearranged and redefined before the final innervation pattern of the cortical plate is achieved by sensory deprivation, cortical, or thalamic lesions. Early manipulations of cortical or thalamic gene expression can also substantially rearrange the thalamocortical topography with an impact on cortical regionalization. However, the mechanisms that guide the thalamic projections and initiate their accumulation below the cortical plate are largely autonomous (Fig. 1.21).

◀ **Figure 1.21** Choreography of the cellular interactions during the development of early thalamocortical connectivity in mouse. (A) Diagrams illustrate three stages of the early establishment of reciprocal connections between dorsal thalamus and cortex in embryonic mouse. At E12–14, the prethalamus and internal capsule contain guidepost cells (*gray*) that already developed projections to the dorsal thalamus and they guide the thalamic projections towards the internal capsule through the diencephalic and telencephalic boundary. Corridor cells (*dark blue*) originate from the lateral ganglionic eminence at embryonic day 12 (E12) and migrate tangentially toward the diencephalon, where they form a permissive "corridor" for the thalamic projections (*red*) to navigate them through the internal capsule.[69] Perireticular cells have been proposed to regulate the entrance of tricarboxylic acids into the subpallium,[15] whereas corridor cells orient the internal pathfinding of tricarboxylic acids inside the medial ganglionic eminence.[69] Subsequently the early corticothalamic projections guide the thalamic projections through the pallial-subpallial boundary as originally proposed in the handshake hypothesis.[5,68,70] (B and C) The thalamocortical projections reach to the pallial subpallial boundary and cross this region by associating to the early corticothalamic projections mostly originating from subplate between E14 and 16. Thalamic projections were labeled with carbocyanine dye DiI (appear red) in a E15 Golli-tau eGFP mouse where the lower layers, layers 6b and some 6a express GFP.[71] The diagrams revealed the co-fasciculation of the thalamic and early corticofugal projections. Boxed area in (A) is presented with high power at (B). Scale 80 and 15 μm, respectively. *(A) From Hanashima C, Molnár Z, Fishell G. Building bridges to the cortex. Cell. 2006;125:8–10.[72] (B and C) From Piñon MC, Jethwa A, Jacobs E, Campagnoni A, Molnár Z. Dynamic integration of subplate neurons into the cortical barrel field circuitry during postnatal development in the Golli-tau-eGFP (GTE) mouse. J Physiol. 2009;587(Pt 9):1903–1915.[71]*

It is now clear that even before birth, our brain is not a blank page. When a newborn child opens their eyes for the first time, starts to cry to attract its mother's attention, the brain is already prepared to process visual, auditory, and somatosensory information. The sensory nervous system feeds into circuits that are preformed to some extent. Although it takes months to reach the full capacity of modality-specific sensory perception, the same neuronal modules and activity patterns are—from in utero stages—transiently present to generate an "operable" cortical network. At these early stages, there are transient neuronal circuits in both the cerebral cortex and the thalamus, which relay and process most of the information from our sensory environment. The early transient circuits both process and produce activity patters that are vital for further maturation of the central pathways. It is vitally important to better understand the transition from early transient to permanent neuronal circuits, as impairments at the earliest stage of brain development can lead to the development of neurological and psychiatric conditions several years or even decades after altered development.[6] These early impairments, such as intrauterine infection, drug exposure, or mild hypoxia, may potentially influence and disturb early brain activity and cause cortical miswiring at later stages of development (Fig. 1.13). The difficulty is that examining brains at the time of diagnosing a neurological or psychiatric disorder does not inform clinicians what happened during development. Therefore it is essential to understand which factors control and modify the early transient neuronal activity in the developing brain.

1.17 Remnants of the developmental scaffold have an important role in the adult, linking brain, body, and behavior

The subplate layer is a highly dynamic zone of the developing cerebral cortex that contains some of the earliest-born neurons of the cortex (Fig. 1.5).[6] The timing of subplate neuron birth, migration, distribution, and degree of programmed cell death has been analyzed in detail in rodents, carnivores, and primates.[16,73] The remaining subplate cells as layer 6b in the adult mouse and interstitial white matter cells in macaque and human were also well documented. Most of these studies examined the distribution of subplate cells labeled with the S-phase DNA replication marker tritiated thymidine ([3H]dT) and revealed the peak of neurogenesis and the period of highest cell death of these neuronal populations. Birthdating studies in macaque revealed that subplate neurons, after having completed their migration, become secondarily displaced inward by the arrival of subcortical and cortical axons.[73] These studies revealed that, although there are some uniform features, subplate shows remarkable variations in width between different cortical areas during development. Moreover, there are great regional variations in the amount of interstitial white matter cells that persist in various cortical areas.[74]

1.17.1 Unique connections formed by layer 6b

First-order thalamic nuclei receive inputs largely from peripheral sense organs and subcortical structures and relay this information to the cerebral cortex. Higher-order nuclei relay information from one cortical area to another and may occasionally receive subcortical input (Fig. 1.22). Layer 5 usually provides input to higher-order thalamic nuclei, via axon collaterals of subcortical projections targeting the superior colliculus and spinal cord, among others. These are powerful and large, feed-forward "driver" inputs for relay to other cortical areas. They may converge onto thalamic cells and also receive driver input from subcortical structures, but they do not form collaterals or synapse in the thalamic reticular nucleus (TRN) (Fig. 1.22). In contrast, layer 6a cells form axon collaterals and synapse with inhibitory neurons in the TRN and this has an impact on the frequency-dependent modulation of thalamic function.

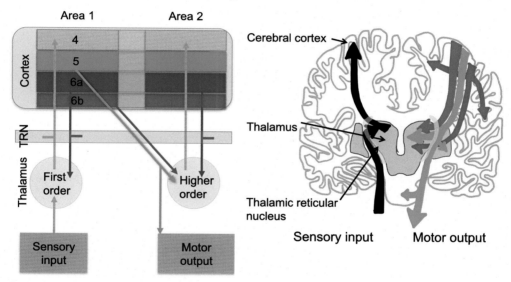

Figure 1.22 The corticofugal projections from layers 5, 6a, and 6b have distinct relationships to first order (also called relay) nuclei and higher-order (also called association) nuclei of the thalamus. Only the first-order thalamic nuclei receive direct input from the sensory periphery. The higher-order thalamic nuclei receive their input from the cortex and relay this back to other cortical areas, providing a route for trans thalamic cortico-cortical communication. Projections from cortical layer 6a provide abundant inputs to all thalamic nuclei and modulate how sensory information is relayed to the cortex. Layer 5 usually provides input to higher-order thalamic nuclei, via axon collaterals of subcortical projections targeting the superior colliculus and spinal cord, among others. These are powerful and large, feed-forward "driver" inputs for relay to other cortical areas. My laboratory discovered that a subgroup of layer 6b neurons (Drd1a-Cre line) preferentially target posterior thalamic nucleus, lateral posterior nucleus of thalamus, and other higher-order and midline thalamic nuclei in the ipsi- and contralateral thalamus, as well as frontal and lateral association cortices. Their axons do not form side branches in (pre)thalamic reticular nucleus. Our work suggests that there is a third type of cortico-thalamic projection in addition to layers 5 and 6a, from layer 6b.[75]

My laboratory has been investigating the role of subplate neurons in the development of the thalamocortical and corticothalamic circuits.[76] Since the remnants of subplate neurons are in layer 6b, I became interested in the differences between the upper and lower parts of layer 6. Layer 6b is very different from layer 6a; they have different connectivity, cell types and gene expression patterns.[16] The morphology of these layers is so different that even Cajal considered them as two separate layers, calling the bottom one layer 7.[14] My laboratory has been examining the input and output characteristics of these neurons using various tracing methods.[75] Our anterograde tracing experiments from a subgroup of layer 6b neurons in mouse revealed that they selectively and specifically target the higher-order thalamic nuclei (Fig. 1.22).[75]

Layer 6a projections have dynamic synapses in thalamus that depend upon their firing frequency.[77] Layer 6a projections stimulate the thalamic projection neurons directly and inhibit them through collaterals that stimulate the thalamic reticular nucleus. Depending to the balance of these influences, they modulate the sensory gating.[77] Layer 6b projections from Drd1a-cre+ neurons do not have side branches or synapses in TRN so do not make connections with inhibitory neurons.[75] These layer 6b projections could act as stimulatory counterparts of zona incerta projections that also selectively target the higher-order thalamic nuclei.[78] However, the layer 6b terminals in posterior thalamic nucleus (PO) are always small in contrast to layer 5 projections, which can be large and small in the Rbp4-cre line.[75] The higher-order thalamic nuclei are involved in local cortical state control. Layer 6b projections to the cortex and to the higher-order thalamic nuclei could form the anatomical substrate of the pathways that regulate which part of the thalamocortical circuits should be active and how transthalamic cortico-cortical communication is regulated.[79] These projections to the higher-order thalamic nuclei could be particularly important when there is contextual conflict, and more attention should be paid to a particular sensory input. For example, if we suddenly realize that something is "just not right" or "novel" in our environment, we must adjust the sensory gain to get better, more precise information.[79] Thus layer 6b projections and contacts with thalamic cells may open up transthalamic cortico-cortical communications especially through the cortical areas involved in higher cognitive processing (Fig. 1.21). These projections to the higher-order thalamic nuclei might enable processing of sensory input in a more global context.[79,80] The layer 6b projections to higher-order thalamic nuclei could be involved in local and global cortical state control.

1.17.2 Unique receptors expressed in layer 6b that can regulate state control

Lateral hypothalamus will be discussed in Tamás' chapter in detail. In lateral hypothalamus, there are neurons that produce a wake-promoting peptide, orexin. This peptide is essential for the stability of our arousal, lacking this peptide or its receptor can lead to sudden loss of arousal and sudden falling asleep. Neurons expressing the neuropeptides orexin-A and orexin-B are exclusively localized to the lateral hypothalamus and perifornical area (10,000−20,000 orexinergic neurons in the human brain) but they have wide projection targets across the central nervous system, including hypothalamus, thalamus, cortex, brain stem, and spinal cord. The projections to other hypothalamic neurons and subcortical arousal centers are important for modulating arousal, appetite, and activity of the hypothalamic−pituitary−adrenal axis. However, the roles of cortical projections remain less well understood.

The adult layer 6b neurons in rodents and the interstitial white matter neurons in primates are the only cortical neurons responsive to orexin[81,82] and layer 6b cells also selectively respond to the neuromodulators neurotensin[83] and cholecystokinin.[84] Layer 6b cells produce slow oscillations that could contribute to slow-wave sleep.[85] Layer 6b has been proposed to function as an orexin-gated feed-forward loop potentiating thalamocortical arousal,[86] based on its intracortical connections. However, there are also significant corticofugal projections from layer 6b, which may be of considerable importance (Fig. 1.23). Layer 6b has considerable similarities with claustrum (Bruguier et al., 2021).

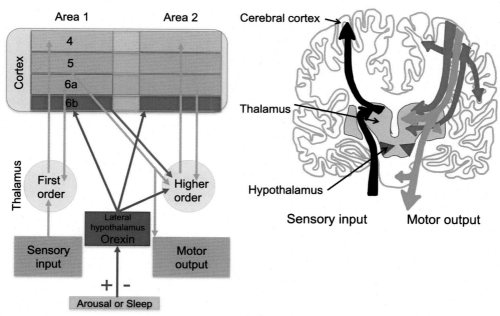

Figure 1.23 A subset of layer 6b cells are intrinsically bursting, responsive to neurotensin, dopamine, histamine, and noradrenaline, have a spiny, nonpyramidal shape, and are depolarized by orexin in a manner that stops intrinsic bursting.[81] It is of significance that both layer 6b neurons and their targets, the higher-order thalamic nuclei, are selectively sensitive to orexin.[81,86] This part of the thalamus falls asleep first when we go to sleep.

My laboratory collected strong evidence that a subtype of layer 6b (Drd1a-Cre+) neurons play a role in the cortical response to stress.[80] We "silenced" Drd1a-Cre+ neurons during postnatal development (P7 onwards) by Snap25 conditional KO (Drd1a-Cre: Snap25-floxed), which abolishes calcium dependent synaptic vesicle release. These mutants displayed normal locomotor activity when exposed to a novel environment, and circadian activity like those of control wild-type mice. However, silencing of Drd1a-Cre+ neurons led to a robust reduction in anxiety-like behavior, as measured in three different behavioral tasks (elevated plus maze, light-dark box, and food neophobia), suggesting that Drd1a-Cre+ neurons may act as a key component in the cortex for anxiety. Similar silencing of layer 5 neuronal population in the cortex (Rbp4-Cre::Snap25-floxed) produced a very different reduced sleep phenotype.[88] The findings in layer 6b "silenced" mice raises the possibility that a small population of cortical neurons that are the remnants of the subplate neurons regulate a range of behaviors dependent on cortical function and provide an important link between body and brain to determine appropriate behavioral response.[79] Orexin receptors have the highest expression in the frontal cortex. The medial prefrontal cortex (mPFC) plays a key role in integrating cognitive and affective information and is acutely sensitive to stress. In mouse experiments, during acute restraint-induced stress, some mPFC neurons show sustained increases in firing rate that outlast the stimulus, which could reflect peptidergic activation. The mPFC has reciprocal projections to the amygdala, hypothalamus, and other brainstem nuclei, that in turn enable top-down modulation of stress responses, and these mPFC circuits are thought to be disrupted in mood and anxiety disorders. The mPFC projections may generally have a net inhibitory effect on downstream targets and lesioning the mPFC leads to reduced measures of anxiety. Therefore the phenotype of reduced anxiety in the Drd1a-Cre::Snap25$^{fl/fl}$ mice is more likely to result from the silencing of Drd1a-Cre+ neurons in mPFC rather than sensory areas.

It is extremely interesting that a neuronal population that has been considered as a developmental vestige in the adult brain could have such a central role in linking autonomic, sensory, and cognitive functions in our brain. Layer 6b neurons are well suited to aid the detection of salient features of our external stimuli. Our brain selects the information that it would like to perceive before it gets to us depending on the state of your body. The circuits that sort foreground from the background are regulated by the autonomic nervous system and our body (Fig. 1.23). It is highly significant that the signals from our body can directly regulate the top-down management of sensory information processing. Our brain directs our eye and directs our hearing, and it is managing our peripheral input. The distribution of the receptors for the signals from the autonomic nervous system and directly from our body is positioned in highly specific locations where they influence the state control. This is not a coincidence, and we should aim to have a more global view of our brain function, including these important control mechanisms. What is surprising for me, for a developmental and evolutionary biologist, is that the elements that mediate these key interactions and hugely important roles in our state control have been largely considered as developmental remnants or left-over elements until very recently. These early generated and largely transient cells in the mammalian cerebral cortex might have conserved evolutionary origin, but they are repurposed and continue to have hugely important roles in the adult to synchronize states between our body and brain to produce the appropriate behavior.

1.18 Summary

Neuronal circuits form and change according to the information they process. There is a large degree of self organization during development. Just as on a computer board, the circuitry shows regional variations. But there is a great difference between how the computer board was put together and how the brain was assembled. The computer was put together by IBM or Apple, the brain was put together by itself based on a program that evolved over millennia and enabled it to integrate the signals provided by the environment. The computer cannot modify its connections dependent on the environment, our brain can. The computer cannot integrate new elements into the circuit, our brain can. In some species, these are very dramatic. One of the greatest tasks in developmental neurobiology is to understand how it is done.

References

1. Molnár Z. On the 400th anniversary of the birth of Thomas Willis. *Brain*. 2021;144(4):1033—1037.
2. Molnár Z. Evolution of cerebral cortical development. *Brain Behav Evol*. 2011;78(1):94—107.
3. Garcia-Moreno F, Molnár Z. Variations of telencephalic development that paved the way for neocortical evolution. *Prog Neurol*. 2020;194:101865.
4. Molnár Z, Pollen A. How unique is the human neocortex? *Development*. 2014;141(1):11—16.
5. Molnár Z. *Development of Thalamocortical Connections*. Springer; 1998. 264 p.
6. Molnár Z, Luhmann HJ, Kanold PO. Transient local and global circuits match spontaneous and sensory driven activity patterns during cortical development. *Science*. 2020;370(6514).
7. Stolp H, Neuhaus A, Sundramoorthi R, Molnár Z. The long and the short of it: gene and environment interactions during early cortical development and consequences for long-term neurological disease. *Front Psychiatry*. 2012;3:06.
8. Vasistha NA, Khodosevich K. The impact of (ab) normal maternal environment on cortical development. *Prog Neurobiol*. 2021. Available from: https://doi.org/10.1016/j.pneurobio.2021.102054.
9. Rakic P. No more cortical neurons for you. *Science*. 2006;313(5789):928—929.
10. Spalding KL, Bhardwaj RD, Buchholz BA, Druid H, Frisen J. Retrospective birth dating of cells in humans. *Cell*. 2005;122:133—143.
11. Salter MW, Stevens B. Microglia emerge as central players in brain disease. *Nat Med*. 2017; 23(9):1018—1027.
12. Molnár Z, Rockland K. Cortical columns. In: 2nd ed. Rakic P, Rubenstein J, eds. *Neural Circuit Development and Function in the Healthy and Diseased Brain: Comprehensive Developmental Neuroscience*. Vol 2. Elsevier; 2020 (Chapter 5).

13. Sherman SM, Guillery RW. The role of the thalamus in the flow of information to the cortex. *Philos Trans R Soc Lond B.* 2002;357:1695—1708.
14. Cajal SR. *Histologie dù Système Nerveux de l'homme et des vertébrés.* Maloine, París; 1909—1911.
15. Molnár Z, Garel S, López-Bendito G, Maness P, Price DJ. Mechanisms controlling the guidance of thalamocortical axons through the embryonic forebrain. *Eur J Neurosci.* 2012;35(10):1573—1585.
16. Hoerder-Suabedissen A, Molnár Z. Development, evolution and pathology of cortical subplate neurons. *Nat Rev Neurosci.* 2015;16(3):133—146.
17. Rakic P. Specification of cerebral cortical areas. *Science.* 1988;241(4862):170—176.
18. Bystron I, Rakic P, Molnár Z, Blakemore C. The first neurons of the human cerebral cortex. *Nat Neurosci.* 2006;9(7):880—886.
19. Molnár Z, Clowry GJ, Šestan N, et al. New insights into the development of the human cerebral cortex. *J Anat.* 2019;235:432—451.
20. Guillemot F, Molnár Z, Tarabykin V, Stoykova A. Molecular mechanisms of cortical differentiation. *Eur J Neurosci.* 2006;23:857—868.
21. Manger P, Slutsky D, Molnár Z. Subdivisions of visually responsive regions of the dorsal ventricular ridge of the iguana (Iguana iguana). *J Comp Neurol.* 2002;453:226—246.
22. Molnár Z, Butler AB. Neuronal changes during forebrain evolution in amniotes: an evolutionary developmental perspective. *Prog Brain Res.* 2002;136:21—38.
23. Molnár Z, Métin C, Stoykova A, et al. Comparative aspects of cerebral cortical development. *Eur J Neurosci.* 2006;23:921—934.
24. Meckel JF. *Beyträge zur vergleichenden Anatomie [Contributions to Comparative Anatomy].* Reclam; 1808.
25. Haeckel E. *Der Monismus als Band zwischen Religion und Wissenschaft, Glaubensbe- kenntniss eines Naturforschers.* Emil Strauss; 1892.
26. Darwin C. *On the Origin of Species by Natural Selection.* John Myrray; 1859.
27. von Baer, KE. Über Entwickelungsgeschichte der Thiere. Beobachtung und reflexion. [On the Developmental History of the Animals. Observations and Reflections]. Königsberg; 1828.
28. Montiel JF, Vasistha N, Garcia-Moreno F, Molnár Z. From Sauropsids to mammals and back: new approaches to comparative cortical development. *J Comp Neurol.* 2016;524(3):630—645.
29. Wolpert L. *The Triumph of the Embryo.* Courier Corporation; 2008:12. ISBN 9780486469294.
30. Silver DL, Rakic P, Grove EA, et al. Evolution and ontogenetic development of cortical structures In: Singer W, Sejnowski TJ, Rakic P, eds. *The Neocortex. Strüngmann Forum Reports.* Vol. 27; 2019.
31. Belgard TG, Marques AC, Oliver PL, et al. A transcriptomic atlas of mouse neocortical layers. *Neuron.* 2011;71(4):605—616.
32. Chodroff RA, Goodstadt L, Sirey TM, et al. Long noncoding RNA genes: conservation of sequence and brain expression among diverse amniotes. *Genome Biol.* 2010;11(7):R72.
33. Tosches MA, Yamawaki TM, Naumann RK, Jacobi AA, Tushev G, Laurent G. Evolution of pallium, hippocampus, and cortical cell types revealed by single-cell transcriptomics in reptiles. *Science.* 2018;360:881—888.
34. Belgard TG, Montiel JF, Wang WZ, et al. Adult pallium transcriptomes surprise in not reflecting predicted homologies across diverse chicken and mouse pallial sectors. *Proc Natl Acad Sci USA.* 2013;110(32):13150—13155.
35. Raff RA. *The Shape of Life. Genes, Development, and the Evolution of Animal Form.* The University of Chicago Press; 1996. ISBN 0-226-70266-9.
36. Lein ES, Belgard TG, Hawrylycz M, Molnár Z. Transcriptomic perspectives on neocortical structure, development, evolution, and disease. *Annu Rev Neurosci.* 2017;40:629—652.
37. Garcez PP, Stolp HB, Sravanam S, et al. Zika virus impairs the development of blood vessels in a mouse model of congenital infection. *Sci Rep.* 2018;8:12774.
38. Barbeito-Andrés J, Pezzuto P, Higa LM, et al. Congenital Zika syndrome is associated 1 with maternal protein malnutrition. *Sci Adv.* 2020;6(2). eaaw6284.
39. Vuong HE, Pronovost GN, Williams DW, et al. The maternal microbiome modulates fetal neurodevelopment in mice. *Nature.* 2020;586:281—286.
40. Wald NJ, Morris JK, Blakemore C. Public health failure in the prevention of neural tube defects: time to abandon the tolerance upper intake level of folate. *Public Health Rev.* 2018;39:2.

41. Steppan CM, Swick AG. A role for leptin in brain development. *Biochem Biophys Res Commun.* 1999;256(3):600—602.
42. Maynard TM, Sikich L, Lieberman JA, LaMantia A-S. Neural development, cell-cell signaling, and the "two-hit" hypothesis of schizophrenia. *Schizophrenia Bull.* 2001;27(3):457—476.
43. Vesalius A, 1543 De Humani Corporis Fabrica Libri Septem, Basel [On the Fabric of the Human Body in Seven Books].
44. Willis T, 1964 Cerebri Anatome cui Accessit Nervosum Descritio et Usus. [Anatomy of the Brain, with a Description of the Nerves and Their Function].
45. Ling ZD, Chang Q, Lipton JW, Tong CW, Landers TM, Carvey PM. Combined toxicity of prenatal bacterial endotoxin exposure and postnatal 6-hydroxydopamine in the adult rat midbrain. *Neuroscience.* 2004;124:619—628.
46. Molnár Z. Thomas Willis (1621—1675), the founder of clinical neuroscience. *Nat Rev Neurosci.* 2004;5(4):329—335.
47. Tyler KL, Malessa R. The Goltz-Ferrier debates and the triumph of cerebral localizationalist theory. *Neurology.* 2000;55(7):1015—1024.
48. Gowers WR, Klein E, Schaefer AE, Langley JN. Preliminary report. In: *Transactions of the International Medical Congress [Seventh Session].* Vol. 1; 1881:242a—242d, 243.
49. Langley JN, Sherrington CS. On sections of the right half of the medulla oblongata and of the spinal cord of the dog which was exhibited by Professor Goltz at the International Medical Congress of 1881 Proc. Physiol. Soc. *J Physiol.* 1885;5:vi.
50. Molnár Z, Brown RE. Insights into the life and work of Sir Charles Sherrington. *Nat Rev Neurosci.* 2010;11(6):429—436.
51. Brodmann K. *Vergleichende Lokalisationslehre der Grosshirnrinde.* Johann Ambrosius Barth; 1909 (in German).
52. von Economo C, Koskinas GN. *The Cytoarchitectonics of the Human Cerebral Cortex.* Oxford University Press; 1929.
53 Roy CS and Sherrington CS. On the regulation of blood-supply of the brain. J. Physiol. 1890;11 (1—2):85—108.
54. Dougherty RF, Koch VM, Brewer AA, Fischer B, Modersitzki J, Wandell BA. Visual field representations and locations of visual areas V1/2/3 in human visual cortex. *J Vis.* 2003;3:586—598. 2003.
55. Ungerleider LG, Mishkin M. Two cortical visual systems. In: Ingle DJ, Goodale MA, Mansfield RJW, eds. *Analysis of Visual Behaviour.* MIT Press; 1982. Chapter 18.
56. Goodale MA, Milner AD. Separate visual pathways for perception and action. *Trends Neurosci.* 1992;15(1):20—25.
57. Elston GN, Benavides-Piccione R, DeFelipe J. A study of pyramidal cell structure in the cingulate cortex of the macaque monkey with comparative notes on inferotemporal and primary visual cortex. *Cereb Cortex.* 2005;15:64—73.
58. Nishimoto S, Vu AT, Naselaris T, Benjamini Y, Yu B, Gallanta JL. Reconstructing visual experiences from brain activity evoked by natural movies. *Curr Biol.* 2011;21(19):1641—1646.
59. Iacoboni M. Who really won the super bowl? The story of an instant-scientific experiment. Edge; 2006. https://www.edge.org/conversation/marco_iacoboni-who-really-won-the-super-bowl.
60. Geschwind N, Levitsky W. Human brain: left-right asymmetries in temporal speech region. *Science.* 1968;161(3837):186—187.
61. Wada J. A new method for the determination of the side of cerebral speech dominance. A preliminary report of the intra-carotid injection of sodium amytal in man. *Igaku Seibutsugaki Tokyo.* 1949;14:221—222.
62. Herculano-Houzel S, Lent R. Isotropic fractionator: a simple, rapid method for the quantification of total cell and neuron numbers in the brain. *J Neurosci.* 2005;25(10):2518—2521.
63. Hart BL, Hart LA, Pinter-Wollman N. Large brains and cognition: where do elephants fit in? *Neurosci Biobehav Rev.* 2007;32(1):86—98.
64. Herculano-Houzel S, Collins CE, Wong P, Kaas JH. Cellular scaling rules for primate brains. *Proc Natl Acad Sci U S A.* 2007;104(9):3562—3567.
65. Krubitzer L, Campi KL, Cooke DF. All rodents are not the same: a modern synthesis of cortical organization. *Brain Behav Evol.* 2011;78(1):51—93.

66. Cadwell CR, Bhaduri A, Mostajo-Radji MA, Keefe MG, Nowakowski TJ. Development and arealization of the cerebral cortex. *Neuron.* 2019;103(6):980—1004.
67. López-Bendito G, Molnár Z. Thalamocortical development: how are we going to get there? *Nat Rev Neurosci.* 2003;4:276—289.
68. Molnár Z. Development and evolution of thalamocortical interactions. *Eur J Morphol.* 2000;38:313—320.
69. López-Bendito G, Cautinat A, Sánchez JA, et al. Tangential neuronal migration controls axon guidance: a role for neuregulin-1 in thalamocortical axon navigation. *Cell.* 2006;125:127—142.
70. Molnár Z, Blakemore C. How do thalamic axons find their way to the cortex? *Trends Neurosci.* 1995;18:389—397.
71. Piñon MC, Jethwa A, Jacobs E, Campagnoni A, Molnár Z. Dynamic integration of subplate neurons into the cortical barrel field circuitry during postnatal development in the Golli-tau-eGFP (GTE) mouse. *J Physiol.* 2009;587(Pt 9):1903—1915.
72. Hanashima C, Molnár Z, Fishell G. Building bridges to the cortex. *Cell.* 2006;125:8—10.
73. Kostovic I. The enigmatic fetal subplate compartment forms an early tangential cortical nexus and provides the framework for construction of cortical connectivity. *Prog Neurobiol.* 2020;194. 101883.
74. Swiegers J, Bhagwandin A, Williams VM, et al. The distribution, number, and certain neurochemical identities of infracortical white matter neurons in a chimpanzee (Pan troglodytes) brain. *J Comp Neurol.* 2021;529(14):3429—3452.
75. Hoerder-Suabedissen A, Hayashi S, Upton L, et al. Subset of cortical layer 6b neurons selectively innervates higher order thalamic nuclei in mice. *Cereb Cortex.* 2018;28(5):1882—1897.
76. Bandiera S, Molnár Z. Development of the thalamocortical systems. In: Halassa M, ed. *The Thalamus*; 2021.
77. Crandall SR, Cruikshank SJ, Connors BW. A corticothalamic switch: controlling the thalamus with dynamic synapses. *Neuron.* 2015;86:768—782.
78. Mitrofanis J. Some certainty for the "zone of uncertainty"? Exploring the function of the zona incerta. *Neuroscience.* 2005;130:1—15.
79. Molnár Z. Cortical layer with no known function. *Eur J Neurosci.* 2019;49(7):957—963.
80. Guidi LG, Korrell KV, Hoerder-Suabedissen A, et al. Functional role of cortical layer VIb in mouse behavior. *Soc Neurosci Abstr.* 2016;634:16.
81. Bayer L, Serafin M, Eggermann E, et al. Exclusive postsynaptic action of hypocretin-orexin on sublayer 6b cortical neurons. *J Neurosci.* 2004;24:6760—6764.
82. Wenger Combremont A-L, Bayer L, Dupré A, Mühlethaler M, Serafin M. Slow bursting neurons of mouse cortical layer 6b are depolarized by hypocretin/orexin and major transmitters of arousal. *Front Neurol.* 2016;7:88.
83. Case L, Lyons DJ, Broberger C. Desynchronization of the rat cortical network and excitation of white matter neurons by neurotensin. *Cereb Cortex.* 2016;27:2671—2685.
84. Chung L, Moore SD, Cox CL. Cholecystokinin action on layer 6b neurons in somatosensory cortex. *Brain Res.* 2009;1282:10—19.
85. Crunelli V, David F, Lőrincz ML, Hughes SW. The thalamocortical network as a single slow wave-generating unit. *Curr Opin Neurobiol.* 2015;31:72—80.
86. Hay YA, Andjelic S, Badr S, Lambolez B. Orexin-dependent activation of layer VIb enhances cortical network activity and integration of non-specific thalamocortical inputs. *Brain Struct Funct.* 2015;220:3497—3512.
87. Bruguier H, Suarez R, Manger P, et al. In search of common developmental and evolutionary origin of the claustrum and subplate. *J Comp Neurol.* 2020;528(17):2956—2977.
88. Krone LB, Yamagata T, Blanco-Duque C, et al. A role for the cortex in sleep—wake regulation. *Nat Neurosci.* 2021;24:1210—1215.

CHAPTER 2

Tamas Horvath: The hunger view on body, brain and behavior

On the following pages, I will give my personal view on brain, body, and behavior that originates from and follows my professional development. My approach is entirely different from the chapters by Zoltán Molnár and Joy Hirsch who have written scholarly and in crystal clarity about brain development and brain—brain interactions, respectively. I came to brain research on a very different trajectory, and I am limited in my ability to speak as scholarly and eloquently as Molnár and Hirsch about the topic that I have pursued, which is the relationship between metabolism and brain functions. This is a relatively new field within neuroscience as the molecular underpinnings of brain control of metabolism started to be discovered less than 30 years ago. Because my work on the specific brain circuits in metabolism regulation predates these discoveries by a few years, I decided to narrate the topic from my point of view, which, by default, is subjective, quasi-autobiographical. My goal here is not to pontificate my wisdom, but rather to provide one conceptual framework out of countless alternatives in a manner that fits my experience. With this, I would also like to provide an alternative example for trainees on the approach to research.

The biggest impact on my professional path came from books I read in high school written by great scientists. One was *The Double Helix: A Personal Account of the Discovery of the Structure of DNA* by James Watson (1968, Atheneum), and the other was from Hans (János) Selye: *The Stress of Life* (New York: McGraw-Hill, 1956). While I learned about scientific information from those books, what affected me more was the way those authors narrated their discoveries and their philosophy of research. From those early years of my life and throughout my journey until today, one of the most important take-home messages has been that there is no blueprint for success in our way of life regarding either meaningful discoveries or professional achievements. That is the "beauty and the beast" of our business. Because of this, I will not try to emulate any of the aforementioned authors or other books on science and research, rather I will write the way it comes from my point of view. I will focus on a very specific aspect of our existence, which is our necessary interest in food. One could make an equally if not more persuasive argument for the need to breed, to drink, to reproduce, to escape from a predator, and so on. The bottom line is that my argument emphasizes the idea to approach the brain and its most sophisticated functions from

Body, Brain, Behavior
DOI: https://doi.org/10.1016/B978-0-12-818093-8.00009-4
67

the outside in and from the bottom up. I am not arguing that this "idea" is a superior approach, rather it is far from it. However, I believe that the door must be open for diverse viewpoints and approaches in the pursuit of the unknown. Otherwise, the last 100 years would have solved much of our problems in biomedicine. Well, it did not. The second part of this book, which highlights many of our conversations from the three different angles of Joy Hirsch, Zoltán Molnár, and me, is a testament to this. Primarily being a researcher, whose fundamental if not sole responsibility is to pursue the unknown, I retain my basic right to be wrong. Thus any declarative sentence you find later, you will have to take with a grain of salt.

2.1 What is the brain?

I take the global point of view that the entire body is a brain and the content of the cerebral cortex, for example, is a component of the mind, brain, and body system. However, the general popular consensus is that it is the brain that makes us who we are. For thousands of years, from the earliest written texts to the societal and scientific views of today, the brain has been viewed as the center of our intelligence and the driver of the development and success of humankind. Over the last century, increasingly sophisticated methods have been brought to bear on the investigation of the inner workings of the brain. The good, the bad, and the ugly of our existence and the interactions amongst ourselves and with our environment have been attributed to the remarkable complexity and beauty of our brains. Undoubtedly, we all feel and believe that our experience in life is coded and decoded by our brains. We are convinced by the fact that we know that we are and that we can conceptualize past, present, and future and recall, at will, episodes from the past because of our brain. This is a reasonable and logical assertion. There is little uncertainty that these attributes, as well as our ability to express ourselves via speech and writing, hinge entirely on our having a central nervous system, and the brain. There is no doubt that the spinal cord as well as the peripheral nervous system are crucial for our existence, but other than writing, they are dispensable for all the other attributes mentioned earlier. It is not that they do not play a role in the complexities of our behaviors, it is that "higher" brain processing and consciousness do not hinge on these parts of the nervous system (see subjects with spinal cord injury).

We learned a great deal about the regional specification of the brain (Chapters 1 & 2). We know how many cells (neurons and glia cells) reside in our brain and how that number differs between us and other animal species with diverse abilities. We also developed a great sense about the cellularity of the various brain regions and how these cells communicate to one another (Chapters 1 & 2). Tools and approaches emerged to interrogate how a behavior is evoked and how the brain impacts peripheral tissue functions. Fundamental principles on how we sense the environment via visual, auditory, olfactory, and tactile stimuli have been uncovered. We made giant steps to comprehend how the central nervous system controls body temperature, pH, nutrient, and hormonal milieu of the organism and keeps them in a relatively steady state (homeostasis). The role of the cerebral cortex in making us who we are is also undisputed. In fact, it is the complex cytoarchitecture, connectivity, and its developmental attributes that may explain the differences in complex behaviors between humans and other species.

Despite all these discoveries, our ability to exploit these advances to better understand how the brain functions, and more importantly, why it dysfunctions remain limited. There is a small chance that the reason for this is that we are looking for "the lost key" at the spot on which the light shines rather than the location of where it was lost. In the following paragraphs, I will attempt to approach the brain and its relationship with the rest of the body in an unorthodox, albeit not entirely new, way with the hope that it can trigger others to also study the brain using an outside-of-the-box approach. There is no attempt here to pretend that a truth will emerge, rather my goal is to suggest that "reboots" might help lead us to new territories of brain research. Let us start with a simple question: what is the brain?

The brain is an organ that processes information from within the body and from the outside world, and with this information, it evokes adaptive movements within the body and the environment. This is the primary function of the central nervous system. What differentiates complex organisms with brains, including humans, from those without brains, such as trees, is the coordinated and adaptive locomotor responses that they can predictably manifest. These movements relate to the pursuit of oxygen, food and water, reproduction and to escape from predators and dangers that jeopardize their survival. While there has been considerable fascination about the brain and its function by scientists and the lay public, this tissue is not superior to any others of the body originally classified by anatomy and then by function. In essence, the brain is not "more" in isolation than the liver, the musculoskeletal system, the white fat, the pancreas, the kidney, or the heart. Removal of any one of these tissues terminates life. From this perspective, the brain is no more and no less than these other organs. Where the brain, as an anatomical entity, differs from the other organs (by the current superficial, state-of-the-art perspective) is its crucial role in coordinating the activity of the various organs to the environment and evoking adequate behavioral (locomotor) responses. This together with the so-called higher brain functions, such as learning, memory, decision-making conceptualization, consciousness puts the brain into a "special" category. It may be, however, that this is a misconception, one that has not led our understanding of the brain much closer than the revolutionary concepts about the central nervous system that emerged in the late 1800s and early 1900s. In the following segments, I will aim to provide an alternative conceptual framework of how the brain integrates with the rest of the body in association with its physiological processes and malfunctions.

2.2 Conceptual framework

Since my early years in veterinary school, I have wondered what makes us different from other animals, specifically in association with our brains and behaviors. For most of my life, I have been under the "impression" that we humans are "special," somehow an advanced product of evolution attested by all the "accomplishments" of human societies. In part because of my work but also due to an appreciation of other single and multicellular organisms, my belief in the superiority of human behavior as the epitome of evolution has been impacted and the notion of the preeminence of the human brain has been challenged. Later, I will go through an arbitrary set of subjects to illustrate how and why these changes occurred in my way of conceptualizing the brain and its role in the promotion of success in the environment. These paragraphs were written over the past 3 years, the last of which was under the "supervision" of the COVID epidemic. That alone has been an amazing experience with all its impact on our lives. To the point of this book, it was remarkable to see how a virus, an organism without a "brain" or anything close to it, could overtake human society around the globe at every level. Despite the potential to restrain this epidemic by man-made inventions (vaccines), it is clearer than ever that biological principles way beyond the realm of any nervous system is capable of organizing and successfully self-propagating at the expense of any "higher" intelligence. Perhaps one day these principles will aid our efforts to better understand ourselves and our brain.

Despite the enormous effort and resources devoted to neuroscience, we know little more about fundamental brain-assigned functions, such as consciousness and schizophrenia, then great minds assumed about them one hundred or even hundreds of years ago. There is always the premise that the next big thing in neuroscience will solve a crucial piece of the puzzle and make breakthroughs that benefit the sick and society at large. For more than 30 years the spending of governments, most notably the US Government, have been exponentially increasing on brain-related issues, be they about how the brain develops, works, what the underlying cause of Alzheimer's disease, dementias, Parkinson's disease, schizophrenia, etc. might be. No breakthrough answers, no breakthrough solutions, and no breakthrough medications.

This lack of progress is accompanied by enormous advances to better understand specific processes and mechanisms related to neurons, neuronal connections, glia cells, endothelial cells, pericytes, and how they may be affected by genetics, epigenetics, the environment, and many other factors. For example, we understand how specific ion channels can affect ion flow in and out of various brain cells affecting many attributes of neurons and glial cells. Cellular biological discoveries advanced our understanding of how cells package, transport, release, and redistribute new and old molecules within and between cells. Through opto- and chemogenetics, we can see how signals can be selectively initiated at any part of the brain with impact on many brain functions, including control of complex behaviors. While these are remarkably elegant advances of fundamental biology of the brain, they did not solve the problem or offer feasible explanations for the aforementioned brain-associated complex functions and disorders. The promise has always been there, and, frequently a 5—10-year timeframe is being offered for the solution to be delivered as long as the money flows. Despite all these hopes and promises delivered with good intentions, sincere beliefs, and enormous "minds," there have been no solutions.

Because of the above-mentioned lack of outcomes regarding "higher" brain functions, I believe that the conceptual framework of how we view the brain will need to change dramatically to deliver fundamental breakthroughs.

A recent book by one of the most ingenious neuroscientists of the past 100 years, György Buzsáki, is entitled *The Brain From Inside Out* (Oxford University Press, 2019). Buzsáki provides an absolute tour-de-force, futuristic take on the inner working of the brain from the frontiers of contemporary neuroscience. His work and concepts on oscillation, their origin and impact on behavioral output is what may provide the closest connection of systems of neuroscience to behavioral psychology. I posit that his views are the culmination of the Neuronal Doctrine most famously popularized by Ramon Y Cajal. Whether such brilliant views are steppingstones to a more insightful comprehension of the brain machinery for the enhancement of solving problems of the mind, or they represent yet another brilliant intellectual advancement that maintains or increases the distance to the "mirage" of "solving" the brain remains to be seen.

2.2.1 Lessons from Cajal and Golgi

To be provocative (and a bit arrogant), I suggest that the Neuronal Doctrine that emerged in the late 1800s and was solidified in the beginning of the 20th century is the underlying cause for the lack of fundamental breakthroughs in brain research over the past 100 years. This doctrine was developed by a handful of scientists, among whom the most well-known and arguably most insightful and influential was Santiago Ramon y Cajal.[1] This Spanish physician—scientist was studying the cytoarchitecture and structure of the brain and the nervous system in general with the use of contemporary and breakthrough technologies, including the modified silver impregnation technique developed by Camillo Golgi (an Italian contemporary of Cajal, an outstanding neuroscientist in his own right).

With Golgi's technique, Cajal analyzed brains, spinal cord, and various tissues to study "nerve" cells. He made critical observations that made him conclude that neurons are individual cells, which receive inputs on one side (dendrites), and then convey the received information to another neuron via another, efferent process (the axons). He concluded that neurons are independent units, which interact with one another via physical contacts (synapses), and that information traveling from one individual cell to another via a complex network of neurons determines the functional output of the brain. In a nutshell, this is the Neuronal Doctrine.

The Neuronal Doctrine is extremely logical and attractive as it explains the functionality of the brain like an engineering diagram. The simplicity and ingenuity of this postulation became the fundamental tenet of contemporary neuroscience and brain research. It was so well received even back then that Cajal and Golgi shared the Nobel Prize for Physiology or Medicine in 1906 for these hypotheses (without proof).

Cajal was an amazing genius: every prediction he made (none of which he provided evidence for) was subsequently proven experimentally by others in the following 100 plus years. It was a remarkable accomplishment of a loner, who lived in a society and environment that placed little value on scientific endeavors. It was a time of difficult communication of views via publications, and in Spanish no less. He traveled Europe and visited noted scientists with his microscope and slides to discuss and convince them of his views. He published papers with excruciating details. Any conceptual breakthrough one attributes to a researcher of any given era subsequent to Cajal can almost always be found in Cajal's own writings and predictions. He foresaw how the Neuronal Doctrine would lead to information flow from point "A" to "B" to "C" to "D" and so on. Every single one of us who was educated in brain research has been indoctrinated by Cajal's works and concepts. We take it for granted that when we study something (within the framework of the Neuronal Doctrine that was ingrained upon us) and find something "new," it will add, even if incrementally, to the solving of the "puzzle" of how the brain works. Well, it has not happened so far. Maybe because we are only at letter "S" or maybe there's another reason.

Alternatively, could it be possible that the fundamental tenet of the Neuronal Doctrine, while an intellectual tour-de-force and a result of ingenious logic, is not the appropriate conceptual framework for the understanding of how the brain works?. Remarkably, this very question was raised more than 100 years ago by Camilo Golgi.

In his Nobel lecture in 1906, Golgi made remarkable (based on today's "state-of-the-art"), stunning comments regarding his view of the state-of-the art regarding the brain: "It may seem strange that, since I have always been opposed to the neuron theory — although acknowledging that its starting-point is to be found in my own work — I have chosen this question of the neuron as the subject of my lecture, and that it comes at a time when this doctrine is generally recognized to be going out of favor." "I admire the brilliancy of the doctrine which is a worthy product of the high intellect of my illustrious Spanish colleague, I cannot agree with him on some points"... "I must declare that when the neuron theory made, by almost unanimous approval, its triumphant entrance on the scientific scene, I found myself unable to follow the current of opinion, because I was confronted by one concrete anatomical fact; this was the existence of the formation which I have called the diffuse nerve network. I attached much more importance to this network, which I did not hesitate to call a nerve organ, because the very manner in which it is composed clearly indicated its significance to me."

In retrospect, with the eye of a neuroscientist, Golgi's conclusions are astonishing, which appear to be defying facts (and the assumed reality). It is important to note that Cajal and Golgi drew their different conclusions looking at and depicting the same things under the microscope (Fig. 2.1). Golgi's notion that the theory was going out of favor back in 1906 is also surprising considering that in the subsequent decades Cajal's postulations were pursued and proven to hold ground establishing an undisputed axiom of today's neuroscience, which is the Neuronal Doctrine. While the facts seemingly favor Cajal's views, as noted before, concepts, strategies, studies, and results built on the Neuronal Doctrine, to date did not lead to a breakthrough in our understanding of how the brain works. While Golgi's fact-based argument may be flawed ("I was confronted by one concrete anatomical fact; this was the existence of the formation which I have called the diffuse nerve network") as electron microscopy eventually showed the physical boundaries between individual neurons, a "truth" may lie in a hybrid of Colgi's and Cajal's views.

As a side note, it is worthy to mention that to date, we still do not understand how the Golgi method works.[2] There are some rudimental explanations, but the fact is that the labeling of a preparation of Golgi staining is seemingly random and does not include most of the neurons in the preparation. The relevance of this hiatus is obvious, when we look at the huge numbers of studies that utilized this technology to compare neurons taken from different animals or postmortem human samples from

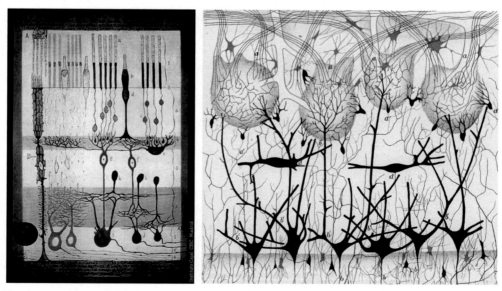

Figure 2.1 *Left:* Outline of the structure of the mammalian retina by Cajal, from *Structure of the Mammalian Retina* (Madrid, 1900) *Right:* Nerve cells in a dog's olfactory bulb (detail), from Camillo Golgi's *Sulla fina anatomia degli organi centrali del sistema nervoso* (Napoli, Milan, Pisa; 1886).

different experimental, physiological, and/or pathological conditions. Many of these studies contributed to the establishment of new theories and even subfields in neurobiology without really having a scientifically sound base. These included analyses of cell size, characterization of dendritic arbors, and synapses as well as axonal projections and axon terminals. The issue with conclusions made using the Golgi approach in these comparative circumstances is that they may be entirely flawed due to the "random" nature of the labeling of neurons by this technique so that potentially different populations of neurons were being compared in the various experimental, physiological, or pathological conditions. The necessity to understand the molecular principles of Golgi staining may seem mute considering technological developments in labeling techniques of populations and subpopulations of neurons in all parts of the brain and spinal cord. However, beyond the intellectually rewarding aspect of understanding of how and why the Golgi silver impregnation labels specific cells, that mechanism may deliver new insights and principles of brain functions.

2.2.2 Regional specification of brain functions

Another fundamental tenet of contemporary neuroscience, which also arises from works of Cajal and others, is the regional specification of the brain regarding behavioral and autonomic control. Regarding this, specific brain regions are assigned to support specific functions. The simplest regional specifications are those of cortical functions relating to sensory inputs and motor control. For example, in the most complex part of the cortex, there are defined regions for decoding visual, auditory, olfactory, and somatosensory inputs (see Chapter 1: The developing brain). The thalamus is divided into regions subserving inputs to and from the cortex and lower areas with emphasis on specific functions. Equally well defined are the cerebellum with its architectural beauty and the basal ganglia both associated with motor control. And there are also specific regions (with much less architectural simplicity) in "ancient" parts of the brain, in the hypothalamus, brain stem, and spinal cord that control various homeostatic functions. These latter functions include the regulation of circadian rhythms, pituitary, temperature, feeding, breathing, heart rate. In between, there are various areas responsible for appropriate information flow between brain regions, proper feedback, and feedforward control regions. Spatial learning and memory processing is associated with the dentate gyrus and hippocampal formation. Reward-like behaviors are tied to the midbrain dopamine system, which is also fundamentally important for fine motor control. The amygdala is the "heart" of emotions, and the prefrontal cortex is tied to working memory, decision-making, and other complex behaviors. Detailed lists of these and other regional specifications can be found in thousands of original contributions and also in hundreds of books written over the past 100 years. At the same time, it is also becoming clear that the boundaries of such arbitrary separations of brain regions and specific, arbitrary defined behaviors, and other functions will not hold the key to a better comprehension of the brain and its disorders. I will take one example, which is the control of eating. This theme is central to my view of the brain as it will become clear from the next segment.

2.3 Eating: linking the environment to the body and brain

Appetite control by the brain has been on the fringe of neuroscience for decades. This is despite the fact that major discoveries about critical aspects of the central nervous system have been more frequently than not connected to appetite (see the work of Pavlov and the vast majority of discoveries in cognitive neuroscience using animal models, including nonhuman primates). By and large, religion as well as intellectualism marginalized eating as a primitive homeostatic "need," which has little to do with the "so superior" process of "thinking" and other "higher functioning" of the brain. Perhaps it was also fueled by the historical dogma of overeating being one of the seven deadly sins. Fast forward to today: appetite and feeding is predominantly regulated by peripheral signals mediated by all brain regions with a dominance of subcortical, ancient parts of the brain. However, interference with almost any part of the brain will have more or less an impact on eating. In the same vein, when a function is dominantly connected to a cortical region (e.g., decision-making by the prefrontal cortex), for a better understanding of physiological and pathological aspects of that behavior, the impact of those "primitive" parts of the brain that have the highest impact on eating will need to be considered. Some examples of this will be provided later. It is more than likely that with a diminished insistence upon regional-specificity of brain functions and dysfunctions, a better comprehension of physiological and pathological brain activities will emerge.

Our main way of fuel intake is via an oral consumption of energy sources, that is, eating food. Lesion studies from more than 50 years ago identified very specific hypothalamic brain areas as positive and negative controllers of eating.[3–8] Lesioning the ventromedial hypothalamic nucleus resulted in massive obesity, while the destruction of the neighboring lateral hypothalamus stopped animals from eating entirely. Consequently, these responses were assigned to those specific brain regions that control feeding. The ensuing five decades using classic neurobiological approaches and transformational genetic discoveries regarding feeding control identified subpopulations of neurons within the hypothalamus that are outside of the ventromedial and lateral hypothalamic nuclei that play indispensable roles in hunger and satiety regulation.[9] Discoveries of the last two decades also provided convincing evidence that areas outside of the hypothalamus, including multiple sites in higher brain regions, including the prefrontal cortex, contribute significantly to how much we and animals eat.[10] At the same time, subpopulations of neurons in the hypothalamus assumed to be involved primarily in eating regulation have been found to significantly contribute to the regulation of other complex behaviors, such as reward seeking, stereotypy, aggression, sexual behaviors, and more. Further details are provided on these and other processes below. However, it is realistic to say that the idea of regional specification of the brain as it was originally conceived is changing and moving to a more dynamic conceptual framework that is less "organized" and more "fluid" than that previously considered. Nevertheless, the hypothalamic centers of eating are a dominant force in behavioral control. The unmasking of fine details of various circuits was the product of decades of granular work boosted by genetic discoveries in the mid-1990s.

For over 30 years, I have been studying two peptidergic neurons in the hypothalamic arcuate nucleus (ARC), one that produces neuropeptide Y (NPY) and agouti-related peptide (AgRP), and the other that produces proopiomelanocortin (POMC)-derived peptides. The involvement of the hypothalamic NPY neurons in the regulation of feeding was originally proposed by the very first study I initiated on June 4, 1990, 4 days after arriving in the United States from Hungary. My supervisor, Dr. Csaba Leranth, a Hungarian neuroanatomist at Yale, gave me the material and said: "do something with it and make a paper out of it."

2.3.1 Hypothalamus, the stepchild of neuroscience and the home of hunger and satiety

The root and conceptual framework of contemporary neuroscience does not originate in the hypothalamus. To simplify my point, the hypothalamus and its related functions has occupied a no-man's land. A territory that was neither prioritized by contemporary neuroscience relating to it is being part of the brain nor by internal medicine, whose focus overlaps the functional relevance of the hypothalamus relating to endocrinology and autonomic control. From the perspective of contemporary, state-of-the-art neuroscience, the structural complexity and functional ambiguity of any given hypothalamic neuronal population or circuitry did not lend itself to easy testing and proof of the tenets of the Neuronal Doctrine. The beautifully organized hippocampus, cortices, and sensory pathways have been much more attractive for those seeking the illusion of an a-to-z interrogation and solution of a neurobiological problem that is so special for mammals, including humans. Not surprisingly, much knowledge of hypothalamic structure and function came from medically oriented endocrinologists, who, unlike traditional researchers, had less innate rigor about the Neuronal Doctrine. The fundamental issue/problem with hypothalamic circuitries from the perspective of a neuroscience "purist" hinging on technology is that the same pathway subserves multiple functions. In fact, the computational principles of hypothalamic circuits are not known. It can be pursued artificially, out of true context, when a single function is investigated (such as feeding), but it will likely provide very little, if any, insights regarding how the hypothalamus can organize predictable and aligned behavioral, endocrine, and autonomic outputs.

My professional development fits perfectly and is benefitted tremendously from this "profile" of the hypothalamus. I am a failed 3rd generation veterinarian (both my father and his father were successful practicing farm veterinarians) whose lifeline was to come to Yale to the Department of Obstetrics and Gynecology to study the neuroanatomy of the hypothalamus in relation to reproductive neuroendocrinology. The project I started just 4 days after arriving at Yale was to explore the anatomical relationship between two subpopulations of neurons in the hypothalamic ARC, one that produces NPY and the other POMC.[11] The premise of the study was that the interplay between these two specific subsets of neurons may be important for the regulation of ovarian cycles as, at that point, the crucial neurons in the control of ovulation, the hypothalamic luteinizing hormone-releasing hormone neurons (LHRH), were believed to be indirect targets of the ovarian hormone, estradiol. The notion was that because NPY and POMC neurons were earlier found to synaptically connect with LHRH cells in the medial preoptic area, it may be that these two ARC peptidergic neurons convey the crucial estrogenic signal for ovulation triggered by a surge of LHRH release into the portal circulation of the median eminence. Thus I examined the synaptic interactions between NPY and POMC neurons.

The reason I was brought to New Haven from Budapest was because I had some training in electron microscopy at the Vet School in Budapest by Ferenc Hajós, a human anatomist himself trained by János Szenthágothai, probably the most influential Hungarian neuroscientist in the world in the 20th century. Hajós had been friends with Csaba Léránth at Yale from their time together in the 1970s and early 1980s in the department of Szenthágothai. Csaba needed reliable and inexpensive labor from a trusted source, hence, my recruitment. Csaba Léránth was an amazing electron microscopist. There were several such Hungarian electron microscopists of his generation, including Péter Somogyi, Miklós Palkovics, József Hámori, István Záborszky, László Seres their offspring, including Tamas Freund. Csaba stood out by his dedication to the quality of the work; he was a true craftsman. He was trained as a dentist, fixed and raced cars, so his approach to details was second to none. Being trained by him and one of his very talented postdocs, Robert Jakab, was one of the luckiest events of my life. It was a classic apprenticeship: he was uncompromising regarding quality and refused to offer shortcuts or to do things for me. I learned the craft by trial and error, something that has lost popularity in the recent past.

I did what was the state-of-the-art at the time regarding investigation of the synaptic interaction between specific populations of neurons: the combination of light and electron microscopy. To be very truthful, this methodology is still one of the most unambiguous ways to conclude that there is indeed synaptic interaction between two neurons. Regardless of the revolutionary improvements of tools, all of which, including electrophysiology, voltage, or calcium flux monitoring even at cellular and subcellular levels, remain to be indirect, relying on interpretation of certain signals. None have replaced electron microscopy despite popular belief. I analyzed the relationship between these NPY and POMC cells (Figs. 2.2 and 2.3).[11] The outcome was very straightforward: NPY neurons heavily innervate POMC cells in the ARC, while POMC neurons do not innervate ARC NPY cells.

While Csaba was a craftsman, I, on the other hand, have very little dexterity. In fact, in hindsight, I find it a quasimiracle that I was able to successfully execute several projects in the 1990s, almost all on my own, that turned out the way they did and ended up being correct in the long run, regarding their functional implications. In other words, while the manual labor gave me little joy, I was very much intrigued and curious about what the findings may mean. In this case, regarding the relationship between NPY and POMC neurons, while I discussed the relevance of the findings in relation to gonadotropin control according to the original plan, I also raised the possibility that the interplay between these two hypothalamic peptidergic systems might be relevant in the control of appetite. This was not a random proposition, however. In 1991, I was approached by a physiologist, Satya Kalra from the University of Florida at Gainesville, who wanted to collaborate on our circuit projects. Satya Kalra, was well-known in both the reproductive neuroendocrinology and feeding fields. With his wife Pushpa Kalra, they made ground-breaking discoveries regarding LHRH control.

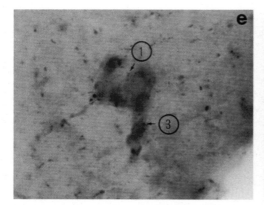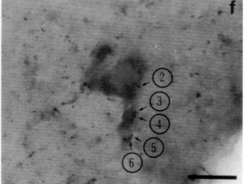

Figure 2.2 Color light micrographs demonstrate vibratome sections of the rat arcuate nucleus double immunolabeled for NPY (black nickel-intensified diaminobenzidine reaction product) and for β-END, the product of the POMC gene, labeled by light brown diaminobenzidine reaction product. e and f show (at two focus planes) one BEND-immunoreactive neuron contacted by multiple, numbered NPY boutons. This material was processed by electron microscopy to discern whether the contacts are synaptic in nature (Fig. 2.3). Scale bar represents 20 μm. NPY, Neuropeptide Y; POMC, proopiomelanocortin. *From Horvath TL, Naftolin F, Kalra SP, Leranth C. Neuropeptide Y innervation of β-endorphin-containing cells in the rat mediobasal hypothalamus. A light and electron microscopic double-immunostaining study.* Endocrinology. *1992;131:2461−2467.*

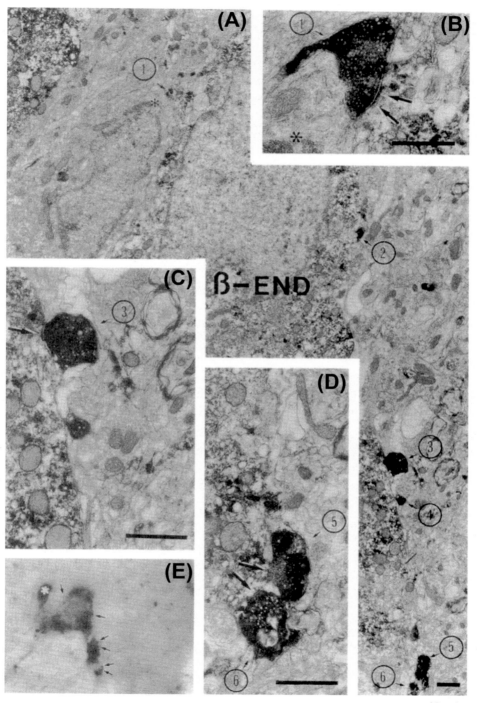

β-END

(Continued)

However, they also made a discovery in 1984, which revolutionized the research of metabolism: they found that NPY is a highly potent promoter of appetite and eating, acting in the brain.[12] Their work was soon replicated by others and NPY emerged as the most potent driver of feeding. It was with this background that in that 1992 paper I proposed that our work on the connectivity of the NPY and POMC neurons was relevant for feeding control. This postulation turned out to be correct through sets of subsequent, unrelated and serendipitous discoveries.

In 1994—95, the laboratory of Jeff Friedman at Rockefeller University cloned the obese gene, leptin, and connected its action to the hypothalamus.[13,14] This discovery, which was the most impactful in the field for over 50 years, revolutionized the entire field and introduced a "new neuroscience" in relation to the periphery and the brain. In 1997, through a chain of coincidental discoveries, the central melanocortin system (to which ARC POMC and AgRP neurons belong) was found to be a crucial determinant of obesity and eating[15,16] and tied to leptin signaling.[17] This was the culmination of the work of Roger Cone and his colleagues, who cloned pigmentation genes in the early 1990s.[18] In 1998, a natural ligand of melanocortin receptors, AgRP, was found to be selectively expressed in ARC NPY neurons by Michael Schwartz's laboratory.[19] The work of Christian Broberger in Tomas Hökfelt's lab provided a full picture how this AgRP system limits the NPY circuitry in various brain regions immediately implicating extrahypothalamic sites in feeding regulation.[20] Finally, in 2001, a blueprint of how leptin acts on ARC POMC cells mediated by GABAergic NPY neurons was unveiled by Michael Cowley, Roger Cone, Malcolm Low, their coworkers, and us.[21] These fundamental observations laid the conceptual framework for the field today, which since has been joined by many talented researchers in the pursuit of the obvious and critical follow-up questions (Fig. 2.4).

2.3.2 Hypothalamic AGRP neurons are mandatory for feeding and life

One such fundamental and important question was to confirm the necessity of these hypothalamic NPY and POMC neurons in hunger and satiety. Richard Palmiter and

◀ **Figure 2.3** Electron micrographs of consecutive ultrathin sections demonstrate the NPY innervation of the β-END-immunoreactive cell seen on the color panels e and f of Fig. 2.1. All of the NPY axon terminals numbered from I to 6 in Fig. 2.1, e and f, were recognized under the electron microscope. Fig. 2.2 shows boutons 1–6 (*short arrows*) and the synaptic contacts (*large arrows*) of boutons 1, 3, 5, and 6. a is a low power magnification of the β-END cell shown on both the color light micrographs and the black and white light micrograph in e. White asterisks in a and e label one of the dendrites of this neuron. b, c, and d are high power magnifications of boutons 1,3, and 5 and 6, respectively, taken from serial sections. Black asterisks on a and b label the same nonimmunoreactive cell used as a landmark to identify bouton 1. Original magnification of e, X250. Scale bars=1 μm. *NPY*, Neuropeptide Y. *From Horvath TL, Naftolin F, Kalra SP, Leranth C. Neuropeptide Y innervation of β-endorphin-containing cells in the rat mediobasal hypothalamus. A light and electron microscopic double-immunostaining study.* Endocrinology. *1992;131:2461–2467.*

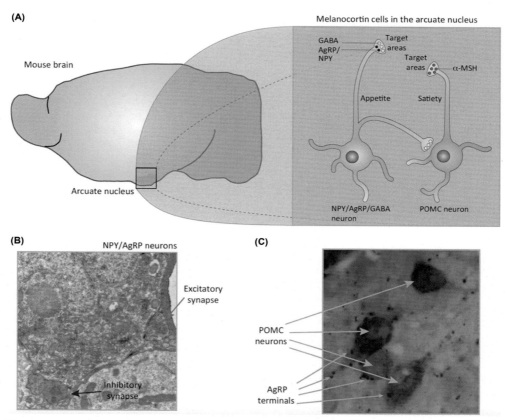

Figure 2.4 Organization of the melanocortin system in the arcuate nucleus. (A) The primary neurons of the melanocortin system are the orexigenic NPY/AgRP/GABA neurons and the anorexigenic POMC neurons (inset). Both populations of cell receive inhibitory and excitatory inputs that control neuronal function. (B) Example of synaptic coverage of the NPY/AgRP neurons in an image obtained using electron microscopy. (C) The NPY/AgRP neurons send inhibitory projections to the neighboring POMC cells, thus regulating the excitability of these cells. α-MSH, α-Melanocyte-stimulating hormone; *AgRP*, arcuate nucleus agouti-related protein; *NPY*, neuropeptide Y; *POMC*, proopiomelanocortin. *From Dietrich MO, Horvath TL. Hypothalamic control of energy balance: insights into the role of synaptic plasticity.* Trends Neurosci. *2013;36(2):65—73. https://doi.org/10.1016/j.tins.2012.12.005. Epub 2013 Jan 12.*[22]

his colleagues lead that quest. Palmiter, an eminent geneticist, neuroscientist, and inventor of technology for genetic modification of mice, set out to interrogate those questions from the perspective of the peptidergic content of NPY/AgRP neurons. His lab knocked out NPY, which did not render animals anorectic.[23] When both NPY and AgRP were knocked out at the same time: no anorexia.[24] I attended an NPY meeting in 1999 at Cayman Islands. Obviously, there were many presentations on NPY and feeding. After each talk, Richard would rise and make the point that his

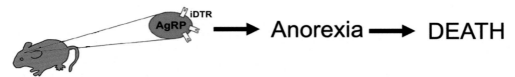

Figure 2.5 AgRP neurons were selectively tagged with diphtheria toxin receptor (wild type mice do not have such receptors). In adulthood, after injection of diphtheria toxin, AgRP neurons were ablated. This, in turn, resulted in cessation of feeding followed by death. *AgRP*, Agouti-related peptide.

Figure 2.6 Ablation of POMC neurons in adult animals resulted in increased eating and elevated body weight. However, these animals did not die. POMC, Proopiomelanocortin.

knockout experiments did not show the mandatory nature of these peptides in feeding control suggesting that these neurons may not be as crucial as people have been thinking. Then, a new tool became available for the examination of such questions more directly and forcefully. This was the initial introduction of diphtheria toxin receptor (DTR) into mouse cells using the cre-lox approach. Mice do not have DTRs; hence, they do not respond normally to diphtheria toxin exposure. However, the induction of DTR into the cell of choice will render these cells vulnerable to exposure to diphtheria toxin. With this new tool, the question could be asked: what happens to feeding when NPY/AgRP neurons are eliminated in the adult animal? Then, Serge Luquet in the Palmiter laboratory, through a set of elegant studies provided the evidence that elimination of AgRP neurons from the adult hypothalamus leads to cessation of eating and loss of weight to the point of death (Fig. 2.5).[25]

At the same time, in a collaborative study between Jens Bruning's lab and mine, we also showed the importance of AgRP neurons in feeding using a similar approach.[26] In addition, we showed that in contrast to AgRP, the elimination of POMC neurons increases feeding and body weight (Fig. 2.6).[26] This latter observation provided direct evidence that POMC neurons are fundamental for satiety and that the anorexia phenotype of AgRP neuron–ablated animals is not because of sickness due to a rapid loss of cells of the hypothalamus but because of the role of these cells in feeding.

2.3.2.1 *The inhibitory neurotransmitter, GABA, drives appetite for life*
Long-projecting principal cells that evoke an activation of other circuits predominantly operate with the excitatory neurotransmitter, glutamate. Indeed, excitation by such a

process is considered a crucial driver of brain activity and related functions. Strikingly, however, it was found that life itself (via eating) is driven by the inhibitory neurotransmitter, GABA.[27] As I noted before, contrary to the effect of ablation of NPY/AgRP neurons in adult mice that caused anorexia and death, knocking out NPY, AgRP, or both of these signaling molecules together from these ARC neurons did not diminish the animals' ability to eat. At the NPY meeting, I referred to the previous note, when Richard Palmiter was discussing these issues, I mentioned that he should also consider that these neurons contain the neurotransmitter, GABA, which we had demonstrated a few years earlier,[28] and in collaboration with Satya Kalra, we showed that this GABA-NPY interplay is relevant for feeding control in the hypothalamus.[30,31] While these observations were confirmed and later elaborated upon by us and others, it was the Palmiter lab that unmasked the crucial involvement of GABA in ARC AgRP neurons in relation to feeding (Fig. 2.7).[32] These studies also confirmed the work of Broberger and Hökfelt that extrahypothalamic sites, such as the parabrachial nucleus and the lateral septum,[20] are sites where these hypothalamic neurons impact feeding. Obviously, these areas and their ascending pathways immediately implicate these hypothalamic neurons in higher brain functions, which makes sense from the perspective of the complex behavioral adaptations needed for feeding. Another finding of theirs was that the elimination of this GABAergic output of the AgRP neurons regarding survival could be rescued if a GABA agonist was provided for several days. This argued for an adult plasticity of the circuitry,[27] a characteristic of hypothalamic feeding circuits uncovered years earlier by a collaborative study between my lab and that of Jeff Friedman.[9,33,34]

2.3.3 Plasticity of hypothalamic circuits and glia cells involved in control of metabolism and homeostasis

In the spring of 2001, Jeff Friedman's laboratory published another landmark paper in which they traced inputs of ARC NPY (AgRP) and leptin receptor-expressing cells.[35] They showed diverse afferents coming to these cells from higher brain regions, including the cortex. They used a new tracing approach, which relied on pseudorabies virus, which was pioneered by Adorján Bartha, a Hungarian veterinarian virologist from my alma mater (Fig. 2.8).[36] By then, I was very much interested in the relationship between the hypothalamus and higher brain regions, in part because of my work on aromatase, the enzyme responsible for conversion of testosterone to estradiol, and hence sexual differentiation of the brain. While most studies on aromatase in the brain until then had concentrated on the hypothalamus and were pioneered by my other mentor at Yale, Frederick Naftolin,[37] then the chair of the Department of Ob/Gyn, I started to explore and uncover the presence of this critical enzyme in brain development in extrahypothalamic regions, including various sensory systems and the cortex.[38] At that time, mainstream neuroscience was not entirely intrigued by parallel discussions on the

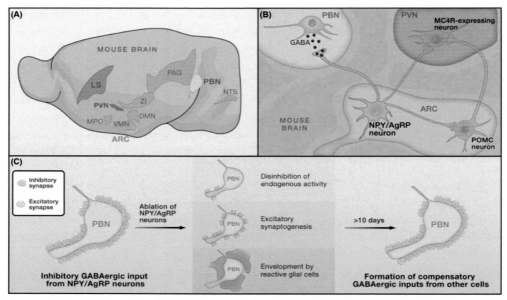

Figure 2.7 The GABAergic output of NPY/AgRP neurons. (A) The NPY/AgRP neurons in the ARC (green) of the hypothalamus send inputs to various nuclei inside and outside the hypothalamus, including the PBN (yellow) and the LS (blue). (B) Metabolic regulation by GABAergic NPY/AgRP neurons is thought to occur through their interactions with the synaptic output of POMC neurons or MC4R-expressing neurons in the PVN. The Wu et al.24 study showed that the GABAergic inputs from the NPY/AgRP neurons to the PBN, which are necessary for maintaining feeding behavior, are independent of POMC neurons. (C) Ablation of NPY/AgRP neurons in the ARC results in the loss of GABAergic inhibitory inputs to target brain regions such as the PBN. One possible mechanistic explanation for the slow adaptive response that occurs after the loss of these neurons is that the ablation first leads to hyperstimulation of the target brain areas due to the disinhibition of the activity of the postsynaptic target neurons, an increase in the excitatory inputs to the target cells, and an increase in the envelopment of the target neurons by reactive glia. After this early adaptation, synaptic plasticity allows for the modification of the neural circuitry, resulting in the recruitment of new inhibitory inputs on the target cells to compensate for the loss of the GABAergic inputs from the NPY/AgRP neurons. AgRP, Agouti-related peptide; ARC, arcuate nucleus; ARC, arcuate nucleus; DMN, dorsomedial nucleus; LS, lateral septum; MPO, medial preoptic area; NPY, neuropeptide Y; NTS, nucleus tractus solitarius; PAG, periaqueductal nucleus; PBN, parabrachial nucleus; POMC, proopiomelanocortin; PVN, paraventricular nucleus; VMN, ventromedial nucleus; ZI, zona incerta. *From Dietrich MO, Horvath TL. GABA keeps up an appetite for life. Cell. 2009; 137:1177–1179.*

hypothalamus and higher brain regions. In 1996, one of our submissions to the *Journal of Comparative Neurology*, the flagship journal of hardcore neuroanatomy, was rejected by the editor-in-chief and not sent out for review because the paper focused on glutamate receptors in neurons and glia cells in relationship to gonadal steroids in the hypothalamus and the hippocampus. He specifically noted that reporting on these two entirely different functional areas of the brain in the same paper is nonsensible, and hence will not be

Figure 2.8 The bust of Adorján Bartha (1923−1996) in garden of the University of Veterinary Medicine in Budapest, Hungary.

considered by JCN. Going back to the DeFalco paper, I vividly remember how excited I was about this work from Friedman's lab, and, I had long conversations with Pasko Rakic and the late Pat Goldman-Rakic at Yale about these potential groundbreaking new directions in neuroscience. This is when I contacted Jeff with an idea of collaboration. He was prompt and welcoming and I visited his lab at Rockefeller in early April of 2001.

By this time, I had been involved in and fascinated by the fact that hypothalamic neurons, specifically in the ARC, showed rapid plasticity in response to changes in the gonadal steroid milieu. The original observation of this synaptic remodeling was described by the laboratories my mentor, Fred Naftolin, and that of my colleague and friend, Luis Miguel Garcia Segura from the Cajal Institute in Madrid.[39] They built on the pioneering work of Matsumoto and Arai[40] who showed a synaptogenic action of estrogen in the ARC. While it was intriguing to explore this issue from the perspective of gonadotropin secretion and estrogen feedback,[28,29] similar to my original work on NPY and POMC connectivity, at that point, I felt that we were running in circles without getting specific answers as to how these changes in synaptology may be relevant to any specific circuits controlling ovarian cycles. Because of our work on the hypothalamic NPY system and feeding, I had also already segue-wayed into the field of central control of metabolism. I realized that Jeff had the model in which we could test the question of plasticity in circuits that can be visualized (AgRP/NPY and POMC neurons) and directly relate to a phenotype, the obese hyperphagia of ob/ob mice. His lab had shown before that leptin administration to these circulating leptin-deficient mice can rapidly restore the feeding and metabolic phenotype of these morbidly obese mice to wild-type controls.[14] In a straightforward approach, we analyzed synaptology of AgRP/NPY and POMC neurons using electron microscopy and electrophysiology. The results were robust and unambiguous: hunger-promoting AgRP/NPY neurons had a domination of excitatory inputs in ob/ob mice, while satiety promoting POMC cells had a domination of inhibitory inputs. Strikingly, exogenous leptin administration rapidly changed this synaptic arrangement to that of wild-type animals in satiety and ob/ob animals started to eat less and lose weight.[33] We also found that these synaptic changes started to emerge preceding the changes in behavior and body composition indicating that these synaptic changes are prerequisite for the altered behavior and autonomic functions (Fig. 2.9).

This idea was further supported by our observation that the gut-derived hunger hormone, ghrelin, induced the opposite changes in synaptology of hypothalamic circuits while promoting eating.[33] By the way, despite my arguments against it, all these observations are actually entirely in-line with the Neuronal Doctrine ... This paper of ours appeared in Science back-to-back with another paper, which focused on the effect of leptin as a growth factor during brain development.[41] Intriguingly, but perhaps not surprisingly, the senior author of that paper, Richard Simerly, like myself, also had his scientific "upbringing" in reproductive neuroendocrinology. And, similar to the impact of estrogens on adult synaptogenesis, estrogen had been found to have a major role as a growth factor in brain development as described by Torran-Allerand in the 1970s.[42] While the growth factor properties of leptin were easily accepted by the field as sensible, the acute and rapid

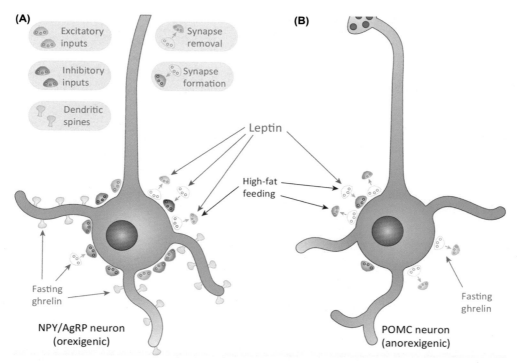

(A)

Excitatory inputs

Inhibitory inputs

Dendritic spines

Synapse removal

Synapse formation

Leptin

High-fat feeding

(B)

Fasting ghrelin

NPY/AgRP neuron (orexigenic)

Fasting ghrelin

POMC neuron (anorexigenic)

Figure 2.9 Schematic illustration of synaptic plasticity of central feeding circuits. Synaptic plasticity in the melanocortin system in the ARC. Summary of synaptic changes and spine formation that occur in NPY/AgRP and POMC neurons in response to different metabolic and hormonal stimuli. (A) Anorexigenic NPY/AgRP neuron that increases its activity in response to hormones released during negative energy balance (e.g., ghrelin). (B) An anorexigenic POMC neuron that is activated during states of satiety or by hormones, such as leptin and insulin. Negative energy balance (e.g., fasting) or ghrelin promotes an increase in the number of excitatory inputs to NPY/AgRP neuron perikarya, while decreasing the number of excitatory inputs to POMC neuron perikarya. Additionally, fasting also leads to an increase in the number of dendritic spines in the NPY/AgRP neurons, in a mechanism dependent on NMDA receptors. Positive energy balance in the form of a high-fat diet leads to a decrease in the number of excitatory inputs to NPY/AgRP neurons and inhibitory inputs to POMC neurons. In addition, high-fat diet increases excitatory inputs to the POMC neuron perikarya. The constellation of changes that occurs during high-fat feeding switches the physiological set point of the melanocortin system in the ARC toward the activation of POMC neurons, while decreasing the excitability of NPY/AgRP neurons.[43] Finally, leptin promotes changes in the synaptic coverage of NPY/AgRP and POMC neurons similar to those in response to high-fat feeding. Specifically, it increases the excitability of POMC neurons by increasing the number of excitatory synaptic inputs to these neurons. In addition, it also increases the number of inhibitory inputs and decreases the number of excitatory inputs onto NPY/AgRP neurons.[44,45] *AgRP*, ARC agouti-related protein; *ARC*, arcuate nucleus; *NPY* neuropeptide Y; *POMC*, proopiomelanocortin. *From Dietrich MO, Horvath TL. Hypothalamic control of energy balance: insights into the role of synaptic plasticity. Trends Neurosci. 2013;36(2):65−73. https://doi.org/10.1016/j.tins.2012.12.005. Epub 2013 Jan 12.*

physical and electrical plasticity we observed in response to leptin and ghrelin was met with skepticism, disbelief, and rumors that perhaps it was made up. I do give tremendous credit to Jeff Friedman, who was open to this idea from the outset and had his own lab members pursue it in follow-up studies,[46] some of whom, notably Scott Sternson, made future, seminal discoveries independently, regarding specific aspects of our original observations.[47] As with every "out-of-the-blue" finding in a field, it is first met with disbelief and ridicule (the operative word in this sentence is "field" because for some, it was not so surprising), then tolerance/acceptance followed by rediscovery of the phenomenon by those specifically who doubted it in the first place presenting it as a new or "real" discovery. With some exceptions, this "time table" is applicable to every unexpected discovery in every field and intellectual endeavor.

It is very difficult to prove causality between synaptic plasticity and brain functions. That difficulty also applies to higher brain functions, such as learning and memory, where a type of synaptic plasticity, long-term potentiation (LTP), was originally described. Nevertheless, it has been powerfully argued that the computational advantages synaptic plasticity offers make it almost inevitable that these processes will be proven to be causally related to behavioral control. Hypothalamic hunger-promoting neurons offer an even better model to test these questions relating to synaptic plasticity as the behavioral output, that is, feeding, is more directly and temporarily tied to activity of these neurons than, for example, learning and memory is tied to higher brain regions. As noted earlier, we observed synaptic changes of feeding circuits to occur preceding the changes that could be observed in behavior suggesting the causality of these cellular changes with behavioral output. In support of this notion, we found that peripheral hormones, of which primary function is not associated immediately with feeding, but has an effect on it, such as estradiol and corticosteroids, induce synaptic changes corresponding to the behavioral outcome.[48,49] To probe this proposition further, we again exploited tenets of the neuronal doctrine. In collaboration with the laboratories of Michael Cowley and Matthias Tschöp, we tested a very simple question: if the qualitative and quantitative synaptic input organizations of hypothalamic neurons are determinants of behavioral response, then the base-wiring of these cells should be predictive of a behavioral and systemic response of an animal to food types that promote eating and obesity. We analyzed such a model developed by Levin.[50] These are outbred rats, which fall into two major categories: rats that resist the development of diet-induced obesity on high fat diet and those that become obese. The fundamental finding of this study was that the synaptic input organization of satiety-promoting POMC neurons was substantially different between these two groups of rats even before they were put on high fat diet when they had similar metabolic profiles (Fig. 2.10).[51]

The base wiring of POMC neurons of those animals who would become obese on high fat diet had higher inhibitory tone than of those animals who were resistant to diet-induced obesity. Thus our data were indicative of the synaptic arrangement being

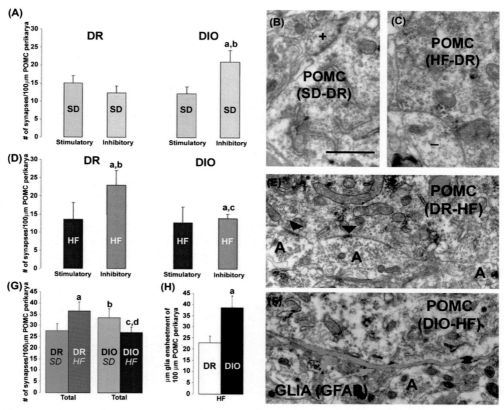

Figure 2.10 Synaptic input organization of Dr and DIO rats. (A) Bar graphs showing the numbers of perikaryal symmetrical and asymmetrical connections on POMC neurons of Dr and DIO animals fed an SD. a and b: $P < .05$. (B and C) Representative electron micrographs showing asymmetrical, putative stimulatory (+; B), and, symmetrical, putative inhibitory ($-$; C) synapses taken from POMC-immunolabeled perikarya of an SD-fed Dr rat (B) and of an SD-fed DIO rat (C). (Scale bar in B: 1 μm for B, C, E, and F.) (D) Bar graphs showing the numbers of perikaryal symmetrical and asymmetrical connections on POMC neurons of Dr and DIO animals fed an high fat diet (HFD). (E and F) Representative electron micrographs showing POMC-labeled perikarya of a Dr HFD mouse (E) and a DIO HFD mouse (F). Note that instead of the axon terminals (A) in the Dr HFD POMC cells, there is glial ensheathment (indicated by green pseudocolor) of the POMC perikarya of the DIO HFD animal. Bar graphs represent mean ± SEM. (G) Bar graphs indicating the total numbers of synapses on POMC perikarya of Dr and DIO rats fed an SD and an HFD. a: $P < .05$ Dr HFD versus Dr SD; b: $P < .05$ DIO SD versus Dr SD; c: $P < .05$ DIO HFD versus DIO SD. (H) Bar graphs indicating glial ensheathment on POMC perikarya of Dr and DIO animals fed an HFD. a: $P < .05$. POMC, proopiomelanocortin. *From Horvath TL, Sarman B, García-Cáceres C, et al. Synaptic input organization of the melanocortin system predicts diet-induced hypothalamic reactive gliosis and obesity. PNAS. 2010;107 (33):14875–14880. Epub 2010 Aug 2. PMCID: PMC 2930476.*

determinant of the so-called set-point of the melanocortin system. We also found that resistant versus vulnerable animals responded differently to exposure to the same high fat diet. While POMC neurons of those animals that remained leaner on the high fat diet gained synapses, POMC neurons in animals that became obese lost synapses. In place of these "missing" neuronal inputs, POMC perikarya of obese mice were covered by processes of astrocytes. Indeed, we observed increased astrogliosis in the hypothalamus of diet-induced obese rats arguing for a hypothalamic inflammatory process triggered by this metabolic shift.[51] These observations raised the possibility that the changes in circuit connectivity in response to metabolic fluctuations may be controlled, at least in part, by nonneuronal brain cells, such as astrocytes.

Over the past decades, our understanding of glial cell function in the brain has evolved significantly. Their roles in various aspects of control of neuronal circuits have been recognized.[52] Astrocytes specifically have been tied to both pre- and postsynaptic control of neuronal transmission via multimodal action. In the hypothalamus, we and others showed the involvement of both astrocytes and microglia in synaptic input organization and activity of neuronal circuits affecting feeding, energy, and glucose metabolism.[51,53–55] In relation to metabolism-related plasticity of hypothalamic circuits, Luis Varela in my lab identified a unique role for astrocytes: he found that AgRP neurons can directly stimulate neighboring astrocytes unmasking a neuron-to-astrocyte communication in the physiological adaptations to changing energy status. Previous reports have described that in hippocampal and cortical areas, neurons communicate with adjacent astrocytes by releasing different neurotransmitters.[52] Among them, GABA has been shown as a main neuron-derived factor that triggers the depolarization of astrocytes. Indeed, we demonstrated that AgRP neuron-released GABA is the main mediator of this cell-to-cell communication in the hypothalamus.[56] Our study also showed that AgRP-activated astrocytes signal back to AgRP neurons in a dual manner. First, as noted before, astrocytic processes replace inhibitory connections on AgRP perikarya: we found that either fasting, ghrelin administration or selective activation of AgRP neurons caused an increase in the glial coverage of AgRP perikarya with a concomitant reduction of the number of inhibitory synapses on their plasma membrane. These observations identify a mechanism whereby the activation of astrocytes by AgRP neuron-released GABA replaces inhibitory inputs from AgRP perikaryal enhancing the activation of these cells by elevated excitatory tone at the time of low-energy availability (Fig. 2.11). Besides this morphological plasticity of AgRP neurons, we also discovered that astrocyte-derived prostaglandins, PGE2, promote the activation of AgRP cells through their actions specifically on EP2 receptors (Fig. 2.11). AgRP neurons are considered "first-order" sensory neurons in the regulation of feeding by mediating information on systemic energy state to a broad array of brain circuits that, in turn, alters complex behaviors.[57] In this regard, the existence of rapid synaptic plasticity of these neurons in response to the changing metabolic environment is practical and essential to assure that peripheral metabolic information can be conveyed to evoke feeding in support of

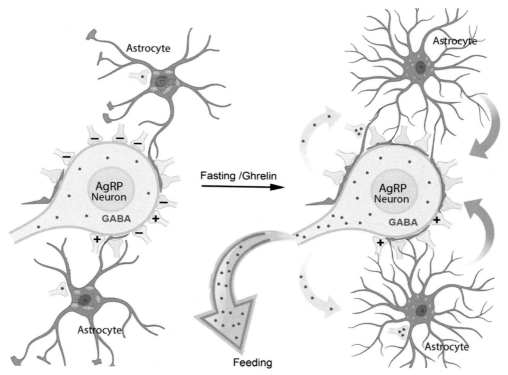

Figure 2.11 Feedforward autoactivation of AgRP neurons via glial mediation. Activation of AgRP neurons by fasting, ghrelin, or chemogenetics engage neighboring glial cells via GABAergic transmission. Activated glial cells have smaller mitochondria, and their processes increase the ensheetment of AgRP neuronal perikaryal resulting in synaptic input organization of these cells that are consistent with their increased excitability and promotion of feeding. In addition, prostaglandin E2 signaling of glial origin excites AgRP neurons. *AgRP*, Agouti-related peptide. *From Varela L, Stutz B, Song JE, et al. Hunger-promoting AgRP neurons trigger an astrocyte-mediated feed-forward auto-activation loop in mice. J Clin Invest. 2021;144239. https://doi.org/10.1172/JCI144239. Online ahead of print.*

survival with limited interfering inhibitory signals arising from other brain areas. In the case of AgRP neurons, the most crucial information flow arises from the periphery to trigger activation of these neurons in support of feeding, which is crucial for survival.[33,34,59] A dominant component of input on peripheral metabolic state of these neurons is in the form of circulating hormone and nutrient signals. Thus the morphological elimination of gating inhibitory synapses by the process described in our studies fits well with the biological necessity of AgRP neuronal activation to support survival. In this regard, AgRP neurons control their own inhibitory input dynamics, in a feedforward manner via glial mediation.

These studies reiterated the critical importance of glial cells in the workings of the hypothalamic machinery in control of feeding and systemic metabolism. We also showed earlier that leptin signaling in glial cells have an impact on the synaptic input organization and functioning of the hypothalamic melanocortin system.[55] Subsequently, Luis and I contributed to studies of the laboratories of Matthias Tschöp and Jens Brüning that revealed the role of insulin- and glucagon–like protein 1 signaling in astrocytes in systemic control of glucose metabolism.[54,58] In addition, Sabrina Diano's lab and that of mine identified an active and acute contribution of microglia to these hypothalamic processes as well.[60,61] Finally, Luis also discovered that the endothelial cells, themselves, have a significant and direct role in control of the hypothalamic machinery that regulates metabolism.[62] This long-winded paragraph is to emphasize that there is much more to the understanding of brain functions than neuronal circuit analyses.

2.4 Plasticity beyond the arcuate nucleus and the hypothalamus triggered by metabolic changes

A recurrent theme in our conceptualization of brain functions is that hunger and satiety impact all functions of the brain, and consequently, all regions of brain. From this assertion comes the notion that if switching functionality of brain circuits is involved in feeding-associated synaptic plasticity as an adaptive cellular response, this process may also be innate to other parts of the brain in response to the changing metabolic environment. From 2005, we undertook several series of studies to interrogate these questions.

2.4.1 Plasticity of lateral hypothalamic hypocretin/orexin cells

First, we turned our attention to the lateral hypothalamic hypocretin/orexin system. These neurons were identified in the late 1990s by two independent groups,[44,63] and they emerged as crucial regulators of arousal and related behaviors. With my long-term collaborator, Xiao-Bing Gao, who is an electrophysiologist involved with the original description of the hypocretin system by de Lecea et al.,[63] we analyzed these neurons in conditions of hunger and satiety.[64] We looked at two parameters: synapses on hypocretin/orexin perikarya using electron microscopy and miniature postsynaptic currents of these cells by slice electrophysiology. In 2004—05, when we conducted these studies, there were no better tools than these to gain insight into synaptic input organization of neurons. I argue that even today, the combination of electron microscopy together with the assessment of miniature postsynaptic events is superior to any other analytic tool of contemporary neuroscience. We complemented these approaches with fluorescence multiple labeling experiments and correlated light- and electron microscopy. Despite the advancement in imaging technology, genetics, and other recently developed "smart" tools, a connection is most unambiguously identified via electron microscopy. While the shortcoming of classical electron microscopy (EM) is its two-dimensional feature, there are ways around that limitation. In 2004, we did not have access to focused ion beam scanning electron microscopy that allows automated 3D reconstruction of samples. We did 3D reconstruction the hard way, which was to serial section entire hypocretin cells and take pictures and analyze every single image of the cell.

I would like to make a note about the remarkable electron microscopical technical support that I have been having the great fortune to be supported by. Most of these individuals were Hungarians, who have learned the craft back in Budapest in the 1970s and 1980s in those out-standing laboratories I referred to before. In the not too distant past, there were three Hungarian electron microscopy technicians working in contiguous hallways here at Yale: Klara Szigeti in the laboratory of Pat Goldman Rakic in the Neurobiology department, Ilona Kovacs in Nihal De Lanerolle's lab at Neurosurgery and Erzsebet Borok in my lab, at the time, in the Ob/Gyn department. Electron microscopy is an art. Each step of the process requires enormous expertise, devotion, diligence, experience, and ingenuity. Each sample behaves differently; nevertheless, the product must be always the highest quality so analyses and comparisons can be made. I would compare the process to the art of molecular cuisine.

The most remarkable aspect of the input organization of these lateral hypotha-lamic hypocretin/orexin neurons was that they were dominated by excitatory inputs, even at the level of the neuronal perikaryal (Fig. 2.12). Gating of neuronal perikarya by inhibitory inputs is crucial to filter excitation-evoked firing of effector neurons, such as the principal neurons of the cortex and hippocampus. This assures appropriate control of "noise" in support of predictable orchestration of behavioral output. The necessity of such gating may be immediately obvious, for example, in the case of motor neurons or pyramidal cells of the cortex, which compute vast amounts of information arising from different brain sites. On the other hand, in the case of hypocretin/orexin neurons, the fundamental role of which is to pro-mote arousal, it is likely, as for the purpose of arousal, "noise," by definition, serves as a crucial signal. We also found that food deprivation, a process invariably promoting arousal, resulted in further recruitment of excitatory inputs onto these lateral hypothalamic neurons.[64] In 1999, in collaboration with Tony van den Pol at Yale, we connected excitatory efferents of this lateral hypothalamic system to hypothalamic hunger-promoting NPY/AgRP neurons[65] as well as to locus coeru-leus arousal-promoting noradrenergic cells.[66] Thus when we found that the dominant excitatory input of hypocretin neurons is further enhanced by food and sleep deprivation,[67] we came to conclude that these neurons, which are also heavily communicating to the cortex, may provide the link between sleep depriva-tion and obesity (Fig. 2.13).[64]

2.4.2 Plasticity of midbrain dopamine neurons

It goes without saying that eating is a "motivated" behavior. The *midbrain dopamine system* is considered a fundamental driver and controller of motivated behaviors. As a

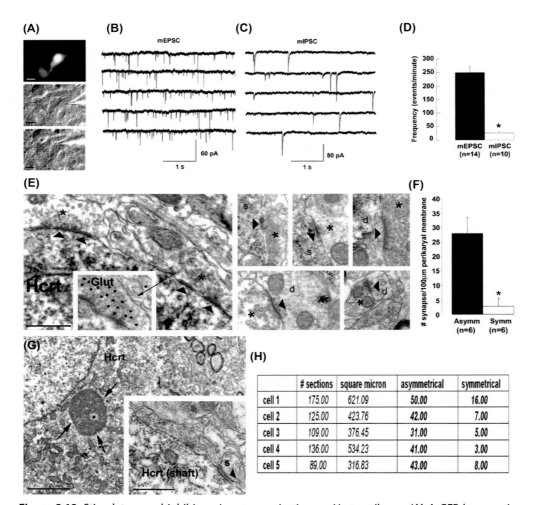

Figure 2.12 Stimulatory and inhibitory input organization on Hcrt perikarya. (A) A GFP-hypocretin-containing neuron was identified and whole-cell patch clamp recording was made on this neuron. Scale bar: 10 μm. (B and C) Raw traces of recorded mEPSCs and mIPSCs are presented. mEPSCs were recorded in the presence of TTX (1 μM) and bicuculline (30 μM), while mIPSCs were recorded in the presence of TTX (1 μM), CNQX (10 μM), and AP-5 (50 μM). (D) Pooled data representing mEPSC and mIPSC frequencies (events recorded per minute) from hypocretin-containing neurons are presented. (E) Electron micrographs showing asymmetric synaptic contacts between unidentified axon terminals (asterisk) and Hcrt somata (s) and dendrites (d). Arrowheads point to synaptic membrane specializations. Arrow on left panel originates on a bouton establishing an asymmetric synapse on an Hcrt perikaryon and points to the inset showing the same bouton in a subsequent section after postembedding immunostaining for glutamate (5 nm immunogold particles). Bar scale on left panel indicates 1 μm for all panels of (E). (F) Statistical analysis revealed that asymmetric, putative excitatory axon terminals dominate inhibitory contacts on Hcrt perikarya. (G) Cytoplasmic

(*Continued*)

further exploration of our proposition that synaptic plasticity is innate to brain circuits involved in feeding control, Alfonso Abizaid in my lab explored this hypothesis regarding the midbrain dopamine system. Through a set of studies building on each other, he found that the gut hormone, ghrelin, promotes the activity of ventral tegmental dopamine neurons, and that this event promotes feeding.[45] The study also described that during this process, ghrelin promoted a rapid synaptic plasticity of these midbrain dopamine neurons as assessed by electrophysiology and electron microscopy, an approach that we used for similar examinations regarding the lateral hypothalamic hypocretin/orexin system[64] and the hypothalamic feeding circuits.[33] We analyzed synapses by electron microscopy (Fig. 2.14) and miniature postsynaptic currents by slice electrophysiology (Fig. 2.15). We substantiated these observations with multiple imaging fluorescent labeling studies (Fig. 2.15).

This discovery was also not a slam-dunk to publish. It was reviewed in three rounds of reviews in 2005—06 at a sister journal of one of the leading scientific journals. The three reviewers were increasingly supportive of the paper as we responded to their comments. In the end, the first two reviewers accepted the paper. The third also agreed with our improvements and had no further substantive comments. However, he or she made a subjective note that in his or her more than 20-year career in neuroscience research, he or she had never seen anything like this before. This comment was in reference specifically to our electrophysiological and electron microscopical observations of rapid synaptic plasticity of midbrain dopamine neurons. Because of this unprofessional and unsubstantiated comment, our paper was rejected. I was furious, livid, and disillusioned. I was stunned that despite two of the reviewers praising our work and the third not finding a flaw and agreeing with the novelty of our results, the paper was rejected because one of the reviewers insinuated that we must have made up our data (as s/he "had not seen anything like this before"). As I noted before, our original description of rapid synaptic plasticity evoked by leptin and ghrelin in the hypothalamus also met with skepticism, but none of those doubts were used directly to confront us. Here, with the midbrain dopamine system, we crossed into a "different" field: neuroscience of motivated behaviors. And in this review

◄ organelles consisting of aggregating transport vesicles (*black arrows*) were frequently found in Hcrt-immunopositive cells. Bar scale represents 2 μm. Inset on (G) shows a dendritic shaft of a hypocretin neuron to be presynaptic to unidentified dendritic spine (s) in an asymmetric synaptic configuration (*arrowhead*). Bar scales represent 3 μm [main panel of (G)] and 1 μm [inset of (G)]. (H) The table showing results of 3D reconstruction of five randomly selected hypocretin-immunopositive neurons from intact, ad libitum—fed animals (*n=5*). Ultrathin sections were 70 nm thick. The surface area of each neuron was determined by multiplying total perikaryal membrane length (from each plane) by 70 nm. *From Horvath, TL, Gao XB. Input organization and plasticity of hypocretin neurons: a possible clues for obesity's association with insomnia. Cell Metab. 2005;1 (4):279—286.*

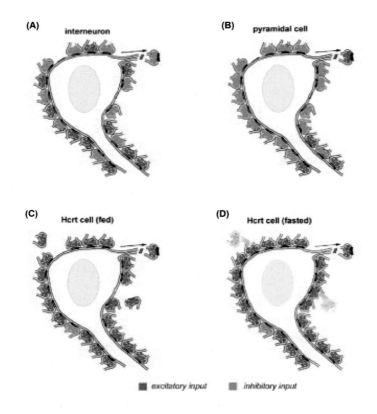

(A) interneuron

(B) pyramidal cell

(C) Hcrt cell (fed)

(D) Hcrt cell (fasted)

■ excitatory input   ■ inhibitory input

(E)

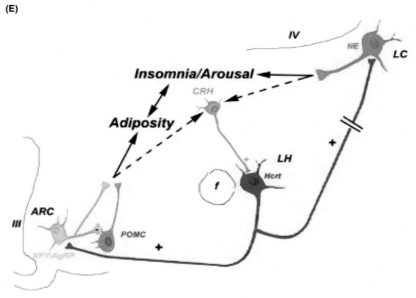

(Continued)

process, we were implicitly told that we must be lying or making it up. To express my anger, I wrote a letter to the editor-in-chief. In that letter, using very simple logic, I pointed out that because they rejected our paper based on the incredulity of a senior reviewer (as was pointed out to me by the editor), they were rejecting our paper because they concluded that our results (at least in part) were the outcome of fraud (scientific misconduct). Because that was the unavoidable conclusion, I demanded that they file a complaint with Yale University with their suspicions. I emphasized that considering these events, it is their duty (and not an option) to file that complaint. They did not. In their reply they assured me that they respect my work (indeed they published papers from my lab subsequently), but they would not publish the paper or file a complaint. We ended up publishing the paper in the *Journal of Clinical Investigations*, where its chief editor at the time was very supportive and empathetic to our case, and we received a fair review and editorial process. Years later, I learned that the senior investigator who raised issues about our original submission was inspired by our work and because of his "curiosity" (disbelief) about it, he pursued and confirmed principles of our findings . . . Again, this experience is not unique, and it is just part of the process of discovery. Originally, Alfonso also investigated whether these ghrelin effects in the ventral tegmental area impact responses to cocaine, a drug known to act via the midbrain dopamine system. In fact, those findings were part of the original submission, and they showed that ghrelin potentiates the effect of cocaine on locomotor responses. Another, otherwise constructive and positive reviewer of the submission that ended up rejected requested that we remove the cocaine data from the submission. He (who later voluntarily informed me of his role in the review process and is a

◀ **Figure 2.13** Input organization of hypocretin neurons and its impact on energy homeostasis and insomnia/arousal. (A—D) Schematic illustration of the inhibitory (*green*) and excitatory (*red*) input organization of various cell types in the brain: interneurons (A), long-projective neurons, such as pyramidal cells (B). Hypocretin cells in the fed state (C) are controlled by excitatory synaptic inputs, which outnumber inhibitory connections 10:1. Food restriction underlies further recruitment (curved, *light red arrows*) of excitatory inputs onto Hcrt perikarya (D). (E) The connectivity and synaptic input organization of lateral hypothalamic hypocretin neurons provide a simple and straightforward explanation for the relationship between insomnia and adiposity: because of the easy excitability of hypocretin neurons, any signal that triggers their activity, regardless of the homeostatic needs, will elevate the orexigenic tone of the arcuate nucleus melanocortin system while also promoting wakefulness through activation of medullary LC NE neurons. Elevated orexigenic output of the melanocortin system as well as sustained arousal at the expense of sleep promote the activity of CRH-producing neurons in the hypothalamus, which, in turn, can further trigger hypocretin neuronal firing. This vicious cycle, in which hypocretin neurons play the role of both trigger and accelerator, can simultaneously promote and worsen adiposity and insomnia. *CRH*, Corticotrophin releasing hormone; *LC*, locus coeruleus; *NE*, noradrenergic. *From Horvath, TL, Gao XB. Input organization and plasticity of hypocretin neurons: a possible clues for obesity's association with insomnia.* Cell Metab. 2005;1(4):279—286.

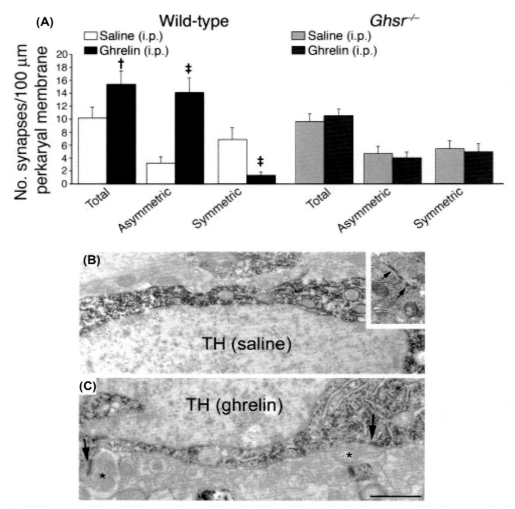

Figure 2.14 Synaptic remodeling induced by peripheral ghrelin in VTA DA cells. (A) Ghrelin increased the number of synapses on VTA DA cells of wild-type mice (*n*=5). Both total and asymmetric synaptic contacts were elevated, while the number of symmetric synapses was decreased. No synaptic changes were observed after peripheral ghrelin injection in *Ghsr*$^{-/-}$ mice (*n*=5). $^{\dagger}P<.05$, $^{\ddagger}P<.01$ versus respective saline-treated controls. (B and C) Electron micrographs showing typical tyrosine hydroxylase (TH)-immunoreactive perikarya of the ventral tegmental area dopamine (VTA) from saline- (B) and ghrelin-treated (C) wild-type mice. Arrows in inset of B indicate a symmetric synapse. Arrows in C indicate asymmetric synaptic contacts. Asterisks indicate unlabeled axon terminals. Scale bar: 1 μm. *From Abizaid A, Liu Z-W, Andrews ZB, et al. Ghrelin modulates the activity and synaptic input organization of midbrain dopamine neurons while promoting appetite. J Clin Invest. 2006;116:3229−3239; Epub Oct 19. PMC 1618869.*

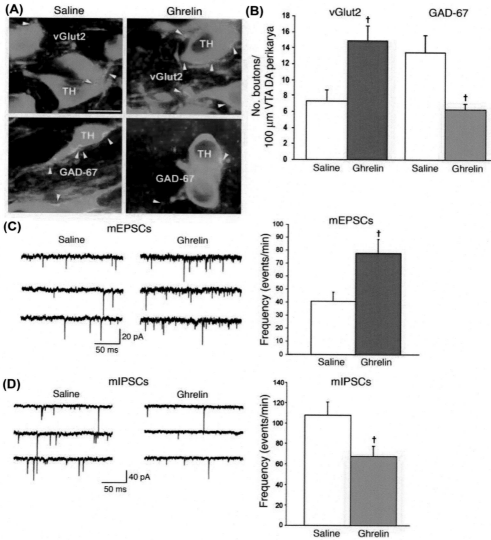

Figure 2.15 Ghrelin alters inhibitory and excitatory inputs of VTA DA cells. (A and B) Appositions between vGlut2-immunoreactive (*red*) and TH-immunopositive (*green*) VTA perikarya were significantly greater in ghrelin-treated animals compared with saline-treated controls. In contrast, appositions between GAD-67—immunolabeled boutons and tyrosine hydroxylase (TH)-immunoreactive ventral tegmental area dopmanine (VTA) perikarya were significantly lower in ghrelin-treated animals compared with saline-treated controls. Scale bars: 10 μm. (C) Corresponding to the observed changes in the number of synapses by light and electron microscopy, ghrelin treatment induced a significant elevation in the frequency of mEPSCs compared with saline controls. (D) Conversely, ghrelin administration triggered a significant decrease in the frequency of mIPSCs that was also in-line with the light and electron microscopy results. †$P < .05$ versus respective saline-treated controls. *From Abizaid A, Liu Z-W, Andrews ZB, et al. Ghrelin modulates the activity and synaptic input organization of midbrain dopamine neurons while promoting appetite. J Clin Invest. 2006;116:3229—3239; Epub Oct 19. PMC 1618869.*

prominent scientist in the field of addictive behavior) made the argument that it confuses people to mix up feeding behavior with that of behavior evoked by drugs of addiction. To have the paper succeed, we reluctantly complied with the request, which today I would not do. This was 2005—06 and that sentiment of separation of these behaviors dominated the state-of-the-art. I did not agree because it was clear to us (and soon to everyone) that mechanisms relating to feeding and addiction overlap and impact one another. A few years later, with Alfonso's lead, we published the ghrelin—cocaine story.[68]

2.4.2.1 The misconception of homeostatic versus hedonic eating

Cocaine induces plasticity of midbrain dopamine neurons that ghrelin seemed to be mimicking.[45] These and other findings argue that the distinction between homeostatic eating (to assure metabolic balance) and hedonic eating is really an artificial and likely unhelpful way to understand issues related to over- or undereating and their disorders, such as obesity and anorexia- or bulimia nervosa. The idea that there are segregated pathways to promote one versus the other is becoming clearly nonsensible. Even the original concept is not about biology, it is about how silos emerged in the field of neuroscience: the midbrain is about reward and addiction (hedonic aspects of life; "more" sophisticated behaviors), while the hypothalamus serves the primitive function of homeostatic control. It is becoming increasingly obvious that such segregation does not exist. From the conceptual perspective of biology, they are the same. The system that controls feeding (i.e., the brain) must assure that there is energy in a sustained manner to support hour-to-hour, day-to-day existence (one would mistakenly call this homeostatic need). It must assure that you eat whenever there is food and that the excess is stored in fat. Fat is your bank savings for survival when food (your currency for survival) does not flow daily. This means that when you eat enough to meet your hour-to-hour or day-to-day needs, you should not stop eating (if food is still available), so that if food becomes unavailable for some time you will not expire. Thus to assure maintenance of homeostasis (survival), you need to overeat and put on fat for reserve if possible. The base circuitry to support hunger is set up to promote that: hunger-promoting NPY/AgRP neurons are GABAergic neurons, which unidirectionally and tonically keep the satiety-promoting POMC neurons silenced. This basic wiring we described in 1992[11] and elaborated later[21] enables that. The fact that there is ongoing plasticity of the system also argues that the system demands soft wiring for rapid adaptations and reorganization of the base wiring. This also enhances the bias in the default wiring, which is to support eating. Finally, the base system in the hypothalamus is so "rigged" to promote feeding whenever it gets the chance that when a subject is completely satiated (supported by POMC neurons and silenced NPY/AgRP cells), it can instantaneously be flipped and satiety can, within minutes, be replaced by ferocious eating. The best example for this is the known effect of cannabis on hunger, especially when consumed after a meal. This is called the "munchies." In this process, after taking cannabis, the subject can develop a ferocious appetite and indiscriminately consume whatever food is around. We found a striking explanation for this phenomenon through a serendipitous chain of events. My friend and colleague in Leipzig, Ingo Bechmann (with whom we identified that NPY/AgRP neurons are GABAergic in 1997 while he was spending time in Csaba Leranth's lab), had a fellow named Marco Koch. Marco was working on the cannabinoid system and wanted to spend a few years with us in New Haven. When Marco arrived, I asked what he wanted to work on. My approach with trainees is that instead of forcing them to work on a specific project, I like them to develop one that best fits their interests and talents, but of course, with some relevance to my overall interest. Marco was persistent that he wanted to work on the cannabinoid system even though I had no real interest. I decided to compromise so I asked him to confirm or disconfirm whether the hypothalamic NPY/AgRP and POMC neurons are involved in the known effect of cannabis on eating when the subject is satiated and otherwise not hungry. The anticipation was that they are either not involved (this is "hedonic" feeding supposedly outside the realm of the hypothalamus), or if they are, we would see that the increased hunger is associated with elevated activity of NPY/AgRP neurons and suppressed activity of POMC cells. Either way, it was going to be something useful that could be published so Marco would not go home empty handed. However, the outcome was stunningly surprising. In a nutshell, Marco uncovered a complete flip in the functionality of these hypothalamic circuits that occurs within minutes! It turned out that the rapid onset of feeding triggered by cannabis was promoted by the same POMC neurons that signaled so that the animal would not eat as they were just fed.[69] The entire fundamentals of this hypothalamic circuitry, on which I had based my career, went up in smoke (pun intended; Fig. 2.16).

It turned out that the mechanism was a primitive one, but with major consequences: activated cannabis receptor on POMC neuronal mitochondria triggered mitochondrial fission mediated by uncoupling protein 2 (UCP2), which, in turn, switched the pool of large core vesicles that were released. The POMC gene product, proopiomelanocortin, is proteolytically processed to several active peptides, including adrenocorticotropic hormone, β-lipotropin, β-endorphin, and α-MSH. α-MSH is the product that is released from large core vesicles by POMC neurons in response to leptin and acts on melanocortin 4 receptors to promote satiety. β-endorphin is packaged in segregated large core vesicles and not coreleased with α-MSH. β-endorphin is a strong promoter of appetite in the short run.[31] Our observations showed that in response to cannabis exposure, at the time of satiety, the release of α-MSH-containing vesicles is switched to the release of β-endorphin-carrying vesicles, which in turn will act on μ-opioid receptors in the paraventricular hypothalamic nucleus to trigger ferocious eating (Fig. 2.17).[69] While these results can explain the action of marijuana

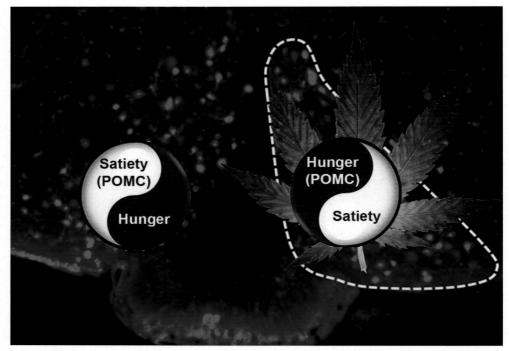

Figure 2.16 Exposure of animals to cannabis mimetics rapidly reversed the function of POMC neurons from supporting satiety and not eating to ferocious eating. *POMC*, Proopiomelanocortin.

POMC efferents

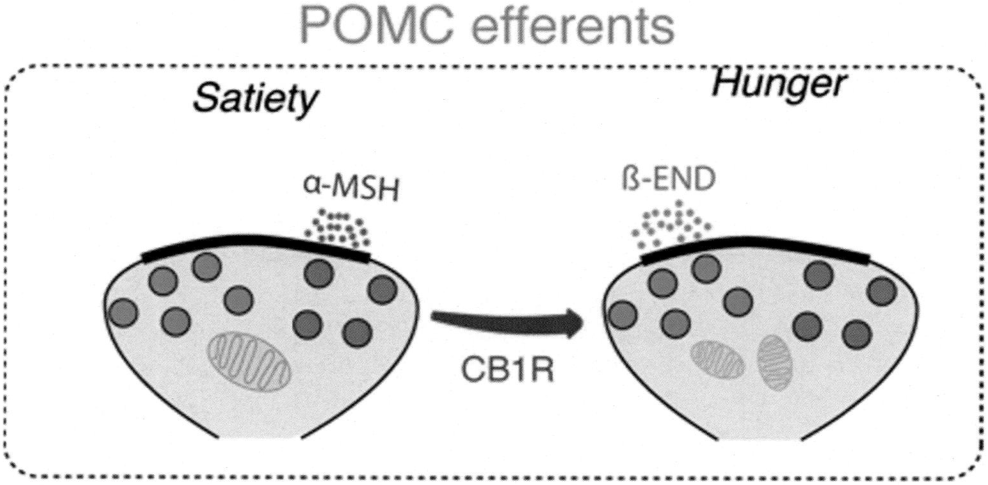

Figure 2.17 Schematic illustration of our findings regarding the cellular adaptations of POMC neurons in response to CB₁R activation, which switches POMC neurons from promotion of satiety to hunger. This process relies on mitochondrial adaptations controlled by UCP2 affecting mitochondria ER interactions and release kinetics of peptide-containing large core vesicles. We hypothesize that this event is relying on cell autonomous expression of CB₁R and UCP2 in POMC neurons. *ER*, Endoplasmic reticulum; *POMC*, proopiomelanocortin; *UCP2*, uncoupling protein 2.

consumption on eating habits, it may also be relevant for physiological control of eating. Most people might have experienced the situation in which one does not eat for a prolonged period. Let us say you skipped breakfast, did not have much time to eat lunch, and it is getting late in the day. You feel tired and low in energy, but you may not feel extraordinarily hungry. This can also occur in people who are food-deprived for prolonged periods. When food is provided, you start to eat, and as you continue eating your appetite really kicks in and you will then become all the hungrier as you eat. This is called rebound feeding. Trying to make up for the lost time when food was not coming into the body. At the time of the first bite of food, after prolonged food deprivation, endogenous cannabinoids are elevated in the brain[70]. When you start to eat, glucose levels will start to rise, and there will be a moment when there is a sufficient level of glucose to drive POMC neurons, but because there is still a high level of local cannabinoids in the environment on these cells, they will release endorphins rather than α-MSH. As you eat more, leptin levels will rise and endogenous cannabinoids will nadir, and from that time on, POMC neurons will release α-MSH and promote cessation of eating.

These surprising observations have implications for other aspects of neuroscience as well. First, it shows that a so-called "hedonic" response is initiated in or controlled from the hypothalamus. This adds to the argument that these artificial and arbitrary word-based categorizations of brain regions and functions are flawed. Obviously, when one goes after munchies after a Thanksgiving dinner for example, they are in an altered mindset. This suggests that these so-called "homeostatic" centers may have much more to do with higher brain functions than the discipline of neuroscience assigns to them. Again, my argument is not that these particular cells in the hypothalamus are "superior" to other parts of the brain, but rather that the historical hierarchical view on brain regions and peripheral signals and their contribution to complex, goal-oriented behaviors is inaccurate. I will illustrate this further with results regarding the effects of the gut hormone ghrelin and hunger-promoting AgRP neurons on processes related to nonclassic eating mechanisms.

2.4.3 Plasticity of the hippocampus (learning, memory, and exercise)

It was soon after our finding of synaptic plasticity of feeding circuits in 2004, when I became interested in pursuing this idea that these "primitive" homeostatic hormones, such as ghrelin, may impact higher brain functions. The study grew out of my then multiple year collaboration with Matthias Tschöp. Matthias was the first to show that ghrelin promotes appetite in his landmark 2000 paper[71], a study that he did and published while at Eli Lilly. In fact, I met Matthias in Indianapolis in 2000, where I gave a talk at Eli Lilly on our emerging findings on mitochondria and neuronal functions. At that time, his paper was not out, but we had a great conversation about science and other topics. Once he exploded in the field, we continued our interactions and started to collaborate on many things, including the issue about ghrelin and the hippocampus. That work was a true broad collaboration between multiple groups and multiple sites. Sabrina Diano and I were coordinating all the complexities.

To briefly summarize, we found ghrelin receptors (growth hormone secretagogue receptor) in the hippocampus and dentate gyrus and that ghrelin binds to these areas.[72] We found that ghrelin can trigger a rapid formation of dendritic spines in the stratum radiatum of the CA1 region of the hippocampus (Fig. 2.18), promotes LTP (Fig. 2.19), and that these changes impact spatial learning and memory (Fig. 2.20). The effect was sufficient to reverse some of the impairments observed in a mouse model of Alzheimer's disease.[72]

Working on this study taught me a lot about behavioral neuroscience. Other than measuring how much an animal eats, I had never been closely involved with the analysis of animal behavior. I did know, however, through conversations that certain people interested in certain brain regions and (assumed) related behaviors usually have their favorite paradigm. It was also clear that there was no agreement among them on which is more informative. Because we had limited experience of our own, Sabrina and I decided that we would collaborate with multiple labs using different paradigms, and, if the outcomes correspond, we would feel comfortable. It turned out that the results supported each other from the cellular level to the behavior of awake, freely moving mice.

These results provided a direct mechanistic explanation for a route of communication directly from the gut to the hippocampus and associated complex behaviors. These findings somewhat explain the term "gut feeling." It also connects the peripheral metabolic state directly to processes relevant to degenerative processes, such as Alzheimer's disease. It also fits the idea of the beneficial effects of calorie restriction on age-related impairments as during that intervention, ghrelin levels are elevated in a sustained manner. The other fundamentally critical message of the observation is that it is more efficient to the acquisition of knowledge when ghrelin levels are high, that is, when one is hungry rather than when ghrelin is low, for example, after a meal. This has multiple implications, most notably for the education and learning environment of trainees, including children. There is a belief that the most important meal of the day is breakfast. That may be so but from the perspective of our findings, it is not about the meal itself or its timing but instead about what would not suppress ghrelin levels for a prolonged period during the time of data acquisition. That coupled with the other finding that ghrelin drives neuronal activity via a lipid utilizing pathway[59] would also be potentially beneficial if during this period lipids are more readily available from the food.

Beyond the alteration of mnemonic functions, ghrelin administration to the hippocampus also promoted feeding,[73] and a recent study by Buzsaki's lab connected hippocampal

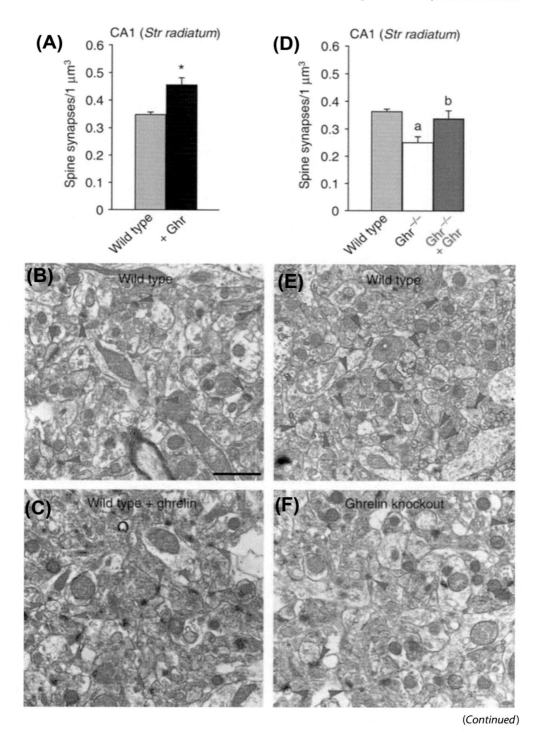

(Continued)

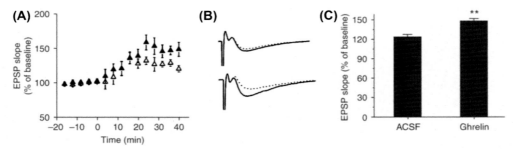

Figure 2.19 (A–C) We tested the effect of ghrelin on generation of LTP in slice preparations. The excitatory postsynaptic potential (EPSP) slope at 35–40 min after 10 Hz stimulation was 124 ± 3.3 (n=12 slices, 10 mice) in artificial cerebrospinal fluid (ACSF)-treated slices and 148.0 ± 3.8 (n=10 slices, 9 mice) in the ghrelin-treated group ($P < .05$), revealing a significant ghrelin-induced change in hippocampal long term potentiation (LTP). *From Diano D, Farr SA, Benoit SC, et al. Ghrelin controls hippocampal spine synapse density and memory performance. Nat Neurosci. 2006;9:381–388. Published online: 19 February 2006 https://doi.org/10.1038/nn1656.*

oscillations as predictors/drivers of fluctuations in peripheral glucose concentrations.[74] These results further emphasize the need to reevaluate regional specifications of the brain and argue for the involvement of broader (if not all) regions of the brain being involved in the control of homeostatic functions relating to metabolism.

Exercise is a known contributor to brain functions at almost all levels with many types of health benefits. For example, voluntary exercise has been shown to promote synapse function of the hippocampal formation in support of enhanced learning and memory performance of animals and humans.[75] In this process, they identified a role for new synapse formations in the dentate gyrus in response to exercise. Marcelo Dietrich (in my laboratory) connected these cellular adaptations to exercise via a mitochondrial lipid metabolizing pathway.[76] Marcelo joined my lab in 2005 as a medical student from Brazil, who interrupted his medical studies to immerse himself in research. He was spending time in Madrid, the Cajal Institute, and he expressed

◀ **Figure 2.18** (A–C) Peripheral ghrelin administration to wild-type mice resulted in an increased density of spine synapses in the stratum radiatum of the hippocampal CA1 subfield (D, a: *$P < .05$). A representative electron micrograph showing spine synapses (*red arrowheads*) in control wild-type (B) and ghrelin-treated wild-type animals (C). (D–F) Ghrelin knockout animals exhibited a lower density of hippocampal spine synapses than wild-type littermates (B: *$P < .05$); however, ghrelin administration to these transgenic animals elevated spine synapse number ($P < .05$) making this parameter indistinguishable from wild-type values. A representative electron micrograph showing spine synapses (red arrowheads) in wild-type (E) and ghrelin knockout animals (F). Scale bar in (B) represents 1 μm for panels (B), (C), (E) and (F). *From Diano D, Farr SA, Benoit SC, et al. Ghrelin controls hippocampal spine synapse density and memory performance. Nat Neurosci. 2006;9:381–388. Published online: 19 February 2006 https://doi.org/10.1038/nn1656.*

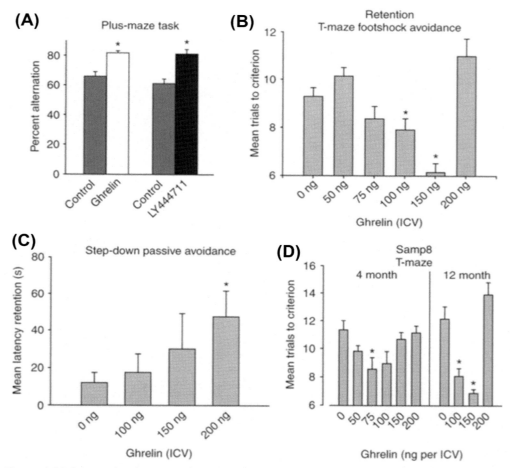

Figure 2.20 Bar graphs showing enhanced performance of animals after ghrelin or ghrelin analog treatment in the plus maze task (A, *P < .05), after ghrelin treatment in the T-maze foot shock avoidance (B, *P < .05) and step-down passive avoidance (C, *P < .05) tasks. (D) Ghrelin also enhanced the performance of SAMP8 mice in the T-maze at 4 months of age (*P < .05) as well as at 12 months (*P < .05). At 12 months of age, these animals showed Alzheimer disease-like symptoms, and a higher dose of ghrelin was required to elicit the same response in these older animals. *From Diano D, Farr SA, Benoit SC, et al. Ghrelin controls hippocampal spine synapse density and memory performance.* Nat Neurosci. 2006;9:381−388. Published online: 19 February 2006 https://doi.org/10.1038/nn1656.

interest to his supervisor, my friend and colleague, Ignacio Torres-Aleman, that he would like to come and work with me at Yale. He was an eager, intelligent fellow who rapidly understood the interest and intent of my laboratory. For the next 9 years, with interruptions so that he could finish his medical training, he became a critical partner in our research endeavors.

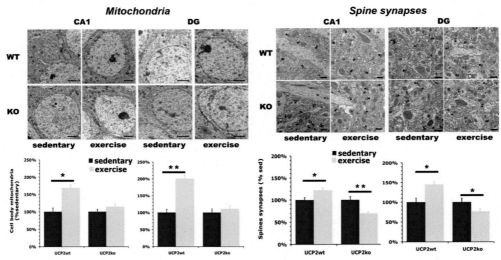

Figure 2.21 Effect of voluntary wheel running on mitochondria and synapses in the hippocampus and dentate gyrus. *Left panels*: Mitochondrial number in the cell body of neurons from the dentate gyrus and CA1 of sedentary (*n*=10 of each genotype) and exercised (*n*=10 of each genotype) UCP2wt and UCP2ko mice. Top, representative electron microscopic images of neuronal cell bodies with mitochondria marked in red. Bottom left, histogram representing a quantitative analysis of mitochondrial number in cell bodies of CA1 pyramidal neurons. Bottom right, histogram representing mitochondrial number in dentate gyrus (DG) neurons. Scale bars, 1 μm. *$P < .05$; **$P < .01$. Error bars represent SEM. *Right panels*: Spine synapse counts of neurons from the dentate gyrus and CA1 of sedentary (*n*=10 of each genotype) and exercised (*n*=10 of each genotype) UCP2wt and UCP2ko mice. Top, representative electron microscopic images of spine synapses marked in blue. Bottom left, histogram representing a quantitative analysis of spine synapse number in CA1 pyramidal neurons. Bottom right, histogram representing spine synapse number in DG neurons. Scale bars, 1 μm. *$P < .05$; **$P < .01$. Error bars represent SEM. *UCP2*, uncoupling protein 2. *From Dietrich M, Andrews ZB, Horvath TL. Exercise-induced synaptogenesis in the hippocampus is dependent on UCP2-dependent mitochondrial adaptation. J Neurosci. 2008;28:10766−10771.*

Marcelo has always been fascinated by the effect of exercise on the brain, in part, because he was (and remains) an avid long-distance swimmer. He set up cohorts of mice on running wheels that were free to run for 4 weeks. He found that hippocampal tissue metabolism was induced by wheel running. It is one thing that whole body, including muscle metabolism, is enhanced by endurance exercise, it was remarkable to find that the respiration of mitochondria in the hippocampus is significantly increased as well. In other words, when you exercise, you also boost the metabolic profile of the hippocampal formation, an area that is crucial for acquiring new knowledge. These metabolic changes came with mitochondrial proliferation (biogenesis or fission) and increased spine synapses of principle cells in the dentate gyrus and the CA1 region of the hippocampus (Fig. 2.21). Marcelo and Zane Andrews (also in my lab) identified a molecule that appeared to be responsible for this metabolic shift in the brain. This molecule was mitochondrial UCP2, which Zane in collaboration with Sabrina Diano's lab had just identified at that time as a key protein in promoting lipid metabolism and the control of reactive oxygen species (ROS) in other neurons (for further details see Section 2.5 and[59]). They ran these exercise studies using both UCP2 knockout and wild-type animals. If this critical protein of lipid metabolism was not present, exercise did not induce metabolic boost of the hippocampal formation, mitochondria could not proliferate, and spine synapses did not increase (Fig. 2.21).

Endurance activity thrives on lipid as fuel. One of my favorite examples for this is the migratory hummingbird (Fig. 2.22). The migration of hummingbirds can involve nonstop, 18–22 hours flying over the Gulf of Mexico. For this trip, they use almost 50% of their body weight in the form of fat as fuel. It is an amazing accomplishment for these tiny (3–4 g body weight) animals.

Beside migrating birds, Marathon runners, long-distance swimmers and other endurance athletes also switch fuel utilization from carbohydrates to lipids to enable sustained muscle activity. During the start of a run, for example, 5 km, there is the first 3–10 minutes, which is the least pleasant as the body is still relying on carbohydrates as fuel for the muscles and during which lactate is released, which is the cause of the unpleasantness (lactate threshold). Once the switch occurs to lipids as the main fuel for muscles, running becomes much more comfortable. The systemic and cellular

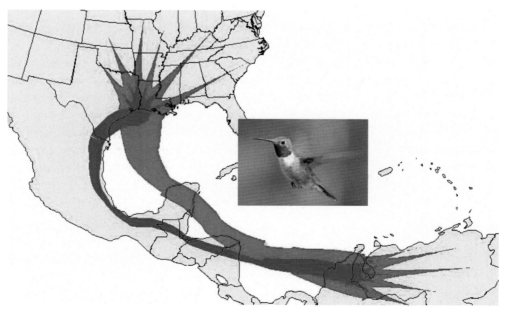

Figure 2.22 Hummingbirds (central picture shows a Ruby-throated Hummingbird) migrate between South-, Central-, and North America spending most of their time in the air and relying almost exclusively on body lipids for fuel for flying.

principles of fat burning (see later) explain why it is beneficial for an organism to utilize fat instead of carbohydrate for endurance activity and perhaps survival in general.

The take-home message of the previous information regarding hippocampus and metabolism is that the structural and functional aspects of promotion of learning by the hippocampus appear to be driven by lipid metabolism (ghrelin and exercise). The physiological
correlates of these events are elevated lipid availability and utilization. This scenario in the
spectrum of hunger and satiety is within the hunger and not the satiety state. It is logical
that when an animal is hungry, it is then that it needs to be most aware of the environment
and learn how survival is best promoted. When it finds food and has plenty of it, satiety will
be reached, and the animal will rest to assure the processing of the acquired fuels. There is
an easy test that can be tried to experience this firsthand. Learn a poem before a huge meal
and then try to learn another one after. It is not a coincidence (although perhaps done without much thought) that most behavioral experiments are done on hungry, unsated animal
subjects. In fact, frequently, they work for food reward.

2.5 Cellular metabolic principles of neuronal responses to the changing metabolic environment

It is one thing to learn something new and another to retain knowledge. For the former, hunger, for the latter, satiety is crucial. They have very different cellular metabolic principles. In the following, I describe cellular events that drive AgRP neurons
under conditions of low-energy availability (to promote hunger), and POMC neurons
that drive satiety under acute energy surplus (after eating).

When AgRP neurons are active, POMC cells are quiescent, and vice versa. Of
course, it makes perfect sense from the perspective of their functional relevance (hunger vs satiety), but how it is regulated at the cellular level has been elusive. When glucose levels decrease and circulating ghrelin increases before meals (or after a prolonged
absence of food), AgRP neurons are activated and POMC neurons are silent, thereby
promoting hunger and lipolysis in the periphery. In the past decade, the idea emerged
that during negative energy balance, pathways promoting lipid metabolism are also
activated in the brain, specifically in the hypothalamus. Indeed, it is lipid utilization
and metabolism that appear to drive AgRP neurons during negative energy balance,[59]
while it is glucose utilization and generation of ROS that seem to drive satiety promoting POMC neurons.[77] Both of these scenarios are likely as circulating lipids are
elevated and glucose is lowered in hunger states, while glucose levels are elevated after
a meal.

2.5.1 Mitochondrial uncoupling protein 2 in neurons

Our interest in the question of cellular metabolic principles of hypothalamic neuronal
functions arose from multiple converging factors in the period between 1994 and

1999: we were working on ARC NPY and POMC cells (1990—); leptin and its receptors were discovered by Jeff Friedman (1994—95) and found to act in the hypothalamus, specifically on NPY and POMC neurons (1997); Sabrina Diano joined our lab at Yale with a thyroid/metabolism research interest and background from Naples, Italy (1994); she and I attended an NPY conference in London, where we learned about the discovery of novel mitochondrial uncouplers, including UCP2[78]; and we localized UCP2 to ARC NPY neurons.[79] Sabrina and I thought it was a very "cool" molecule, which, if it worked as predicted, might "heat" up the synapse, and thus, we proposed that there may be "thermal synapses" in the brain (Fig. 2.23).[79,80] This principle was later strengthened by the work that Sabrina's lab lead, in which a glia—neuronal interplay utilizing locally formed thyroid hormone was identified as a putative driver of such a thermogenic mechanism.[81]

From this point on, our interest turned to neuronal mitochondria and their potential role beyond its classically accepted the function of ATP generation and cell death. The name *mitochondrion* was introduced in 1898 by Carl Benda, and studies covering the past one-hundred-some years have demonstrated the fundamental role of these organelles in providing energy to the cells. Mitochondria are the primary energy source for cellular function by converting nutrient flux into ATP. Products of glucose and fatty acids are actively transported across the inner mitochondrial membrane and into the matrix where they enter the tricarboxylic acid cycle and are oxidized to carbon dioxide. In the process, there is a production of guanosin triphosphate (GTP) and reduced cofactors, nicotinamid adenin dinucleotid (NADH) and FADH2. These cofactors are then oxidized by the electron transport chain. This process is accompanied by a transfer of protons (H^+ ions) across the inner membrane that creates an electrochemical proton gradient, providing energy to power ATP synthase and the generation of ATP. However, protons can reenter the mitochondrial matrix through facilitated diffusion, a process called mitochondrial uncoupling. This proton leak is, in part, mediated by inner membrane proteins encoded by the nuclear DNA, the UCPs. There is also a leak of electrons from the respiratory chain leading to the generation of ROS.

Sabrina and I were intrigued by the mitochondria and by UCPs, UCP2 in particular. There was and has been a considerable amount of debate regarding the role of UCPs other than UCP1 is a bona fide uncoupler. There were and continue to be disagreements about the physiological and pathological roles of UCP2. People have been questioning the ability of UCP2 to transfer protons through the inner membrane of the mitochondria, which is the innate function of UCP1 located in the brown adipose tissue and responsible for thermogenesis. I had the great privilege to meet and discuss matters relating to these processes with pioneers of this field. In 2002, at a meeting organized by Sabrina's graduate school supervisor, Fernando Goglia, we met Vladimir Skulachev, a Russian scientist who made fundamental discoveries regarding mitochondrial uncoupling back in the 1960s in the Soviet Union.[43] We had a fascinating conversation about these emerging new UCPs

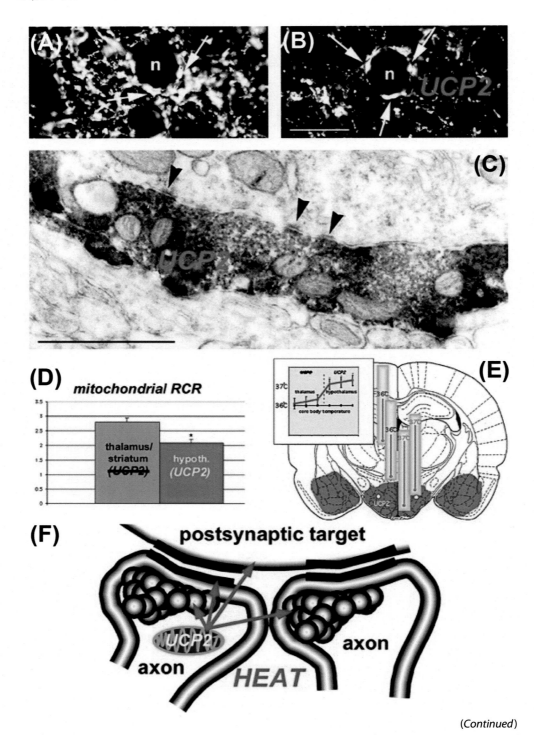

(Continued)

and how and under what circumstance they might function. We thought about various possibilities how UCP2 might be a conditional uncoupler, for example, controlled by voltage. Around the same time, a very attractive concept was put forward by Brand et al. (in which UCP2 is a conditional uncoupler when it is interacting simultaneously with oxygen–free radicals in the mitochondrial matrix that are derived from lipid peroxidation products).[82] The outcome of this process is neutralization of ROS, a function that had been assigned to UCP2 earlier and pursued in the quest for an understanding of neurode-generation.[83] For example, we observed that UCP2 expression is induced in the hippo-campus after lesions or in models of epilepsy, and, in the substantia nigra in models of Parkinson's disease.[84–86] In all these cases, UCP2 was found to control ROS generation and diminish degenerative processes. Similar observations were made regarding ischemic brain injury.[87] We also found that the engagement of UCP2 promotes proliferation of the mitochondria, a process considered by most as a beneficial adaption for cellular integrity and function. Thus, in 2005, we proposed that UCP2 is an important regulator of neuro-nal functions by controlling cellular activity, plasticity, and integrity (Fig. 2.24).[83]

2.5.2 Fat metabolism drives hunger promoting NPY/AgRP neurons

By 2006, the concept of UCP2 being involved in neuronal ROS control under con-ditions of lipid metabolism was very attractive to us in pursuit of the molecular drivers of hunger promoting AgRP neurons at a time of food scarcity (when blood levels of fat are elevated and glucose is diminishing). We started to do this in earnest in collabo-ration with Sabrina's lab 2 years before, when Zane Andrews joined my lab as a post-doctoral fellow. Zane was from New Zealand, and he threw himself at the task with a fearless, Kiwi attitude. By 2006, we made strides in this endeavor and submitted a paper to Neuron, which was reviewed, but roundly rejected. That fall, I attended a

◀ **Figure 2.23** Brain UCP2 in axons predicts increased mitochondrial proton leak and heat production. Light (A, B) and electron (C) micrographs demonstrate the abundant expression of UCP2 in presynaptic axon terminals. Arrows on the light micrographs of (A) and (B) point to UCP2-containing presynaptic terminals in the central amygdaloid nucleus that establish symmetrical synapses (C, *arrowheads*) on the postsynaptic target. (D) The measurement of mitochondrial respiration in brain regions (left panel) where UCP2 is present (hypothalamus) and where no UCP2 was detected (thalamus/striatum) showed a lower mitochondrial phosphorylation level [lower respiratory control ratio (RCR)] in the hypothalamus that was caused exclusively by an increased proton leak of mitochondria in UCP2-containing regions. (E) In agreement with this increased mitochondrial uncoupling activity, brain temperature in the UCP2-expressing hypothalamus was significantly higher than that of the thalamus (right panel) and the core body temperature. (F) The presence of UCP2 in axon terminals together with the significant, positive correlation between UCP2 and mitochondrial uncoupling and local brain temperature suggests that heat produced presynaptically by UCP2 may have a direct influence on axonal temperature leading to the modulation of presynaptic and postsynaptic events. *UCP2*, Uncoupling protein 2. *From Horvath TL, Warden CH, Hajos M, Lombardi A, Goglia F, Diano S. Brain UCP2: uncoupled neuronal mitochondria predict thermal synapses in homeostatic centers. J Neurosci. 1999;19(23):10417–10427.*

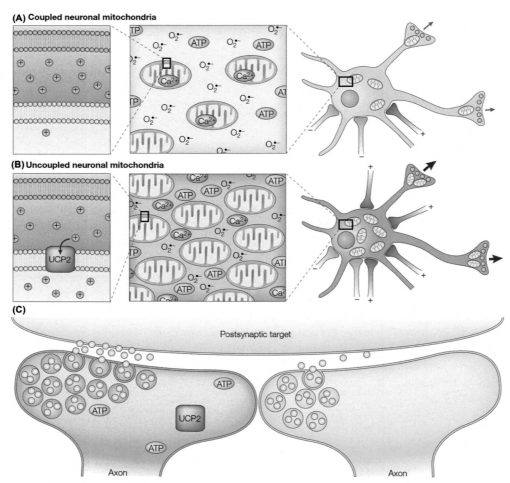

Figure 2.24 Proposed mechanism through which neuronal uncoupling proteins can regulate neuronal function. (A) Coupled mitochondria have a large mitochondrial membrane potential across the inner mitochondrial membrane. A large membrane potential promotes strong proton drive through ATP synthase and, as a consequence, enhances ROS production and mitochondrial calcium influx, both of which are known to promote neuronal dysfunction. Eventually, this coupled mitochondrial state limits synaptic plasticity and neurotransmission. (B) Increased UCP activity allows controlled proton reentry (leak) into the mitochondrial matrix without affecting ATP synthase activity. Consequently, ROS production is diminished and mitochondrial calcium efflux is increased. Although acute uncoupling reduces mitochondrial ATP production due to decreased proton drive through ATP synthase, chronic uncoupling leads to mitochondrial proliferation and greater ATP production per cell. Therefore uncoupled neurons are readily amenable to dynamic fluctuations in neuronal activity and adapt rapidly and efficiently through enhanced synaptic plasticity and neuronal transmission. (C) Shows UCP-induced mechanisms that lead to enhanced neurotransmission (left). By dissipating the mitochondrial membrane potential, uncoupling leads to local heat

(Continued)

special workshop organized by the National Institute of Aging (organized by Ron Kohanski) titled Uncoupling Proteins and Aging. There I had the good fortune to meet and have dinner with great minds in biochemistry of mitochondria: Martin Brand, David Nicolls, and Daniel Ricquiert. The depth and breadth of information these people had on mitochondria was amazing. Equally remarkable to me was their mutual respect for one another, even though they disagreed on almost every detail that was discussed around the dinner table. I enjoyed watching the banter between these great scientists and was actually very encouraged by their disagreements. During the meeting, in reference to our work, one of them noted that we have no "right" to study UCP2 in the absence of understanding its function. That statement was transformative for me, because it highlighted and reinforced (as it was not the first nor the last such comment about our work) that within the boundaries of reason, you are doing something novel and relevant if prominent people start to pick on you and disparage your work. We did reconsider our work on UCP2 and hypothalamic NPY/AgRP neurons and eventually published it in *Nature*.[59] The work had a very simple message: in response to hunger signals (ghrelin), AgRP neurons are driven by an intracellular long-chain fatty acid metabolizing pathway, and UCP2 serves as a critical enabler of that process, in part, by maintaining ROS control (Fig. 2.25).[59]

This scenario is sensible from multiple perspectives. First, during food deprivation (hunger), lipids released from the fat stores become the dominant fuel for cells. Thus, for a neuron that really needs to function during this time (to promote the need to eat), it makes perfect sense that they would use the available fuel, lipids. Second, these neurons need to function for prolonged periods until food is found. Considering that our current human condition of food abundance is a fluke in evolution, these periods could last for a day, weeks, or even longer. Thus it is also reasonable that the cellular energetic process that "feeds" these neurons is with the least amount of ROS generation. Sustained ROS levels can have massive damage inflicted on multiple cellular compartments, including the nucleus (DNA) and membranes (lipid peroxidation). From a broader conceptual perspective, the previous characteristics of lipid utilizing pathways are also in-line with the known life-enhancing effect of calorie restriction,[88] a nutritional intervention that will propagate almost continuous activation of AgRP neurons and lipid utilization as long as the intervention lasts (days, weeks, years, or decades).

◀ generation. It is therefore proposed that local temperature gradients increase neurochemical diffusion through extracellular compartments to postsynaptic neuronal targets. In addition, elevated cellular ATP promotes active processes—such as vesicle formation, transportation, and exocytosis—at the presynaptic nerve terminal, thereby promoting neurotransmission. *ROS*, Reactive oxygen species; *UCP*, uncoupling protein. *From Andrews ZB, Diano S, Horvath TL. Mitochondrial uncoupling proteins in the central nervous system: in support of function and survival. Nat Rev Neurosci. 2005;6 (11):829—840.*

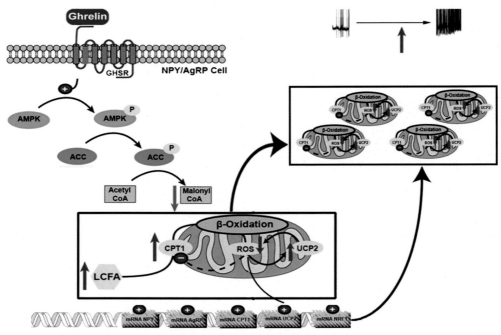

Figure 2.25 Schematic diagram showing the proposed sequence of intracellular events leading to ghrelin-induced activation of NPY/AgRP neurons. Ghrelin binds to its cognate membrane bound receptor, the growth hormone secretagouge receptor (GHSR) on NPY/AgRP neurons leading to immediate induction of action potentials concomitant with phosphorylation and activation of AMP kinase (AMPK). Phosphorylation of AMPK inhibits the activity of acetyl-coenzyme A carboxylase (ACC), reduces malonyl CoA levels, and thereby disinhibits carnitine palmitoyltransferase 1 (CPT1). Increased activity of CPT1 utilizes more long chain fatty acid (LCFA) for mitochondrial beta oxidation and subsequently generates ROS production. Fatty acids and ROS increase UCP2-dependent uncoupling activity and UCP2 mRNA, which subsequently reduces the overproduction of ROS in a feedback manner and thus reduces the inhibitory effect of ROS on CPT1 activity (as indicated by the dashed line with negative symbol) and contributes to the maintenance of increased action potentials. Buffering of ROS production by UCP2 allows appropriate ghrelin-dependent gene transcription events to occur. Increased mitochondrial biogenesis in wild-type mice enables sustained neuronal firing over time. AgRP, Agouti-related peptide; NPY, neuropeptide Y; ROS, reactive oxygen species; UCP2, uncoupling protein 2. *From Andrews ZB, Liu Z-W, Wallingford N, et al. UCP2 mediates ghrelin's action on NPY/AgRP neurons by lowering free radicals. Nature. 2008 Jul 30.*

2.5.3 Glucose utilization and ROS drive satiety-promoting POMC neurons

When glucose, leptin, and insulin levels rise in the circulation following eating, POMC neurons are activated and promote cessation of feeding. There is a cascade of signaling events triggered by leptin binding to leptin receptors in POMC neurons that Elmquist,[89] Myers,[90] and others elegantly described. However, glucose is critical for POMC neuronal activation. It has been argued that glucose-dependent POMC activity

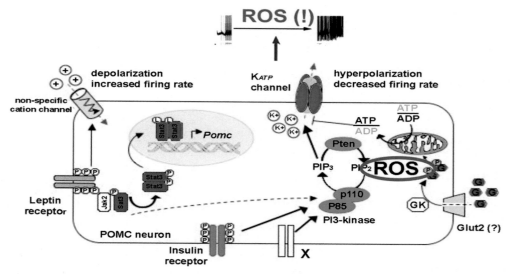

Figure 2.26 Cellular attributes of activation of POMC neurons after a meal. After eating, glucose leptin and insulin levels rise in the circulation. Each of these metabolic signaling molecules impacts POMC neurons at the level of signaling and at the level of mitochondrial fuel utilization. Leptin acts through the Jak—Stat signaling pathway to affect the transcription of various genes. Insulin receptors act on multiple signaling systems, including the PI2 kinase pathway. Glucose enters the cell and it goes through metabolism leading to ROS generation. ROS is also promoted by many other pathways, including PI2 kinase—initiated signaling events. The generated ROS has an important role in the electric activation of POMC cells that is also enabled by K_{ATP} potassium channels. *POMC*, Proopiomelanocortin; *ROS*, reactive oxygen species.

is promoted, in part, by the activation of ATP-sensitive potassium (KATP) channels, which leads to the depolarization of the cells. In conditions of elevated glucose and leptin levels, increasing activity of POMC neurons is also associated with an increase in ROS levels, the by-product of substrate utilization by the mitochondria.[77] In fact, we showed that selective sequestration of ROS in the hypothalamus promoted hunger,[59] while increasing hypothalamic ROS levels induced POMC neuronal activation and promoted satiety.[77] In addition, we unmasked that impairment of POMC neuronal firing during diet-induced obesity, in the face of elevated leptin levels (a state referred to as "leptin resistance"), can be reversed by elevation of ROS levels.[77] Thus the physiological promotion of cessation of eating after a meal is supported by ROS (Fig. 2.26).

2.5.4 Fission and fusion of mitochondria in hunger and satiety

Because hunger and satiety follow each other in a continuum, adaptation of the intracellular site where fuels are utilized, the mitochondrion is crucial to make proper switches in fuel utilization to support behaviors and autonomic functions in the face of the changing peripheral metabolic environment.

The role of mitochondria is known to extend far beyond the sole role of supplying energy for cellular function. Over the past few decades, mitochondria have been shown to be involved in signaling through ROS, regulation of the membrane potential, apoptosis-programmed cell death, calcium signaling (including calcium-evoked apoptosis), and regulation of cellular metabolism. In 1914, Lewis and Lewis were already describing mitochondrial dynamics in living cells "The mitochondria are almost never at rest, but are continually changing their position and also their shape".[91] Mitochondria continually adapt their shapes through fusion, fission, and motility in response to mitochondrial life cycle and changes in energy demand and supply balance. Mitochondrial dynamics are essential for the proper function of the cell including viability, apoptosis, but also bioenergetic adaptation.

Fission is thought to occur when mitochondria are damaged or accumulate deleterious components and in response to high levels of cell stress. Fission is thought to segregate damaged mitochondria and can lead to autophagy. However, fission also correlates with the changing availability of fuels. Specifically, we observed that fission is triggered in HeLa cells[92] and AgRP neurons[93] when fat utilization is increasing at the expense of carbohydrates (Fig. 2.27).[93] In contrast, we found that mitochondrial fusion is critical for the activation of POMC neurons to promote satiety at a time of increasing carbohydrate availability (Fig. 2.27).[92] Fusion of mitochondria can compensate for each other's defects by sharing RNA or proteins. Fusion also enhances communication with the endoplasmic reticulum and this process is important in control of intracellular calcium homeostasis. The fundamentals of these principles can readily be disturbed in both AgRP and POMC neurons in response to continuous overnutrition,[93,94] which is, in part, the consequence of ROS (Fig. 2.27).

I believe that the cell biological principles of mitochondrial dynamics in relation to hunger and satiety, and fat and carbohydrate utilization, are not unique to these hypothalamic neurons. Rather, it is likely that these fission and fusion "cycles" in neurons and other cells within and outside of the brain are reflective of the given metabolic milieu. It is also reasonable to suggest that these cycles in fission and fusion, and fat and carbohydrate availability, determine cellular functions matching the ever-changing function of cells and tissues to most successfully adapt to the environment. This also indicates that those molecules that govern fission and fusion of mitochondria, including dynamin-related protein 1, UCP2, mitofusins, OPA1., and their eventual impairments may play fundamental roles in multiple disorders, including those associated with brain functions.

2.6 The conundrum of interventions based on metabolic principles to fight obesity and aging at the population level

The signaling modality necessary for the activation of POMC neurons and to reduce eating is ROS based. This process is crucial for reaching satiety after a meal and for the

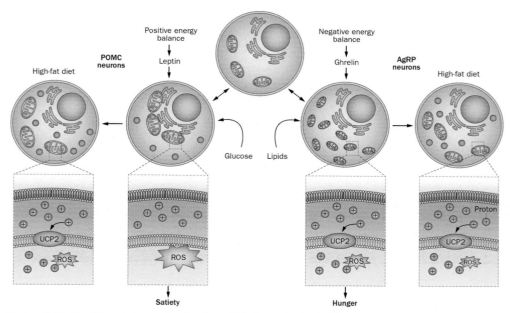

Figure 2.27 Feeding states and mitochondrial dynamics in NPY—AgRP and POMC neurons. The activity of POMC neurons promotes satiety associated with glucose utilization, elevated ROS levels, increases in mitochondrial fusion, and contacts with the ER. By contrast, the activity of NPY—AgRP neurons during fasting or caloric restriction promotes the utilization of lipids and hunger, which is associated with low levels of ROS and increases in mitochondrial fission. Exposure to a high-fat diet promotes the inhibition of mitochondria—ER interactions, increases in peroxisome proliferation and UCP2 activity, which leads to increased proton leak and reduced levels of ROS in both POMC and NPY—AgRP neurons. *AgRP*, Agouti-related protein; *ER*, endoplasmic reticulum; *NPY*, neuropeptide Y; *POMC*, proopiomelanocortin; *ROS*, reactive oxygen species; *UCP2*, uncoupling protein 2. *From Nasrallah C, Horvath TL. Mitochondrial dynamics in the central regulation of Metabolism. Nat Rev Endocrinol. 2014;10:650—658. https://doi.org/10.1038/nrendo.2014.160. Epub 2014 Sep 9.*

promotion of the clearance of glucose from the circulation. Indeed, impaired functioning of the POMC system during diet-induced obesity not only interferes with control of eating, but it can also contribute to the development of type 2 diabetes. To successfully promote decreased eating and clearance of glucose, ROS must be in an elevated state in the hypothalamus. Because clearance of glucose means that sugars are loaded to various peripheral tissues, the satiety process, by default, will come with elevated ROS levels in peripheral tissues as well (Fig. 2.28). Thus, if a medication successfully propagates satiety by decreasing hunger via the hypothalamus, it will result in sustained elevation of ROS in the brain and the periphery. This will result in loss of weight and correction of diabetes, but, due to the multitude of harmful effects of chronic ROS exposure, it is equally likely to cause long-term unwanted tissue impairments and potentially premature death. A tissue that is most vulnerable to overload of glucose is the heart. Cardiomyocytes, similar to

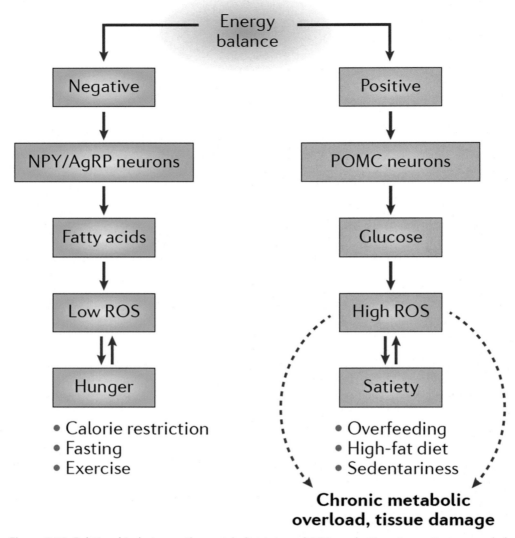

Figure 2.28 Relationship between the metabolic state and ROS production. A negative energy balance—a metabolic state that is characteristic of chronic calorie restriction—activates NPY/AgRP neurons in the hypothalamus. These neurons preferentially utilize free fatty acids as the main source of fuel to sustain firing. Low levels of ROS production are maintained in these neurons, thereby allowing them to fire at higher rates. Increases in ROS levels in the NPY/AgRP neurons have deleterious effects and prevent increases in firing rate. On the other hand, a positive energy balance—a state that is characteristic during periods of overfeeding—promotes the activity of POMC neurons in the hypothalamus. These cells mainly utilize glucose for energy metabolism, which leads to a high production of ROS. Chronic promotion of these mechanisms will lead to an accumulation of ROS and consequent damage in different tissues. *AgRP*, Agouti-related peptide; *NPY*, neuropeptide Y; *POMC*, proopiomelanocortin; *ROS*, reactive oxygen species. *From Dietrich MO, Horvath TL. Limitations in anti-obesity drug development: the critical role of hunger-promoting neurons in integrative physiology.* Nat Rev Drug Discovery. 2012;11(9):675—691. *https://doi.org/10.1038/ nrd3739. [Epub ahead of print].*[95]

AgRP neurons, work very efficiently and without significant damage by preferentially utilizing fat as fuel for their continuous function. A shift toward increased satiety and POMC tone may result in cardiomyopathies and heart failure.

Late onset chronic diseases, including neurodegenerative disorders such as Alzheimer's disease, dementias, and Parkinson's disease are leading causes of mortality and morbidity creating enormous emotional and financial burden on the society. As the aging population grows, it is expected that neurodegenerative disorders will further impact our society. Guided by the idea that late onset chronic diseases are a consequence of prolonged overworking of tissues that have certain vulnerabilities (e.g., genetic, epigenetic), it is reasonable to assert that the cellular energy metabolism of brain cells will determine the health and longevity of brain functions. The robust effects of calorie restriction on chronic diseases lend support to the argument that late onset disorders are the consequence of sustained high levels of substrate oxidation by the various tissues.

Fig. 2.27 as well as the description of cellular principles of hunger previously indicate the favorable bioenergetic profile of native energy balance, which can be accomplished by, among other means, calorie restriction. In-line with this, during the past decades, it became increasingly popular to argue and promote the notion that controlled calorie restriction results in the avoidance of age-related illnesses (such as diabetes, Alzheimer's disease, and cancers) and will actually add years to your life.[88] The argued beneficial effects of calorie restriction are based on many animal studies and also some preliminary human experiments. Clearly, the cellular principles of calorie restriction should immediately predict a beneficial outcome at every level: cell, tissue, and organism. However, the picture might not be that simple. One issue is with ROS itself. While sustained ROS levels are considered by almost everyone as a negative force on health, ROS provide a crucial signal for fundamentally important cellular and tissue events. Thus indiscriminately suppressing ROS (by sustained calorie restriction), by default, will interfere with numerous important physiological processes. Another issue is that you need energy surplus to build structures in cells and tissues. Yes, overgrowth of cells can lead to cancers, but to properly address cellular, tissue, and organismal needs, cells must be able to assemble and disassemble organelles (all membranous structures such as endoplasmic reticulum, Golgi apparatus, cell membranes) and build with proteins. When you are in constant negative energy balance, you cannot accomplish many of these essential processes. Here is a nonscientific example. I was invited to give a talk at a university. At the hotel, I waited for my host in the lobby while enjoying the happy hour offerings. I had not met my host before but had seen his portrait on the web, so I somewhat knew who to anticipate. My host showed up in what I perceived to be very bad shape: skinny, hair very thin and patchy, skin wrinkled way beyond his chronological age. I was startled and had a sympathetic outburst of empathy: I was sure that he had a terminal illness. I was tempted to ask how he was doing, but I did not. Lucky for me: he told me he had been calorie restricting himself for 20 years.

While looks do not matter, his outward appearance assumingly due to the calorie restriction took me aback, thus reinforcing the point that other essential processes may suffer.

One study that was done some years ago looked at how various strains of mice respond to calorie restriction with regard to their lifespan.[96] While this paper has recently been scrutinized, I believe that its outcome is fundamentally critical to consider for the human condition. They looked at dozens of mouse strains and found that one group responded to calorie restriction with increased lifespan, one group had no change in lifespan, and another group had shorter lifespan. The crucial message of this paper is that it cannot be predicted for any given individual whether they will benefit or not from long-term calorie restriction. To put it differently, you are your own experiment; the odds are even in that there is a 30% chance to add some years to your life, 30% chance of no impact on your lifespan, and a 30% chance of a shorter lifespan compared to having no calorie restriction. No one can definitively know the answer regarding the benefits/detriments of calorie restriction. One thing is almost for certain that decades of calorie restriction will result in a number of negative attributes, including hair loss, poor skin tent, weak bones/muscles, elevated stress hormones, and mood alterations. In my view, it is probably not the panacea that it has been purported to be.

Overall, it is prudent to suggest that body weight and lifespan control, like the determinants of height, hinge critically on genetics, developmental and environmental forces. Furthermore, individual variability dominates the biological entity. With regard to individual variability, I suggest that the level of excitability of the AgRP neurons (level of hunger) may play a previously unsuspected but critical role.

2.7 Hunger promoting NPY/AgRP neurons affect higher brain functions and brain disorders

Similarly to ghrelin, the adipose hormone, leptin, that promotes satiety has also been shown to impact these structures and associated brain functions, such as motivated behavior, learning, and memory. Functional imaging studies of the human brain confirm that these effects of the peripheral metabolic hormones do target the same brain sites that were unmasked in experimental animal models. All and all, these seemingly novel and interesting findings simply emphasize the relevance of the assertion that the function of all aspects of the central nervous system is under the control of the metabolic needs of the body and that peripheral tissues exert their requirements by shifting the activity of various brain regions, in part, by humoral signals. Thus it is plausible to assume that the modulation of hormonal or nutrient signals important for metabolic shifts will affect higher level brain functions and, consequently, generate psychiatric and neurological responses with likely implications to the etiology of correlated disorders. Later, I will discuss the specific impact of AgRP neurons on various brain functions not classically associated with eating.

2.7.1 AgRP neurons regulate development of midbrain dopamine neurons and their impact on behavior

As discussed previously (Section 2.4.2), midbrain dopamine neurons mediate ghrelin-induced feeding.[45] While that study was focused, in part, on ghrelin's action directly in the ventral tegmental area, the most robust effect of ghrelin on metabolism is mediated by the ARC NPY/AgRP neurons.[97] Around 2010, after Marcelo Dietrich returned to New Haven from Brazil, he started to interrogate the impact of AgRP neurons more broadly on complex behavior. First, while I was on sabbatical in Madrid, he generated a model, in which he knocked out sirtuin 1 selectively from AgRP neurons. I was furious when I first learned about it during one of my trips back to New Haven. Sirtuins were "hot" in those days as potential magic bullets to stop or reverse aging (this idea still lingers today ...). Marcelo, being young and very enthusiastic, wanted to study sirtuins even though I was dead-set against it because hundreds of other well-meaning and able people were already working on the topic. Early on in my career, I decided that for me to do something that many others are doing is not worth the effort, because they will do a great job, likely better and faster than me. I would be wasting my time and energy, the time of my trainees and my precious resources. Marcelo understood and agreed. He moved the project forward, not from the perspective of sirtuins, but by how the intervention that he did impacted AgRP neurons. In brief, he showed that knocking out sirtuin 1 from AgRP neurons made these cells less responsive to the critical peripheral hunger signal, ghrelin (Fig. 2.29).[98]

Not surprisingly, this impaired responsiveness of AgRP neurons to peripheral metabolic signals made animals less hungry with an overall elevated POMC activity and a leaner phenotype.[98] Marcelo started to test these animals in paradigms unrelated to feeding *per se* but related to motivated behaviors. He found a remarkable set of behavioral outcomes in these mice. He observed in an open-field test that these animals, which were less interested in food, were much more intrigued by nonfood-related novelty in the environment than their wild-type littermates (Fig. 2.30). This elevated "motivation" of AgRP-Sirt1 KO animals was also reflected in their heightened locomotor response to cocaine, and they also preferred the chamber in which they received the cocaine injection in a conditioned place preference test (Fig. 2.30).

We were stunned by these phenotypes. We altered one signaling molecule in the AgRP system, which did not dramatically change the functionality of these neurons from a broad neurobiological perspective. This alteration made them less responsive to ghrelin, a gut signal of hunger. Here was a phenotype where the animal is less interested in food, but more interested in behaviors, including both natural and drug-evoked novelty seeking. In a way, impaired AgRP neuronal function specifically related to peripheral metabolism uncoupled food seeking from novelty seeking. Of course, our immediate thought was that the midbrain dopamine neurons must be

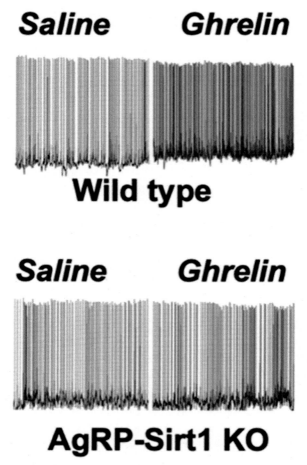

Figure 2.29 Firing frequency of AgRP neurons in wild-type and AgRP-Sirt1-KO mice. Slice electro-physiology showed that in wild-type mice, bath application of ghrelin increases firing frequency of AgRP neurons (upper panel). In contrast, ghrelin did not alter the firing of AgRP neurons in AgRP-Sirt1-KO mice. *AgRP*, Agouti-related peptide. *Modified from Dietrich MO, Antunes C, Geliang G, et al. AgRP neurons mediate SirT1's action on the melanocortin system and energy balance: roles for SirT1 in neuronal firing and synaptic plasticity. J Neurosci. 2010;30(35):11815–11825.*

involved—and they were. We found that in early developmental stages, AgRP neurons innervate midbrain dopamine neurons, and that dopamine cells of AgRP-Sirt1 KO animals are hyperexcitable to the extent that LTP can be evoked in these cells, which could not be accomplished in wild-type animals (Fig. 2.31).

The robust presence of AgRP fibers in the midbrain was during a critical early perinatal period of mice. This period in mice corresponds to the transition between the second and third trimester of human pregnancy. This robust effect of a slight

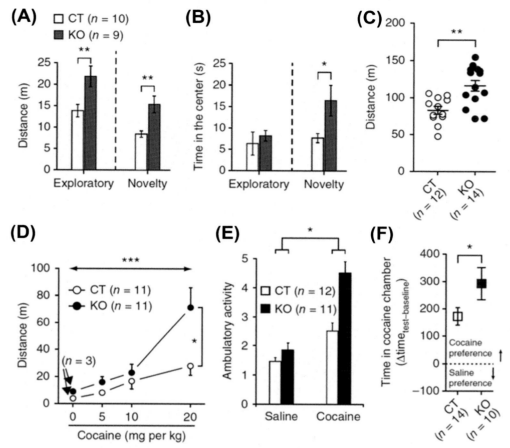

Figure 2.30 AgRP neurons determine the behavioral response to novelty and cocaine. (A and B) Exploratory activity (A) and time spent in the center in the open-field test (B) comparing AgRP-Sirt1 (KO) and littermates CT. (C) Exploratory activity in a different open-field test during 60 min. (D and E) AgRP-Sirt1 mice displayed increased locomotor response to acute cocaine (D) and in a cocaine-sensitization procedure (E) compared with control mice. (F) In the conditioned place preference (CPP) test, AgRP-Sirt1 mice had increased response compared with control mice. $*P < .05$, $**P < .01$, $***P < .001$. Data are presented as mean $\pm$ SEM. *AgRP*, Agouti-related peptide; *CT*, control mice. *From Dietrich MO, Bober J, Ferreira JG, et al. AgRP neurons regulate the development of dopamine neuronal plasticity and non food-associated behaviors. Nat Neurosci. 2012;15:1108–1110.*[99]

alteration in AgRP neuronal activity in response to peripheral metabolic signals on the developing midbrain reward circuitry has far-reaching implications for the impact of peripheral metabolism (including that associated with pregnancy) on complex behaviors and on their impairments. In fact, many if not most psychiatric disorders have important developmental correlates, including anorexia nervosa and schizophrenia.

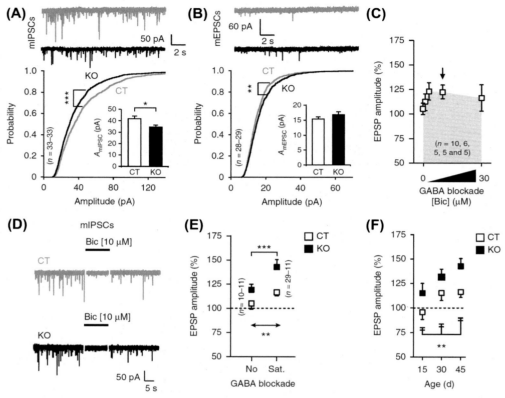

Figure 2.31 AgRP neuronal excitability determines the connectivity of ventral tegmental area (VTA) dopamine neurons. (A and B) Representative traces, probability plots, and average peak amplitude (insets) of miniature postsynaptic currents in control and AgRP-Sirt1 mice. (C) Dose−response of GABA blockade on long term potentiation (LTP) propagation in control. (D) Saturating doses of GABA blockade [10 μM bicuculline (Bic), arrow in (C)] eliminated mIPSCs in both control and AgRP-Sirt1 mice. (E) AgRP-Sirt1 mice had increased LTP either in the absence of GABA blockade (No) or in the presence of saturating doses of GABA blocker (Sat). (F) Quantification of LTP during development in control and AgRP-Sirt1 mice. $^*P < .05$, $^{**}P < .01$, $^{***}P < .001$. Data are presented as mean ± SEM. *AgRP*, Agouti-related peptide. *From Dietrich MO, Bober J, Ferreira JG, et al. AgRP neurons regulate the development of dopamine neuronal plasticity and non food-associated behaviors. Nat Neurosci. 2012;15:1108−1110.*

2.7.2 AgRP neurons trigger stereotypic behaviors

One surprising aspect of the findings on AgRP neurons and the midbrain dopamine system was the dissociation between interest in food and interest in novelty and response to cocaine. I would have predicted that the activity of AgRP neurons parallel interest in novelty and the rest of the word as hunger drive all those various behaviors to succeed in the environment. However, those findings appeared to relate to a specific developmental period during which AgRP efferents shape the emerging midbrain dopamine circuitry. Thus that study did not address the overall impact of the adult AgRP circuitry on behavior. To interrogate that, Marcelo developed a model of acute control of AgRP neurons using a channel that can be artificially driven by exogenous capsaicin, the active component of chili pepper.[57] These receptors, called vanilloid receptor subtype 1 (TRPV1), are expressed in multiple sites of the body, most notably sensory afferents (hence the feeling of hot chili peppers). They are not expressed in AgRP neurons. Marcelo made an animal, in which he selectively knocked in TRPV1 in AgRP neurons in a TRPV1 whole-body knockout background. In this model, he could activate AgRP neurons with an injection of capsaicin, which resulted in ferocious eating by animals if food was available. Marcelo's goal, however, was to see how these animals behave if food is not available and their AgRP neurons are activated. He showed that if AgRP neurons are activated in adult animals, there was a positive correlation between interest in food and interest in novelty.[57] AgRP neuron-activated animals were also less anxious and more inclined to move outside of their comfort zone. These are all logical responses since to survive, a hungry animal must explore everything outside its comfort zone.

An intriguing phenotype that we observed was that of stereotypy, in which animals would start to engage in compulsive behaviors. If it was the home cage, they seemed to start compulsively digging. One could say that they are searching for food. However, when these animals were provided small marble pieces, they would actually start to bury them compulsively (Fig. 2.32). Mice do like to engage in the act of burying marbles, but AgRP neuron-activated mice would compulsively and relentlessly bury these marble balls. If a time limit was set, it was obvious that they are much more preoccupied with these marbles than those animals whose AgRP system was not activated, so much so, that they seemed to be addicted to this behavior as revealed by a modified conditioned place preference test (Fig. 2.32).

Repetitive behaviors may serve as positive adaptations in many circumstances. Obviously, regarding the need for air, water, food, or reproduction, repeated failed attempts should not discourage any living organism from trying. Giving up could mean death or species eradication. To have hope win over experience is a fundamentally critical trait in biology. We all have it in various levels. Those of us who pursue the unknown must have it or be able to develop it. Hunger-promoting AgRP neurons appear to be a key part of our "design" to promote those traits, especially if food is not immediately and easily available.

2.7.3 AgRP neurons drive endurance exercise and control death in a model of anorexia nervosa

Stereotypy, obsessive, and repetitive behaviors also contribute to many psychiatric conditions, including schizophrenia and eating disorders. Schizophrenia, as well as eating disorders with life-long consequences, have important developmental components. Frequently, it is the period of periadolescence when many of these disorders emerge. Eating disorders, such as anorexia nervosa, have broadly overlapping symptomatology

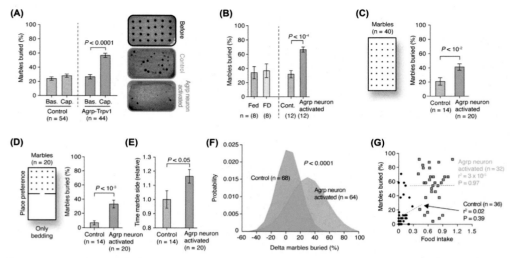

Figure 2.32 Repetitive behaviors after Agrp neuron activation. (A) marbles buried after Agrp neuron activation. (B) Marble buried in fed, FD, control, and Agrp-neuron-activated mice. (C) Marble buried in the modified marble-burying test. (D) Marble buried in the modified place-preference test. (E) Time animals spent in the marble side relative to control animals. (F) Normal distribution fitted to pooled experimental data [delta marbles buried (capsaicin injection—baseline)]. *P* value was calculated using an unpaired *t*-test with Welch's correction. (G) Linear regression analysis correlating marble-burying behavior and food intake. Each data point represents one mouse. Female mice were used in this study. Error bars represent mean ± SEM, and *P* values were calculated using the *t*-test. *AgRP*, Agouti-related peptide; *FD*, food-deprived.

with schizophrenia. For example, anorexia nervosa patients also have delusions (regarding their body), disorganized thoughts as well as emotional disturbances. Anorexia nervosa is a multifaceted and debilitating illness characterized by starvation, persistent fear and anxiety of weight gain, and preoccupation with body image and maladaptive food choices. There is a disruption in homeostatic energy balance mechanisms, where the persistence of homeostatic hunger is overridden by dysfunctional self-regulatory and reward pathways that drive food aversion and severely restrict food intake. One of the most devastating aspects of the illness is its impact on young people, with an early onset during adolescence that sets the course of high comorbidity and with the highest mortality among psychiatric illnesses. Critically, there are no effective pharmacological treatments approved for the illness and psychosocial interventions are inadequate and associated with high relapse rates.

There are multiple coinciding events that appear to be indispensable for the onset of anorexia nervosa. First, it is very likely that genetic predisposition is necessary for symptom development. However, genetics, on its own, is unlikely to be a determinant of eating disorders. This argument is best illustrated by the lack of association of single gene mutations with anorexia nervosa and by the very limited penetrance of chromosomal segments that were associated with the disease.[100] On the other hand, in almost all cases, the onset of the disease emerges in adolescence in temporal association with stress (predominantly psychological) and dieting (food restriction). Frequently, obsessive endurance exercise is also part of the behavioral profile of subjects with anorexia nervosa. Of course, millions of people during adolescence can be under various stressors or engage in diet and exercise, nevertheless, 99.9% of these women and men do not develop life-threatening psychiatric conditions. This argues for the importance of genes. In addition, there may also be a component related to sex hormones, because the vast majority of subjects with anorexia- and bulimia nervosa are women. How sex hormones or sex-related genes play a role in the etiology of these disorders is entirely unknown at present. In animal (mouse) models, it was conclusively shown that the constellation of genetics, stress, obsessive exercise, and restricted feeding are all crucial components to the development of anorexia nervosa. Sometimes there is a misconception that anorexia nervosa is a form of anorexia where people are not hungry. It is actually the exact opposite: elevated hunger becomes rewarding for these individuals as this is the sure sign that they are in control of the desired change in their body. While many periadolescent women with anorexia nervosa had started dieting and exercise because of the peer pressure-associated desire to lose weight, large numbers of people engage in these behaviors to be 100% in control. These individuals are frequently high achievers, perfectionists. When they realize that it is impossible to be in total control of the environment, they turn to their bodies of which they can be in total control.

As I described before, hunger-promoting AgRP neurons are involved in developmental attributes of the reward system, and, they also promote ambulatory locomotor activity and obsessive compulsive behaviors. A few years ago, Marcelo's now-independent lab also showed that AgRP neurons control learning and memory processes.[101] These are all hallmarks of anorexia nervosa. Given this information and because of her insatiable curiosity and drive to pursue the unknown, Maria Miletta (in my laboratory) asked the question whether AgRP neurons may have something to do with attributes of anorexia nervosa. She exploited two mouse models with altered AgRP circuit integrity. One allows for the elimination of AgRP neurons early postnatally without any death of the animals. This model was developed by Richard Palmiter's lab[25] and builds on the remarkable plasticity of the hypothalamus early postnatally. The other model is the one I described before, in which AgRP neurons can be selectively and remotely activated by administration of capsaicin.[57] She studied these animals in a mouse model of activity-based anorexia.[102] Maria found that regardless of AgRP circuit integrity, food restricted, periadolescent animals engaged in compulsive exercise if a running wheel was available, a characteristic behavior of subjects with anorexia nervosa. She also observed a positive correlation between AgRP circuit activity and exercise volume of these food-restricted mice. Strikingly, animals with impaired AgRP circuit integrity had abolished food intake while exercising and died of exhaustion after a few days of compulsive running. On the other hand, those animals in which AgRP neurons were activated daily had significantly increased the endurance of compulsive exercise without lethality. In-line with the role of AgRP neurons peripheral fuel partitioning, AgRP circuit-impaired animals were unable to mobilize fat during food restriction during exercise, but when provided elevated fat content through the restricted diet, the death of AgRP circuit impaired animals was completely prevented (Fig. 2.33).

Another remarkable finding of Maria was that similar to the human condition, engagement of calorie restriction, and exercise-induced long-lasting behavioral alterations in mice way beyond the presentation of anorexia nervosa. This strengthens the validity of animal models of the human condition with obvious caveats.[103] Strikingly, however, these long-lasting effects were also abolished if animals were given the restricted diet in the form of high fat (Fig. 2.34)[104].

It may sound counterintuitive that fat benefits anorexia nervosa subjects. They dread even the thought of fat as it is assumed by most that you gain weight by eating fat. First, that assertion is incorrect. Fat alone does not induce obesity. Ketogenic diet, which is almost all fat, induces weight loss. The cellular principles of hunger and calorie restriction are also based on lipid metabolism. When combined with carbohydrates, there is weight gain. But fat alone is not obesogenic. To support this idea of fat potentially being beneficial for anorexia nervosa, Barbara Scolnick recently reported a case, in which remission of anorexia nervosa occurred in a patient on ketogenic diet in combination with ketamine.[104] Overall, Maria Miletta's observations shed new light on a previously unsuspected organizational role of AgRP neurons in the regulation and dysregulation of complex behaviors and survival in an animal model of anorexia nervosa with immediate translational implications.

2.7.4 A thought on schizophrenia and metabolism regulation

Let us take a look at schizophrenia from a "metabolic" angle. I tend to bring up the subject of schizophrenia when I attempt to make my points about the hypothalamus and higher brain functions. The reason I do is because neuroscientists working on cortical and other higher brain regions do not respond well to Socratic reasoning for the explanation of how hypothalamic AgRP neurons impact higher brain functions. But if I start out stating that schizophrenia is a metabolic disorder whose etiology hinges on metabolism, they may give me 5 minutes.

Schizophrenia is a debilitating mental disorder that affects 1% of the population regardless of geographical or socioeconomic status. There is no cure for schizophrenia and there is no clear understanding of how symptoms associated with schizophrenia emerge.

During the last three decades, significant investments have been made in schizophrenia research by academia as well as the drug industry, with only modest progress.

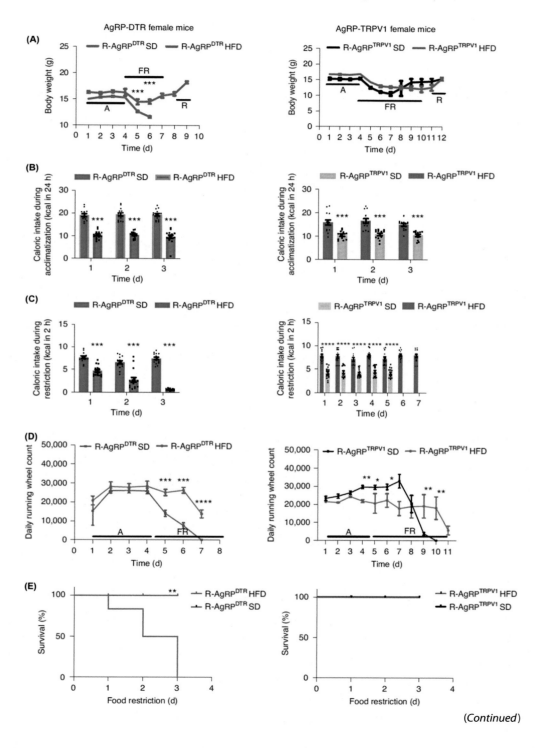

(Continued)

In clinical practice, a wide array of safe antipsychotics exists with various side-effect profiles. However, many critical symptoms of the disease are still not impacted by the current treatments, including cognitive impairment associated with schizophrenia. Therefore it has been questioned whether the right therapeutic targets are pursued, and whether the current preclinical animal models capture the most critical mechanisms of the disease.

An abnormality in dopamine neurotransmission has been well demonstrated in schizophrenic patients, although it has been recognized that dopaminergic hyperactivity is seen only in a subset of patients during the acute stages of the illness. In addition to dopamine, practically all known classic neurotransmitters, including glutamate, GABA, serotonin, norepinephrine, acetylcholine, and a variety of neuropeptides and endogenous neuromodulators, have been tied to schizophrenia. Several affected genes may indicate a dysfunctional NMDA receptor-mediated glutamate transmission. However, it can be argued that alteration in any of the main brain neurotransmitter systems would subsequently influence other neurotransmitters as well.

Schizophrenia is a genetically predisposed disease, and like other complex disorders, it is characterized by the contribution of multiple risk genes in combination with epigenetic and environmental processes leading to the development of the disease. However, it is still unclear whether a relatively few highly penetrant but rare major genetic deficits would determine the disease, or a large number of genes would

◀ **Figure 2.33** High fat diet (HFD) restores normal survival in adolescent mice lacking AgRP neurons and exposed to ABA, while it prevents the development of activity-based anorexia (ABA) phenotype in mice with activated AgRP neurons. (A–E) All studies were conducted in female control and Neonatally diphtheria toxin-deleted AgRP (AgRPDTR) mice (left) and AgRPTRPV1 and control mice (P36–P50; right). Only data for female mice are shown. Data are expressed as the mean ± SEM R-AgRPDTR SD, $n=16$; R-AgRPDTR HFD, $n=15$. R-AgRPTRPV1 SD, $n=13$; R-AgRPTRPV1 HFD, $n=14$. (A) Body weight changes during ABA. Day 5: R-AgRPDTR SD versus R-AgRPDTR HFD, ***$P=.0004$. Day 6: R-AgRPDTR SD versus R-AgRPDTR HFD, ***$P=.0001$. (B) Caloric intake during acclimatization (kcal in 24 h). Days 1–3: R-AgRPDTR SD versus R-AgRPDTR HFD, ***$P=.0001$; R-AgRPTRPV1 SD versus R-AgRPTRPV1 HFD, ***$P=.0001$. (C) Caloric intake during food restriction (kcal in 2 h). Days 1–3: R-AgRPDTR standard diet (SD) versus R-AgRPDTR HFD, ***$P=.0001$; R-AgRPTRPV1 SD versus R-AgRPTRPV1 HFD, ****$P<.0001$. (D) Daily running wheel activity during acclimatization and food restriction. Day 4: R-AgRPTRPV1 SD versus R-AgRPTRPV1 HFD, **$P=.0077$. Day 5: R-AgRPDTR SD versus R-AgRPDTR HFD, ***$P=.0009$; R-AgRPTRPV1 SD versus R-AgRPTRPV1 HFD, *$P=.0498$. Day 6: R-AgRPDTR versus R-AgRPDTR HFD, ****$P<.0001$; R-AgRPTRPV1 SD versus R-AgRPTRPV1 HFD, *$P=.034$. Day 7: R-AgRPDTR SD versus R-AgRPDTR HFD, ****$P<.0001$. Day 9: R-AgRPTRPV1 SD versus R-AgRPTRPV1 HFD, **$P=.0065$. Day 10: R-AgRPTRPV1 SD versus R-AgRPTRPV1 HFD, **$P=.0062$. (E) Survival during food restriction. A log-rank test was used for comparison of survival curves. R-AgRPDTR SD versus R-AgRPDTR HFD, **$P=.0071$; R-control versus R-AgRPTRPV1, no statistical difference ($P \geq .9999$). AgRP, Agouti-related peptide. *From Miletta MC, Iyilikci O, Shanabrough M, et al. AgRP neurons control compulsive exercise and survival in an activity-based anorexia model. Nat Metab. 2020;2:1204−1211. https://doi.org/10.1038/s42255-020-00300-8.*

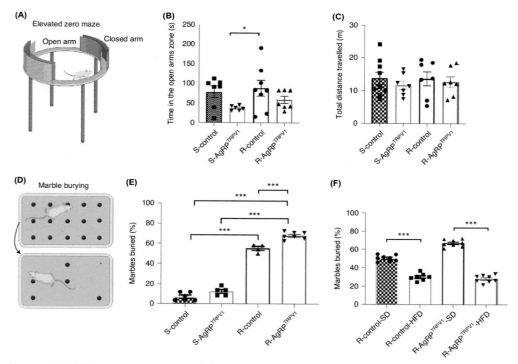

Figure 2.34 HFD prevents repetitive behavior in mice with activated AgRP neurons and exposure to ABA. (A) Elevated zero maze apparatus. (B) Time (s) spent in the open arms in the elevated plus test (S-control, $n=7$; S-AgRPTRPV1, $n=6$; R-control, $n=8$; R-AgRPTRPV1, $n=7$) in mice fed with SD. R-control versus S-AgRPTRPV1, $P=.0406$. (C) Total distance traveled (m) in the elevated plus test (S-control, $n=9$; S-AgRPTRPV1, $n=7$; R-control, $n=7$; R-AgRPTRPV1, $n=7$) in mice fed with SD. No statistical significance was found. (D) Graphical representation of the marble-burying test. (E) The percentage of marbles buried after exposure or not to activity-based anorexia (ABA) in AgRPTRPV1 and control mice fed with SD. S-control, $n=6$; S-AgRPTRPV1, $n=5$; R-control, $n=4$; R-AgRPTRPV1, $n=7$. S-control versus S-AgRPTRPV1, $P=.1742$; S-control versus R-control, $P<.0001$; S-AgRPTRPV1 versus R-AgRPTRPV1, $P<.0001$; S-control versus R-AgRPTRPV1, $P<.0001$; R-control versus R-AgRPTRPV1, $P<.0001$. (F) The percentage of marbles buried after exposure to ABA in AgRPTRPV1 and control mice. Mice were fed with a 65% HFD and SD. R-control-SD, $n=8$; R-control-HFD, $n=7$; R-AgRPTRPV1-SD, $n=7$; R-AgRPTRPV1-HFD, $n=8$. R-control-SD versus R-control-HFD, $P<.0001$; R-control-SD versus R-AgRPTRPV1-SD, $P<.0001$; R-control-SD versus R-AgRPTRPV1-HFD, $P<.0001$; R-control-HFD versus R-AgRPTRPV1-SD, $P<.0001$; R-control-HFD versus R-AgRPTRPV1-HFD, $P<.0001$; R-AgRPTRPV1-SD versus R-AgRPTRPV1-HFD, $P=.7686$. All studies were conducted in control and AgRPTRPV1 female mice between P53 and P56, at least 1 week after recovering from ABA. Data are expressed as the mean ± SEM. Illustrations were created with BioRender (https://biorender.com). *AgRP*, Agouti-related peptide. *From Miletta MC, Iyilikci O, Shanabrough M, et al. AgRP neurons control compulsive exercise and survival in an activity-based anorexia model. Nat Metab. 2020;2:1204–1211. https://doi. org/10.1038/s42255-020-00300-8.*

contribute although each gene accounting for only a small increment in risk. Interestingly, a genome-wide search of copy number variants (CNVs) revealed an association between CNVs of certain high penetrant genes not only with schizophrenia, but also other psychiatric illnesses, such as autism and bipolar disorder.[105] These findings might point to the fact that most psychiatric diseases share a number of symptoms; therefore these genetic abnormalities might correlate better with symptoms of various psychiatric disorders rather than with a singular psychiatric disease itself. The concept and application of endophenotypes better define this relationship: endophenotypes capture and measure pathophysiological signals contributing to the clinical manifestation of the disease. Neurophysiological abnormalities associated with schizophrenia could be considered endophenotypes; their values in genetic studies of schizophrenia have already been demonstrated.[106]

Arguing for the relevance of metabolism regulation in relation to schizophrenia is supported by the following facts. For decades, insulin-induced coma was used as a standard treatment of schizophrenics in Europe and in the United States[107–111] (https://en.wikipedia.org/wiki/Insulin_shock_therapy). Insulin is an essential regulator of metabolism through the regulation of glucose metabolism, which is the fundamental driver of brain functions. While it could be said that in insulin-induced coma, it is the "coma" component that matters, as if a computer were rebooted. Alternatively, it is the increased glucose load of vulnerable circuits and brain areas by insulin that makes a long-standing impact on the elimination of positive symptoms of schizophrenia. In support of this latter notion, intriguingly, contemporary atypical psychotropic drugs that are effective in suppressing the positive symptoms cause massive obesity and diabetes.[112,113] Another, very simple argument in support of the idea that schizophrenia might be the outcome of systemic metabolic impairments is the fact that positive symptoms, such as delusions, hallucinations, racing thoughts, as well as disorganized thoughts and emotional shifts, all of which are characteristics of schizophrenia are considered absolutely normal states of mind during sleep. These "symptoms" are normal elements in dreams while sleeping. Among many different homeostatic parameters, a crucial difference between wakefulness and sleep is systemic fuel utilization. While during wakefulness, the body is dominantly glucose utilizing, during sleep, fuel preference moves from glucose to fat and core body temperature declines. So, from a perspective of fuels, lower glucose availability appears to be permissive of "symptoms" of schizophrenia. This is in-line with the putative impact of insulin and psychotropic drugs on increased glucose utilization—associated diminishment of positive symptoms of patients with schizophrenia. This implies that obesity and diabetes triggered by these drugs may not be side-effects, rather they are on target effects. This also suggests that while the brain and cortical regions, in particular, are essential for the symptomatology of schizophrenia, the underlying cause for the emergence of this disease may be found in metabolic principles rather than the Neuronal Doctrine and the focus points it delivered in neuroscience.

2.8 The metabolic concept of higher brain functions

In cortico-centric neuroscience, the distinction between sensory-, motor-, and somato-sensory areas is a sensible simplification of the ins and outs of the most sophisticated aspects of the brain. While it allowed a logical and deductive way to conceptualize the putative functional principles of the brain, it has not yet answered the most fundamental issues relating to brain functions. From a very simplistic point of view of classic brain research, the value of understanding how the brain works would lie in the utilization of that knowledge to impact maladaptive behaviors and other functions impacted by the central nervous system. Impaired behaviors may be relevant to identified and character-ized psychiatric and neurological disorders, as well as behaviors that are most effective to adapt to changing environment (natural, societal, etc.). For example, from the perspec-tive of successful elimination of delusions in the case of schizophrenia by any means, may or may not rely on the understanding of the inner workings of the central nervous system of the subject. If the successful treatment is based on contemporary understand-ing of neuroscience, that is a great achievement. However, if it is the outcome of a ser-endipitous discovery of an intervention without the understanding of its mechanism of action within the conceptual framework of contemporary neuroscience, it would be equally impactful. It is the outcome that we strive to achieve (i.e., to impact perception and behavior) regardless of the how. In this regard, and as an example, one of the intriguing successful, albeit controversial, treatments of schizophrenia's positive symp-toms was accomplished by the peripheral pancreas-derived hormone, insulin. The dis-covery of insulin just had its 100-year anniversary. Insulin-induced coma can eliminate positive symptoms of schizophrenia without our understanding of how it is accom-plished. If we knew how it worked (which we do not; see later), it might actually aid our quest to understand the brain as well. But if we do not know the mechanisms, it still can be used as long as it does more-good-than-harm. It is perception and output over which we would like to have control. Holistically, we wish people to perceive their environment with contentment, which they may accomplish by dynamically adjusting (adapting) their behavior. One specific angle on this relationship between the environ-ment and the organism is the need to gather fuels to sustain life.

No cell functions in our body without appropriate fuels. These fuels or their pre-cursors come from the environment via consumption of foods. This chapter engaged the topic that feeding behavior governed by hypothalamic NPY/AgRP neurons is a crucial part of our existence. Hence, everything the brain "does" must relate to eating and processing of food. My argument is that this very simple act and related processes, some of which I described before, play much more fundamental roles in controlling all brain functions than contemporary neuroscience considers.

The oxidation of cellular substrates determines life. While cellular substrate oxidation is evidently critical for life, how orchestration of dynamically changing fuel availability is matched with cellular needs of different tissues in support of successful homeostasis of the entire organism is less clearly understood. As discussed before, the hypothalamus and the AgRP neurons, in particular play a significant role in these processes.

For the adequate coordination of tissue functions, it is obvious that the various tissues must communicate with one another and with the brain to make a proper adjustment in fuel recruitment (eating), fuel distribution, and fuel utilization. For instance, when there is insufficient food available, that information needs to be communicated by peripheral tissues, such as the fat, muscle, liver, and pancreas, so that adaptive behaviors initiated by the brain can be mounted to locate and eventually consume nutrients from the environment. But before that pursuit of food can occur, the body needs to adjust to be able to function with declining glucose levels, which is the hallmark of food deprivation. For example, sensing the declining glucose levels, the pancreas will lessen the secretion of insulin and the fat tissue will lessen the release of the hormone leptin. On the other hand, release of glucagon from the pancreas, corticosteroids from the adrenal glands and ghrelin from the stomach will rise. These events will contribute to the production and release of glucose by the liver and kidney into the circulation, so in the absence of carbohydrate intake, the crucial fuel for life, glucose, can be maintained at some level. At the same time, fat stores of the body, mainly the white adipose tissue, will release lipid, which will become the alternative fuel for the body until a new source of food is found and then consumed. These changes in the periphery will signal to the brain to adjust all the behaviors of the organism to engage in the pursuit of food.

There are multiple ways via which the periphery communicates with the brain. Hormones can signal to various parts of the brain, most notably a subcortical region called hypothalamus, which will trigger a cascade of ascending brain pathways eventually affecting the cortex. Signals from the periphery can also travel to the brain via sensory neuronal afferents of the autonomic nervous system impacting the brain stem. From there, again, ascending signals by various routes can reach the cortex. In addition, the changing circulating fuel milieu (declining glucose and elevating lipid levels) will affect what fuel a neuron can use, including those in the cortex.

It is becoming clear that fuel availability on its own can have a major impact on the functionality of cells, including brain cells. In a simplified way, the metabolic environment of the periphery determines how the cortex must function to adjust behavior for survival. Thus cortical functions and the resultant complex behaviors are downstream to the function of peripheral tissues (Fig. 2.35). Contemporary neuroscience, by and large, does not take this into account when trying to understand how the brain works or how malfunctions of the brain emerge.

All complex behaviors governed by diverse brain regions are aligned either to hunger or to satiety. There is no exception. For example, general arousal, sleep/wake cycles, interest in the environment, decision-making, learning and memory processing, and locomotion all need to serve the purpose of responding either to hunger or

State-of-the-Art

Proposed View

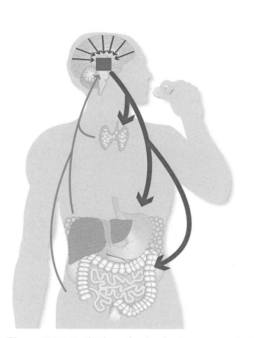
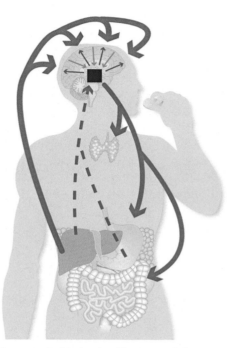

Figure 2.35 Rethinking body—brain communications in control of behavior. The state-of-the-art on human behavior asserts that it is driven by signals arising from the cerebral cortex. There are signals from these areas that also travel via lower brain regions, such as the hypothalamus (*blue box*), to affect peripheral tissues via the autonomic nervous system and the pituitary. In turn, these peripheral tissues provide some feedback to the brain. This chapter provides an alternative conceptual framework. In the *proposed view*, the cortex remains to be an effector in behavioral control, with the caveat that the effector function is downstream to multitude of signals coming from the periphery via humoral and neuronal signals. In this, the hypothalamus plays a crucial organizational role to align peripheral tissue functions with the output of the brain to adequately execute behavior. In this scheme, the peripheral organs, such as the liver, adipose tissue, pancreas, bone, and muscle, play a critical regulatory role in determining what behavioral adaptations need to ensue.

to satiety. These behaviors must be different under these two metabolic conditions otherwise the goals of each of these states will not be fulfilled. The objectives of these metabolic states are distinctly different. In the case of hunger, the primary goal is to identify, pursue, and consume food. When satiety emerges, the primary aim is to process, store, and utilize nutrients for maintenance, growth and reproduction. Of course, adjustments in various complex behaviors during hunger or satiety do not mean that these behaviors make 180-degree switches. For example, learning can occur both during hunger or satiety, but this process is more efficient when the subject is hungry. The situation is different when it comes to decision making regarding a third party. A study was done a few years ago on Israeli parole judges.[114] They analyzed how judges decide the fate of parolees with similar pedigrees. Seemingly astonishingly what they found was that with identical pedigree, those parolees who came before a judge after breakfast or lunch break were more likely to get paroled than those who came in front of the judge before these breaks. The deciding difference was not the paperwork, but potentially the metabolic status of the judges. Were the judges more lenient when satiated or more balanced and objective? The answer is not critical; the point is that systemic metabolism had a profound impact on a complex process of decision-making. Anyone can test this simple metabolic principle by having a conversation with a spouse, child, parent, or coworker on an issue of contention just before lunch or an hour after lunch. It will be blatantly clear to most that compromise and comprehension are tremendously enhanced by satiety. Thus, if we are looking for a resolution to benefit both parties, it might be better to negotiate during the state of satiety.

It is possible to dissect any complex or simple behavior from the perspective of a systemic metabolic state (hunger or satiety). It also goes without saying that extreme states of hunger and satiety are rare. But it can be inferred that any behavior we manifest at any given time is affected by the metabolic state.

Examples of this abound: for you to properly eat, you need to be awake. Therefore a relationship must exist between these circuits in the hypothalamus and those brain regions that control sleep and wakefulness; you need to be able to remember and locate food, which relates to short- and long-term memory as well as reward and their support circuits in various brain regions; you have to make a decision when and how it is best to pursue food, which relates to brain regions involved in decision making; and finally you have to physically pursue the food, which involves locomotion, chewing, and swallowing and their respective brain circuits. I not only contend that hypothalamic hunger-controlling neurons affect these circuits and behaviors under normal conditions, but also that these hypothalamic cells may have relevance in the emergence of dysfunctions of these brain areas, such as sleep disorders, dementias, mood and addictive disorders, depression, schizophrenia, Parkinson's disease, and other "classical" brain disorders.

2.9 Summary

In summary, this chapter has made the argument that complex, goal-oriented behaviors, even those we attribute to higher mammals and humans, have a direct relationship with hunger and, hence, with the machinery that controls hunger in the hypothalamus (Fig. 2.35). Hypothalamic hunger-promoting AgRP neurons are uniquely positioned to coordinate peripheral tissue functions with that of the output of the brain in control of behavior. They are situated outside of the blood—brain—barrier having the ability to receive unfiltered, blood—born signals from the periphery. They also receive ascending information from the periphery via the autonomic nervous system. They, in turn, send out signals through neuronal and glial connections reaching both higher brain regions and as peripheral organs via the autonomic nervous system and the endocrine hypothalamus. The combination of signals controlled by AgRP neurons and their mitochondrial within and outside of the brain will then "meet up" in the cortex to evoke or suppress a behavior. The routes are infinite and involve all the cell types of the brain and beyond. Interrogating these communications with tools developed on the conceptual basis of the neuronal doctrine to better understand, body, brain, and behavior will deliver pieces of a puzzle that may or may not belong to the puzzle of the comprehension of brain functions and dysfunctions. I look at hunger promoting AgRP neurons as the "heart" of our brain. Our lives end with the stoppage of the heart, but the heart will stop and life will end if AgRP neurons and, hence, hunger are removed.

References

1. Santiago Ramon y Cajal, "Structure of the Mammalian Retina," Madrid, 1900.
2. Nicholls J.G. From neuron to brain. *Sinauer Associates*. pp. 5. ISBN 0878934391; 2001.
3. Hetherington AW. The relation of various hypothalamic lesions to adiposity and other phenomena in the rat. *Am. J. Physiol.* 1941;133:326.
4. Hetherington AW, Ranson SW. The relation of various hypothalamic lesions to adiposity in the rat. *J. Comp. Neurol.* 1942;76:475.
5. Hetherington AW. Non-production of hypothalamic obesity in the rat by lesions rostral or dorsal to the ventromedial hypothalamic nuclei. *J. Comp. Neurol.* 1944;80:33.
6. Brobeck JR, Tepperman J, Long CNH. Experimental hypothalamic hyperphagia in the albino rat. *Yale J. Biol. Med.* 1943;15:831.
7. Brobeck JR. Mechanism of the development of obesity in animals with hypothalamic lesions. *Physiol. Rev.* 1946;26:541—559.
8. Anand BK, Brobeck JR. Hypothalamic control of food intake in rats and cats. *Yale J. Biol. Med.* 1951;24:123—146.
9. Horvath TL, Diano S. The floating blueprint of hypothalamic feeding circuits. *Nat Rev Neurosci.* 2004;5:662—667.
10. Waterson MJ, Horvath TL. Neuronal regulation of energy homeostasis: beyond the hypothalamus and feeding. *Cell Metab.* 2015;22(6):962—970. Available from: https://doi.org/10.1016/j.cmet.2015.09.026. Epub 2015 Oct 22.
11. Horvath TL, Naftolin F, Kalra SP, Leranth C. Neuropeptide Y innervation of β-endorphin-containing cells in the rat mediobasal hypothalamus. A light and electron microscopic double-immunostaining study. *Endocrinology.* 1992;131:2461—2467.
12. Clark JT, Kalra PS, Crowley WR, Kalra SP. Neuropeptide Y and human pancreatic polypeptide stimulate feeding behavior in rats. *Endocrinology.* 1984;115:427—429.
13. Zhang Y, et al. Positional cloning of the mouse obese gene and its human homologue. *Nature.* 1994;372:425—432.
14. Halas JL, et al. Weight-reducing effects of the plasma protein encoded by the obese gene. *Science.* 1995;269:543—546.

15. Fan W, Boston BA, Kesterson RA, Hruby VJ, Cone RD. Role of melanocortinergic neurons in feeding and the agouti obesity syndrome. *Nature*. 1997;385:165–168.

16. Huszar D, et al. Targeted disruption of the melanocortin-4 receptor results in obesity in mice. *Cell*. 1997;88:131–141.

17. Seeley RJ, et al. Melanocortin receptors in leptin effects. *Nature*. 1997;390:349.

18. Cone RD, Mountjoy KG, Robbins LS, et al. Cloning and functional characterization of a family of receptors for melanocortin peptides. *Ann N Y Acad Sci*. 1993;680:342–363. Available from: https://doi.org/10.1111/j.1749-6632.1993.tb19694.x.

19. Hahn TM, Breininger JF, Baskin DG, Schwartz MW. Coexpression of Agrp and NPY in fasting-activated hypothalamic neurons. *Nat Neurosci*. 1998;1:271–272.

20. Broberger C, Johansen J, Johansson C, Schalling M, Hökfelt T. The neuropeptide Y/agouti-related protein (AgRP) brain circuitry in normal, anorectic and monosodium glutamate-treated mice. *Proc Natl Acad Sci U S A*. 1998;95(25):15043–15048. Available from: https://doi.org/10.1073/pnas.95.25.15043.

21. Cowley MA, Smart JL, Rubinstein M, et al. Leptin activates anorexigenic POMC neurons through a neural network in arcuate nucleus. *Nature*. 2001;411:480–484.

22. Dietrich MO, Horvath TL. Hypothalamic control of energy balance: insights into the role of synaptic plasticity. *Trends Neurosci*. 2013;36(2):65–73. Available from: https://doi.org/10.1016/j.tins.2012.12.005. Epub 2013 Jan 12.

23. Erickson JC, Clegg KE, Palmiter RD. Sensitivity to leptin and susceptibility to seizures of mice lacking neuropeptide Y. *Nature*. 1996;381(6581):415–421. Available from: https://doi.org/10.1038/381415a0.

24. Qian S, Chen H, Weingarth D, et al. Neither agouti-related protein nor neuropeptide Y is critically required for the regulation of energy homeostasis in mice. *Mol Cell Biol*. 2002;22(14):5027–5035. Available from: https://doi.org/10.1128/MCB.22.14.5027-5035.2002.

25. Luquet S, Perez FA, Hnasko TS, Palmiter RD. NPY/AgRP neurons are essential for feeding in adult mice but can be ablated in neonates. *Science*. 2005;310:683–685. Available from: https://doi.org/10.1126/science.1115524.

26. Gropp E, Shanabrough M, Borok E, et al. Agouti-related peptide-expressing neurons are mandatory for feeding. *Nat Neurosci*. 2005;8(10):1289–1291 [Epub ahead of print]. *: Co-senior authors of the paper.

27. Dietrich MO, Horvath TL. GABA keeps up an appetite for life. *Cell*. 2009;137:1177–1179.

28. Horvath TL, Bechmann I, Kalra SP, Naftolin F, Leranth C. Heterogeneity in the neuropeptide Y-containing neurons of the rat arcuate nucleus: GABAergic and non-GABAergic subpopulations. *Brain Res*. 1997;756:283–286.

29. Horvath TL, Garcia-Segura LM, Naftolin F. Control of gonadotrophin feedback: the role of estrogen-induced hypothalamic synaptic plasticity. *Gynec Endocrin*. 1997;11:139–143.

30. Pu S, Jain MR, Horvath TL, Diano S, Kalra PS, Kalra SP. Morphological and pharmacological evidence of synergistic interactions between neuropeptide Y and γ-aminobutyric acid in stimulation of feeding. *Endocrinology*. 1999;140:933–940.

31. Kalra, SP, Xu B, Dube MG, Pu S, Horvath TL, Kalra PS. Interacting appetite regulating pathways in the hypothalamic regulation of body weight. *Endocr. Rev*. 1999;20:67–100.

32. Wu Q, Boyle MP, Palmiter RD. Loss of GABAergic signaling by AgRP neurons to the parabrachoal nucleus leads to starvation. *Cell*. 2009;137(7):1225–1234. Available from: https://doi.org/10.1016/j.cell.2009.04.022.

33. Pinto S, Liu H, Roseberry AG, et al. Rapid re-wiring of arcuate nucleus feeding circuits by leptin. *Science*. 2004;304(5667):110–115.

34. Horvath TL. The hardship of obesity: a soft-wired brain to feed and retain. *Nat Neurosci*. 2005;8(5):561–565.

35. DeFalco J, Tomishima M, Liu H, et al. Virus-assisted mapping of neuronal inputs to a feeding center in the hypothalamus. *Science*. 2001;291(5513):2608–2613. Available from: https://doi.org/10.1126/science.1056602.

36. Bartha A. Experiments to reduce the virulence of Aujeszky's virus. *Magy Allatorv Lapja*. 1961;16:42–45.

37. Naftolin F, Ryan KJ, Petro Z. Aromatization of androstendione by the diencephalon. *J Clin Endocrinol Metab*. 1971;33(2):368–370. Available from: https://doi.org/10.1210/jcem-33-2-368.

38. Horvath TL, Wikler KC. Aromatase in developing sensory systems of the rat. *J Neuroendocrinology.* 1999;11(2):77—84.
39. Garcia-Segura L, Baetens D, Naftolin F. Synaptic remodelling in the arcuate nucleus after injection of estradiol valerate in adult female rats. *Brain Res.* 1986;366:131.
40. Matsumoto A, Arai Y. Synaptogenic effect of estrogen on the hypothalamic arcuate nucleus of the adult female rat. *Cell Tissue Res.* 1979;198:427—433.
41. Bouret SG, Draper SJ, Simerly RB. Trophic action of leptin on hypothalamic neurons that regulate feeding. *Science.* 2004;304(5667):108—110. Available from: https://doi.org/10.1126/science.1095004.
42. Toran-Allerand CD. Sex steroids and the development of the newborn mouse hypothalamus and preoptic area in vitro: implications for sexual differentiation. *Brain Res.* 1976;106:407—412.
43. Liberman EA, Topaly VP, Tsofina LM, Jasaitis AA, Skulachev VP. Mechanism of coupling of oxidative phosphorylation and the membrane potential of mitochondria. *Nature.* 1969;222 (5198):1076—1078. Available from: https://doi.org/10.1038/2221076a0.
44. Sakurai T, Amemiya A, Ishii M, et al. Orexins and orexin receptors: a family of hypothalamic neuropeptides and G protein-coupled receptors that regulate feeding behavior. *Cell.* 1998;92:696.
45. Abizaid A, Liu Z-W, Andrews ZB, et al. Ghrelin modulates the activity and synaptic input organization of midbrain dopamine neurons while promoting appetite. *J Clin Invest.* 2006;116:3229—3239, Epub Oct 19. PMC. Available from: 1618869.
46 Sternson SM, Shepherd GMG, Friedman JM. Topographic mapping of VMH-arcuate nucleus microcircuits and their reorganization by fasting. *Nat. Neurosci.* 2005;8:1356—1363. Available from: https://doi.org/10.1038/nn1550.
47. Yang Y, Atasoy D, Su HH, Sternson SM. Hunger states switch a flip-flop memory circuit via a synaptic AMPK-dependent positive feedback loop. *Cell.* 2011;146(6):992—1003. Available from: https://doi.org/10.1016/j.cell.2011.07.039.
48. Gao Q, Mezei G, Rao Y, et al. Anorexigenic estradiol mimics leptin's effect on re-wiring of melanocortin cells and Stat3 signaling in obese animals. *Nat Med.* 2007;13(1):89—94. Epub 2006 Dec 31.
49. Gyengesi E, Liu ZW, D'Agostino G, et al. Corticosterone regulates synaptic input organization of POMC and NPY/AgRP neurons in adult mice. *Endocrinology.* 2010;151:5395—5401. Epub 2010 Sep 15.
50. Levin BE, Dunn-Meynell AA, Balkan B, Keesey RE. Selective breeding for dietinduced obesity and resistance in Sprague—Dawley rats. *Am J Physiol.* 1997;273:R725—R730.
51. Horvath TL, Sarman B, García-Cáceres C, et al. Synaptic input organization of the melanocortin system predicts diet-induced hypothalamic reactive gliosis and obesity. *PNAS.* 2010;107 (33):14875—14880, Epub 2010 Aug 2. PMCID: PMC. Available from: 2930476.
52. Araque A. Astrocytes process synaptic information. *Neuron Glia Biol.* 2008;4(1):3—10. Available from: https://doi.org/10.1017/S1740925X09000064. Epub 2009 Feb 27.
53. Fuente-Martín E, García-Cáceres C, Granado M, et al. Leptin regulation of glutamate and glucose transporters in hypothalamic astrocytes. *JCI.* 2012;122(11):3900—3913. Available from: https://doi.org/10.1172/JCI64102 (Epub ahead of print). Pii64102.
54. Garcia-Caceres C, Quarta C, Varela L, et al. Astrocytic insulin signaling couples brain glucose uptake with nutrient availability. *Cell.* 2016;166:867—880. Available from: https://doi.org/10.1016/j.cell.2016.07.028.
55. Kim JG, Suyama S, Koch M, et al. Leptin signaling in atsrocytes regulates hypothalamic neuronal circuits and feeding. *Nat Neurosci.* 2014;17(7):908—910. Available from: https://doi.org/10.1038/nn.3725. Epub 2014 Jun 1.
56. Varela L, Stutz B, Song JE, et al. Hunger-promoting AgRP neurons trigger an astrocyte-mediated feed-forward auto-activation loop in mice. *J Clin Invest.* 2021;144239. Available from: https://doi.org/10.1172/JCI144239. Online ahead of print.
57. Dietrich MO, Zimmer MR, Bober J, Horvath TL. Hypothalamic Agrp neurons drive stereotypic behaviors beyond feeding. *Cell.* 2015;160:1222—1232. Available from: https://doi.org/10.1016/j.cell.2015.02.024. Published online March 4 2015.
58. Timper K, Del Río-Martín A, Cremer AL, Bremser S, Alber J, Giavalisco P, et al. GLP-1 receptor signaling in astrocytes regulates fatty acid oxidation, mitochondrial integrity, and function. *Cell Metab.* 2020;31(6):1189—1205. Available from: https://doi.org/10.1016/j.cmet.2020.05.001.

59. Andrews ZB, Liu Z-W, Wallingford N, et al. UCP2 mediates ghrelin's action on NPY/AgRP neurons by lowering free radicals. *Nature*. 2008;454(7206):846–851. Epub 2008 Jul 30.

60. Jin S, Kim JG, Park JW, Koch M, Horvath TL, Lee BJ. Hypothalamic TLR2 triggers sickness behavior via a microglia-neuronal axis. *Sci Rep*. 2016;6:29424. Available from: https://doi.org/10.1038/srep29424.

61. Kim JD, Yoon NA, Jin S, Diano S. Microglial UCP2 mediates inflammation and obesity induced by high-fat feeding. *Cell Metab*. 2019;30(5):952–962. Available from: https://doi.org/10.1016/j.cmet.2019.08.010. Epub 2019 Sep 5.

62. Varela L, Suyama S, Huang Y, et al. Endothelial HIF-1a enables hypothalamic glucose uptake to drive POMC neurons. *Diabetes*. 2017;66(5):1511–1520. Available from: https://doi.org/10.2337/db16-1106 [Epub ahead of print]. pii: db161106.

63. de Lecea L, Kilduff TS, Peyron C, et al. The hypocretins: hypothalamus-specific peptides with neuroexcitatory activity. *Proc Natl Acad Sci USA*. 1998;95:322–327.

64. Horvath TL, Gao XB. Input organization and plasticity of hypocretin neurons: a possible clues for obesity's association with insomnia. *Cell Metab*. 2005;1(4):279–286.

65. Horvath TL, Diano S, van den Pol AN. Synaptic interaction between hypocretin (orexin) and NPY cells in the rodent and primate hypothalamus — a novel hypothalamic circuit implicated in metabolic and endocrine regulations. *J Neurosci*. 1999;19:1072–1087.

66. Horvath TL, Peyron C, Diano S, et al. Hypocretin (orexin) activation and synaptic innervation of the locus coeruleus noradrenergic system. *J Comp Neurol*. 1999;415:145–159.

67. Rao Y, Liu ZW, Borok E, et al. Prolonged wakefulness induces experience-dependent synaptic plasticity in mouse hypocretin/orexin neurons. *J Clin Invest*. 2007;117(12):4022–4033, PMC. Available from: 2104495.

68. Abizaid A, Mineur YS, Roth RH, et al. Reduced locomotor responses to cocaine in Ghrelin deficient mice. *Neuroscience*. 2011;192:500–506. 2011 June 12. [Epub ahead of print].

69. Koch M, Varela L, Kim JG, et al. Hypothalamic POMC neurons promote cannabinoid-induced feeding. *Nature*. 2015;519(7541):45–50. Available from: https://doi.org/10.1038/nature14260. Epub 2015 Feb 18.

70. Di Marzo V, Goparaju SK, Wang L, Liu J, Bátkai S, Járai Z, Fezza F, Miura GI, Palmiter RD, Sugiura T, Kunos G. Leptin-regulated endocannabinoids are involved in maintaining food intake. *Nature*. 2001;410:822–825. Available from: https://doi.org/10.1038/35071088.

71. Tschop M, Smiley DL, Heiman ML. Ghrelin induces adiposity in rodents. *Nature*. 2000;407:908–913.

72. Diano D, Farr SA, Benoit SC, et al. Ghrelin controls hippocampal spine synapse density and memory performance. *Nat Neurosci*. 2006;9:381–388. Available from: https://doi.org/10.1038/nn1656. Published online: 19 February 2006.

73. Carlini VP, Varas MM, Cragnolini AB, Schiöth HB, Scimonelli TN, de Barioglio SR. Differential role of the hippocampus, amygdala, and dorsal raphe nucleus in regulating feeding, memory, and anxiety-like behavioral response to ghrelin. *Biochem Biophys Res Commun*. 2004;313(3):635–641. Available from: https://doi.org/10.1016/j.bbrc.2003.11.150.

74. Tingley D, McClain K, Kaya E, Carpenter J, Buzsáki G. A metabolic function of the hippocampal sharp wave-ripple. *Nature*. 2021;597(7874):82–86. Available from: https://doi.org/10.1038/s41586-021-03811-w. Epub 2021 Aug 11.

75. Cotman CW, Berchtold NC. Exercise: a behavioral intervention to enhance brain health and plasticity. *Trends Neurosci*. 2002;25:295–301.

76. Dietrich M, Andrews ZB, Horvath TL. Exercise-induced synaptogenesis in the hippocampus is dependent on UCP2-dependent mitochondrial adaptation. *J Neurosci*. 2008;28:10766–10771.

77. Diano S, Liu Z-W, Jeong JK, et al. Peroxisome proliferation—associated control of reactive oxygen species sets melanocortin tone and feeding in diet-induced obesity. *Nat Med*. 2011;17:1121–1127. Available from: https://doi.org/10.1038/nm.2421. Published online: 28 August 2011.

78. Fleury C, Neverova M, Collins S, et al. Uncoupling protein-2: a novel gene linked to obesity and hyperinsulinemia. *Nat Genet*. 1997;15:269–272.

79. Horvath TL, Warden CH, Hajos M, Lombardi A, Goglia F, Diano S. Brain UCP2: uncoupled neuronal mitochondria predict thermal synapses in homeostatic centers. *J Neurosci*. 1999;19(23):10417–10427.

80. Horvath TL, Stachenfeld NS, Diano S. A temperature hypothesis of hypothalamus-driven obesity. *Yale J Biol Med*. 2014;87:149−158.

81. Coppola A, Liu ZW, Andrews ZB, et al. A central thermogenic-like mechanism in feeding regulation: an interplay between arcuate nucleus T3 and UCP2. *Cell Metab*. 2007;5(1):21−33, MID: 17189204; PubMed Central PMCID: PMC. Available from: 1783766.

82. Brand MD, Esteves TC. Physiological functions of the mitochondrial uncoupling proteins UCP2 and UCP3. *Cell Metab*. 2005;2(2):85−93. Available from: https://doi.org/10.1016/j.cmet.2005.06.002.

83. Andrews Z, Horvath B, Barnstable CJ, et al. UCP2 is critical for nigral dopamine cell survival in a mouse model of Parkinson's disease. *J Neurosci*. 2005;25:184−191.

84. Bechmann I, Diano S, Warden CH, Bartfai T, Nitsch R, Horvath TL. Brain mitochondrial uncoupling protein 2 (UCP2): protective stress signal in neuronal injury. *Biochem Pharm*. 2002;64:369−374.

85. Diano S, Matthews RT, Patrylo P, et al. Uncoupling protein 2 prevents neuronal death including that occurring during seizures: a mechanism for pre-conditioning. *Endocrinology*. 2003;144 (11):5014−5021. Epub 2003 Aug 21.

86. Andrews ZB, Diano S, Horvath TL. Mitochondrial uncoupling proteins in the central nervous system: in support of function and survival. *Nat Rev Neurosci*. 2005;6(11):829−840.

87. Deierborg T, Wieloch T, Diano S, Warden CH, Horvath TL, Mattiasson G. Overexpression of UCP2 protects thalamic neurons following global ischemia in the mouse. *J Cereb Blood Flow Metab*. 2008;28(6):1186−1195 [Epub ahead of print] PMC. Available from: 2642535.

88. Fontana L, Partridge L. Promoting health and longevity through diet: from model organism to humans. *Cell*. 2015;161(1):106−118. Available from: https://doi.org/10.1016/j.cell.2015.02.020.

89. Gautron L, Elmquist JK. Sixteen years and counting: an update on leptin in energy balance. *J Clin Invest*. 2011;121(6):2087−2093. Available from: https://doi.org/10.1172/JCI45888. Epub 2011 Jun 1.

90. Myers Jr. MG. Leptin receptor signaling and the regulation of mammalian physiology. *Recent Prog Horm Res*. 2004;59:287−304. Available from: https://doi.org/10.1210/rp.59.1.287.

91. Lewis MR, Lewis WH. Mitochondria in Tissue Culture. *Science*. 1914;39:330−333.

92. Song JE, Alves TC, Stutz B, Sestan-Pesa M, Kilian N, Diano S, et al. (2020). Mitochondrial fission governed by Drp1 regulates exogenous fatty acid usage and storage in HeLa cells. *Metabolites*. 2021 May 2021;11(5):322. Available from: https://doi.org/10.3390/metabo11050322.

93. Dietrich MO, Liu Z-W, Horvath TL. Mitochondrial dynamics controlled by mitofusins regulate Agrp neuronal activity and diet-induced obesity. *Cell*. 2013;155(1):188−199. Available from: https://doi.org/10.1016/j.cell.2013.09.004.

94. Schneeberger M, Dietrich MO, Sebastián D, et al. Mitofusin 2 in POMC neurons connects ER stress with leptin resistance and energy imbalance. *Cell*. 2013;155(1):172−187. Available from: https://doi.org/10.1016/j.cell.2013.09.003.

95. Dietrich MO, Horvath TL. Limitations in anti-obesity drug development: the critical role of hunger-promoting neurons in integrative physiology. *Nat Rev Drug Discovery*. 2012;11(9):675−691. Available from: https://doi.org/10.1038/nrd3739 [Epub ahead of print].

96. Liao CY, Rikke BA, Johnson TE, Diaz V, Nelson JF. Genetic variation in the murine lifespan response to dietary restriction: from life extension to life shortening. *Aging Cell*. 2010;9(1):92−95. Available from: https://doi.org/10.1111/j.1474-9726.2009.00533.x. Epub 2009 Oct 30.

97. Cowley MA, Smith RG, Diano S, et al. The distribution and mechanism of action of ghrelin in the CNS demonstrates a novel hypothalamic circuit regulating energy homeostasis. *Neuron*. 2003;37:649−661.

98. Miletta MC, Iyilikci O, Shanabrough M, et al. AgRP neurons control compulsive exercise and survival in an activity-based anorexia model. *Nat Metab*. 2020;2:1204−1211. Available from: https://doi.org/10.1038/s42255-020-00300-8.

99. Dietrich MO, Antunes C, Geliang G, et al. AgRP neurons mediate Sirt1's action on the melanocortin system and energy balance: roles for SirT1 in neuronal firing and synaptic plasticity. *J Neurosci*. 2010;30(35):11815−11825.

100. Dietrich MO, Bober J, Ferreira JG, et al. AgRP neurons regulate the development of dopamine neuronal plasticity and non food-associated behaviors. *Nat Neurosci*. 2012;15:1108–1110.

101. Watson HJ, et al. Genome-wide association study identifies eight risk loci and implicates metabopsychiatric origins for anorexia nervosa. *Nat Genet*. 2019. Available from: https://doi.org/10.1038/s41588-019-0439-2.

102. Zimmer MR, Schmitz AE, Dietrich MO. Activation of AgRP neurons modulates memory-related cognitive processes in mice. *Pharmacol Res*. 2019;141:303–309. Available from: https://doi.org/10.1016/j.phrs.2018.12.024. Epub 2019 Jan 2.

103. François M, Fernández-Gayol O, Zeltser LM. A framework for developing translationaly relevant animal models of stress-induced changes in eating behavior. *Biol Psychiatry*. 2021. Available from: https://doi.org/10.1016/j.biopsych.2021.06.020. S0006-3223(21)01428-1. Online.

104. Scolnick B, Zupec-Kania B, Calabrese L, Aoki C, Hildebrandt T. Remission from chronic anorexia nervosa with ketogenic diet and ketamine: case report. *Front Psychiatry*. 2020;11:763. Available from: https://doi.org/10.3389/fpsyt.2020.00763. eCollection 2020.

105. Lee KW, Woon PS, Teo YY, Sim K. Genome wide association studies (GWAS) and copy number variation (CNV) studies of the major psychoses: what have we leartn? *Neurosci Biobehav Res*. 2012;36(1):556–571. Available from: https://doi.org/10.1016/j.neubiorev.2011.09.001. Epub 2011 Sep 17.

106. Javitt DC, Spencer KM, Thaker GK, Winterer G, Hajós M. Neurophysiological biomarkers for drug development in schizophrenia. *Nat Rev Drug Discov*. 2008;7(1):68–83. Available from: https://doi.org/10.1038/nrd2463.

107. Sakel MJ. *The great physiodynamic therapies in psychiatry: an historical reappraisal*. In F. Marti-Ibanez et al., eds. New York; 1956:13–75.

108. Doroshov DB. (2007) Performing a cure for schizophrenia: insulin coma therapy on the wards. 2007 J Hist Med Allied Sci. Apr;62(2):213–43. Available from: https://doi.org/10.1093/jhmas/jrl044. Epub 2006 Nov 14.

109. Jones K. Insulin coma therapy in schizophrenia. *J R Soc Med*. 2000;93(3):147–149. Available from: https://doi.org/10.1177/014107680009300313.

110. Kalinowsky LB. The discoveries of somatic treatments in psychiatry: facts and myths. *Compr Psychiatry*. 1980;21(6):428–435. Available from: https://doi.org/10.1016/0010-440x(80)90044-9.

111. Neustatter WL. *Modern psychiatry in practice*. London: J. & A. Churchill 1948:224.

112. Citrome L, Holt RIG, Walker DJ, Hoffmann VP. Weight gain and changes in metabolic variables following olanzapine treatment in schizophrenia and bipolar disorder. *Clin Drug Investig*. 2011;31(7):455–482. Available from: https://doi.org/10.2165/11589060-000000000-00000.

113. Das C, Mendez G, Jagasia S, Labbate LA. Second-generation antipsychotic use in schizophrenia and associated weight gain: a critical review and meta-analysis of behavioral and pharmacologic treatments. *Ann Clin Psychiatry*. 2012;24(3):225–239.

114. Danzinger S, Levav J, Avnaim-Pesso L. Extraneous factors in judicial decisions. *PNAS*. 2011;108:6889–6892.

Further reading

Andrews AB, Horvath TL. Uncoupling protein 2 regulates lifespan in mice. *Am J Physiol Endo Met*. 2009;296(4):E621–E627, Epub 2009 Jan 13. PMC. Available from: 2670629.

Andrews ZB, Erion D, Beiler R, et al. Ghrelin promotes and protects nigrostriatal dopamine function via a UCP2-dependent mitochondrial mechanism. *J Neurosci*. 2009;29(45):14057–14065, PubMed Central PMCID: PMC. Available from: 2845822.

Berger JM, Singh P, Khrimian L, et al. Mediation of the acute stress response by the skeleton. *Cell Metab*. 2019;30(5):890–902.e8. Available from: https://doi.org/10.1016/j.cmet.2019.08.012. 2019 Sep 12. pii: S1550-4131(19)30441-3.

Diano S, Horvath TL. Mitochondrial uncoupling protein 2 in glucose and lipid metabolism. *Trends Mol Med*. 2012;18(1):52–58. 2011 Sep 12. [Epub ahead of print].

Flynn RA, Belk JA, Qi Y, et al. Systematic discovery and functional interrogation of SARS-CoV-2 viral RNA-host protein interactions during infection. *Cell*. 2020;184(9):2394−2411.e16. Available from: https://doi.org/10.1016/j.cell.2021.03.012. Epub 2021 Mar 11.

Gao Q, Horvath TL. The neurobiology of feeding end energy expenditure. *Annu Rev Neurosci*. 2007;30:367−398.

García-Cáceres C, Balland E, Prevot V, et al. Role of astrocytes, microglia and tanycytes in brain control of systemic meyabolism. *Nat Neurosci*. 2019;22(1):7−14. Available from: https://doi.org/10.1038/s41593-018-0286-y. Epub 2018 Dec 10.

Golgi C *Sulla fina anatomia degli organi centrali del sistema nervosa*. Napoli, Milan, Pisa; 1886.

Halene, et al. *Genes Brain Behav*. 2009;8(7):661−675.

Hermes G, Nagy D, Waterson M, et al. Role of mitochondrial uncoupling protein-2 (UCP2) in higher brain functions, neuronal plasticity and network oscillation. *Mol Metab*. 2016;5:415−421. Available from: https://doi.org/10.1016/j.molmet.2016.04.002. eCollection 2016 Jun1.

Horvath TL. Suprachiasmatic efferents avoid phenestrated capillaries but innervate neuroendocrine cells including those producing dopamine. *Endocrinology*. 1997;138:1312−1320.

Horvath TL. An alternate pathway for visual signal integration into the hypothalamo-pituitary axis: retinorecipient intergeniculate neurons project to various regions of the hypothalamus and innervate neuroendocrine cells including those producing dopamine. *J Neurosci*. 1998;18:1546−1558.

Horvath TL, Bruning JC. Developmental programming of the hypothalamus: a matter of fat. *Nat Med*. 2006;12:52−53.

Horvath TL, Diano S, Miyamoto S, et al. Uncoupling proteins-2 and 3 influence obesity and inflammation in transgenic mice. *Int J Obes*. 2003;27:433−442.

Horvath TL, Andrews ZB, Diano S. Fuel utilization by hypothalamic neurons: roles for ROS. *Trends Endocrinol Metab*. 2009;20(2):78−87. Epub 2008 Dec 10.

Jin S, Yoon NA, Liu Z-W, et al. Drp1 is required for AgRP neuronal activity and feeding. *eLife*. 2021;10:e64351. Available from: https://doi.org/10.7554/eLife.64351.

Keshavan MS, Nasrallah HA, Tandon R. *Schizophr Res*. 2011;127(1-3):3−13.

Kim JG, Sun B-H, Dietrich MO, et al. AgRP neurons regulate bone mass. *Cell Rep*. 2015;13(1):8−14. Available from: https://doi.org/10.1016/j.celrep.2015.08.070. Epub 2015 Sep 24.

Matarese G, Procaccini C, Menale C, et al. Hunger-promoting hypothalamic neurons modulate effector and regulatory T-cell responses. *PNAS*. 2013;110(15):6193−6198. Available from: https://doi.org/10.1073/pnas.1210644110. Epub 2013 Mar 25.

Nasrallah C, Horvath TL. Mitochondrial dynamics in the central regulation of Metabolism. *Nat Rev Endocrinol*. 2014;10:650−658. Available from: https://doi.org/10.1038/nrendo.2014.160. Epub 2014 Sep 9.

Onorati O, Li Z, Liu F, et al. Zika virus disrupts phospho-TBK1 localization and mitosis in human neuroepithelial stem cells and radial glia. *Cell Rep*. 2016;16(10):2576−2592. Available from: https://doi.org/10.1016/j.celrep.2016.08.038.

Oury F, Khrimian L, Gardin A, et al. Maternal and offspring-derived osteocalcin influences brain development and functions. *Cell*. 2013;155(1):228−241. Available from: https://doi.org/10.1016/j.cell.2013.08.042.

Plum L, Ma X, Hampel B, et al. Enhanced PIP3 signaling in POMC neurons causes neuronal silencing via KATP channel activation and leads to diet-sensitive obesity. *J Clin Invest*. 2006;116(7):1886−1901, Epub 2006 Jun 22 PMC. Available from: 1481658.

Rash BG, Micali N, Huttner AJ, Morozov YM, Horvath TL, Rakic P. Metabolic regulation and glucose sensitivity of cortical radial glial cells. *PNAS*. 2018;115(40):10142−10147. Available from: https://doi.org/10.1073/pnas.1808066115. Epub 2018 Sep 17.

Romanov RA, Zeisel A, Bakker J, et al. A novel organizing principle of the hypothalamus reveals 2 molecularly segregated periventricular dopamine neurons. *Nat Neurosci*. 2017;20(2):176−188. Available from: https://doi.org/10.1038/nn.4462. Epub 2016 Dec 19.

Ruan HB, Dietrich MO, Liu ZW, et al. O-GlcNAc transferase-controlled Agrp neurons suppress browning of white fat. *Cell*. 2014;159:306−317.

Ryu S, Shchukina I, Youm YH, et al. Ketogenic diet restrains aging-induced exacerbation of coronavirus infection in mice. *Elife*. 2020;10:e66522. Available from: https://doi.org/10.7554/eLife.66522.

Sambataro, et al. *Neuropsychopharmacology*. 2013;38(5):846−853.

Shadel GS, Horvath TL. Mitochondrial ROS signaling in organismal homeostasis. *Cell*. 2015;163:560−569. Available from: https://doi.org/10.1016/j.cell.2015.10.001.

Simon-Areces J, Dietrich MO, Hermes G, Garcia-Segura LM, Arevalo M-A, Horvath TL. Ucp2 regulates neuronal differentiation of the hippocampus and related adult behavior. *PLoS One*. 2012;7(8): e42911. Epub 2012 Aug 8.

Song JE, Alves TC, Stutz B, et al. Mitochondrial fission governed by Drp1 regulates exogenous fatty acid usage and storage in HeLa cells. *Metabolites*. 2020;11(5):322. Available from: https://doi.org/10.3390/metabo11050322.

Taylor-Giorlando M, Scheinost D, Ment L, Rothman D, Horvath TL. Prefrontal cortical and behavioral adaptations to surgical delivery mediated by metabolic principles. *Cereb Cortex*. 2019;29 (12):5061−5071. Available from: https://doi.org/10.1093/cercor/bhz046.

Vacic, et al. *Nature*. 2011;471(7339):499−503.

Varela L, Horvath TL. AgRP neurons: a switch between peripheral carbohydrate and lipid utilization. *EMBO J*. 2012;31:4252−4254. Available from: https://doi.org/10.1038/emboj.2012.287.

Varela L, Horvath TL. Leptin and insulin pathways in POMC and AgRP neurons that modulate energy balance and glucose homeostasis. *EMBO Rep*. 2012;13:1079−1086. Available from: https://doi.org/10.1038/embor.2012.174.

Varela L, Schwartz M, Horvath TL. Mitochondria controlled by UCP2 determines hypoxia-induced synaptic remodeling in the cortex and hippocampus. *Neurobiol Dis*. 2016;90:68−74.

Vogt MC, Paeger L, Hess S, et al. Neonatal insulin action impairs hypothalamic neurocircuit formation in response to maternal high fat feeding. *Cell*. 2014;156:495−509. Available from: https://doi.org/10.1016/j.cell.2014.01.008. Epub 2014 Jan 23.

Yasumoto Y, Stoiljkovic M, Kim JD, et al. Ucp2-dependent microglia-neuronal coupling control ventral hippocampal circuit function and anxiety-like behavior. *Mol Psychiatry*. 2021;20:1−13. Available from: https://doi.org/10.1038/s41380-021-01105-1. Online ahead of print.

Yoon, et al. *Biol Psychiatry*. 2013.

Zimmer MR, Fonseca AHO, Iyilikci O, Pra RD, Dietrich MO. Functional ontogeny of hypothalamus AgRP neurons in neonatal mouse behaviors. *Cell*. 2019;178(1):44−59.e7. Available from: https://doi.org/10.1016/j.cell.2019.04.026. Epub 2019 May 16.

CHAPTER 3

Joy Hirsch: Brain-to-Brain

3.1 Full circle

The year of "conversational boot camp" with my colleagues Tamas and Zoltán resulted in the 19 conversations included in this book and has inspired each of us to place our disciplines into a context that includes the other as well as a much broader range of "real-world" topics. This is quite an accomplishment because our disciplines are conventionally so different, and our domains of expertise are so specialized that often relevance to life issues such as education, general health, and well-being across the life span, general medicine and disease, psychiatric and developmental disorders, and social organization remains outside the bounds of what we do. However, over the course of our 19 conversations we discovered that, actually, they are not. Extending the parable of the Blind Men and the Elephant (Zoltán, Chapter 1: The developing brain), we can conceptualize a grand circle that encloses a central space. The circle is formed by the disciplines that we represent, human development, human neural physiology, and social connections between humans, and suggests a cycle from procreation and human development to human physiology and neuroendocrinology, and onto social interactions (Fig. 3.1). These disciplines relate many common questions and aims that are enclosed within this conceptual framework, including developmental disorders, psychiatric disease, general medicine, individual health and well-being, education, and social organization. Extending this conceptual framework, one can imagine that the circle can include many other disciplines but with similar topics and questions included in the center. Our wondering conversations in the aggregate addressed them all by infusing contributions from each of our points of view. Being challenged within this relatively undirected and intense experience of just talking and listening to Zoltán and Tamas has shaped and expanded my appreciation of these "bigger pictures." Perhaps together we have discovered a partial solution to the "elephant dilemma" which is to connect each of the interrogators combining their experiences. In the case of the elephant, combining the experiences of the wall, the spear, the snake, the tree, the fan, and the rope could jointly lead to the construction of an elephant. By analogy, the emerging inclusion of the "center realities" enclosed within our "circle" has enriched and extended each of our representations of our respective disciplines.

3.1.1 What makes us social?

This chapter begins where the previous two chapters have left off and starts with the self-evident axiom that humans are a profoundly social species. From the beginning of our

Body, Brain, Behavior
DOI: https://doi.org/10.1016/B978-0-12-818093-8.00008-2

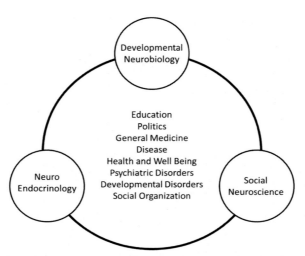

Figure 3.1 Insights from the conversations. The central space within the fusion of developmental neurobiology, neuroendocrinology, and social neuroscience includes many common questions and topics related to developmental disorders, psychiatric disease, general medicine, individual health and well-being, education, social organization, politics, etc. This framework can be generalized to include many conventional disciplines on the outer ring that enclose a wide range of common questions within.

postnatal lives to the end of our lifespans and at all points in between, humans gravitate toward each other. Isolation from human contact is one of the most severe and potentially damaging conditions to health and well-being. "Socialness" is conserved across most of the animal kingdom and further highlights the relevance of social interactions for humans. Humans are naturally drawn to other humans and are endlessly fascinated by watching other humans. Sitting on a park bench and watching other humans is an endlessly fascinating pastime. Our attention systems seem to be tuned to seek information from other human beings, and the transfer of emotional experience by nonverbal cues such as facial expressions and acoustical signals is a well-known phenomenon. Given the accelerated interest in and relevance of prosocial interactions, empirical approaches and theoretical frameworks for dyadic interactions have not advanced accordingly. The neural mechanisms that underlie live social interactions are poorly understood, in part, due to the paucity of noninvasive neural technologies capable of providing objective neural data that represent the fundamental principles.

Social behaviors are the core of our behavioral repertoire. Primary social cues consist of facial expressions, eye-to-eye contact, affective touch, acoustical expressions, and spoken language, all of which include behaviors conventionally outside the domains of neural imaging technologies when they are dynamically and reciprocally connected to other brains. Fortunately, with the emergence of head-mounted hemodynamic detectors and technologies such as functional near-infrared spectroscopy (fNIRS), pairs of individuals can be imaged during live and social interactions, and the underlying neural systems can be exposed. These technical advances serve to reduce the knowledge gap between single brain neural processes and those during dyadic interactions. Pioneering opportunities to investigate and understand human brains connected with other human brains during various live social interactions are now in their early days. This chapter presents some of this emerging foundation.

3.1.2 The social brain starts with the holes in your head

Zoltán and Tamas point out in previous chapters that social interactions between humans are modified (1) by our individual differences as part of our developmental sequences that tune us for socialization and also by (2) internal signaling from body sources outside of the brain. However, from the point of view of this chapter, social interactions between individuals start with the sensory systems that send and receive relevant information to and from other individuals. These sensory systems include the holes in your head such as eyes, ears, mouth, and nose that are specialized for the detection of information about the external world. Other somatosensory information such as touch is acquired from external sources via a distributed sensory system that covers the entire body. These specialized mechanisms transduce external signals into meaningful information that is encoded by the nervous system. Here, at the point of entry, the principle of functional specificity is indisputable. That is, each of these systems is specialized to detect a specific form of environmental energy, that is, wavelengths of light, air pressure waves, chemical particles, and mechanical pressure, and tasked with transducing these stimulus domains into decodable neural signals processed by primary and secondary sensory systems. These systems occupy separate brain real estate as they enter the brain and are functionally and spatially separate.

3.1.3 The "black box" model of the brain: sensory input versus behavioral output

Psychophysics. Models that relate stimulus properties to human sensation, perception, and action introduced empirical approaches to brain research. For example, Weber's Law, proposed in 1834 by Weber and De Pulsu[1], a German physiologist, showed that any noticeable change in a stimulus is a constant ratio of the original stimulus. For example, if I is the intensity of a light, then a perceptible change in light intensity by the human brain, delta I, is predicted by delta I/I. This general relationship was found to apply to most human sensory experiences and provided an objective basis for the future development of a science referred to as "psychophysics." "Psychophysical" models link the sensory behavior of the animal (usually human) to the properties of the stimulus and consider the brain as a "black box" that performs the transduction (Fig. 3.2). For example, the relationship between light intensity and the perception of brightness can be

The "Black Box" Brain Model

Figure 3.2 The "black box" model of the brain is based on comparisons of stimulus properties that are detected (responses) in relation to those that are presented as stimuli. These "stimulus/response" functions are referred to as psychometric functions and describe the transduction functions that are performed by the brain.

quantified as a specific "stimulus/response" function that is often referred to as the psychometric function of human brightness perception. Psychometric functions were pioneered by Helmholtz[2] and sensory physiologists such as Graham[3] followed the tradition. The shape and characteristics of the functions are assumed to be the consequence of ensembles of neural operations that are performed within the "black box" of the brain. That is, these neural operations were not directly observed, but rather considered a part of a multicomponent system consisting of a stimulus such a light that was input to a transducer device (the brain) where it was coded and integrated with relevant information that led to responsive behavior such as a perception or an action such as deciding if the light was different from a background intensity. Neural operations, in this model framework, were thought to be tools that transduced input signals to output events and the entire system was taken as a single "psychometric" unit.

3.1.4 What's in the black box?

The quest to probe what is in the black box is rooted in discoveries related to the anatomy of the neuron and enabled by the development of microscopes and staining techniques. A flurry of discoveries starting in the late 1890s and early 1900 illustrated the varied and branching structures of nerve cells and led to a new science of cytology and onto a long-standing scientific enterprise to understand the organization and operations of the nervous system. These discoveries include the famous drawings of Cajal and Golgi as well as the comparative neuroanatomy of Brodmann. These advances are discussed in previous chapters by Zoltán and Tamas. This neuron journey also included the seminal advances of Hodgkin and Huxley in 1952[4] with the discovery that the neuron was a physiological unit and a generator of an impulse due to exchanges of sodium and potassium ions across a membrane. The significance of this touchstone event lies in the connection of the

The Single-Neuron Brain Model

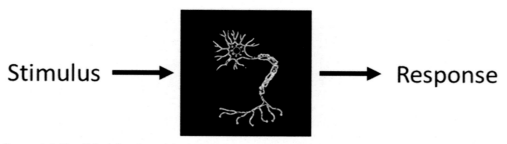

Stimulus ⟶ Response

Figure 3.3 The "black box" model of the brain is conceptualized to include the properties of neurons that perform the transduction functions. These advances introduce the concept of linking hypotheses that connect behavior to the actions of neurons.

neuron to specific stimulus parameters and behavioral responses. Thus function was merged with structure and raised questions like, "how does it work" (Fig. 3.3).

3.1.5 How does black box machinery work?

Studies of the human visual system led to a series of general structure-function insights, including the discovery by Hartline[5] that the excitement of photocells in the eye generated electrical impulses in the optic nerve. He was awarded the 1967 Nobel Prize in Physiology or Medicine for this and the discovery of both excitatory and inhibitory neural responses to light. However, the science of neurons, their anatomy and mechanisms of operation, did not fully extend to the behavior of the organisms until Hubel and Wiesel[6] recorded signals from neurons in the visual system of a cat. They related neural firing rates to the motion and orientation of a line stimulus in the visual field. Together they received the 1981 Nobel Prize in Physiology for Medicine for these discoveries of information processing in the brain. These and continuing discoveries of neural encoding related to external stimuli became a pivotal influence on the emerging discipline of "functional neuroanatomy."

3.1.6 Complex structural machinery and black box operations

Surprise! There is a lot more squishy stuff in the "black box." Contemporary neuroscience has discovered a treasure trove of neural machinery with principles of organization and structure/function relationships that make up our brains. The advantage of objective measures of neural responses and the compelling need for cures of brain diseases continues to drive modern brain science. Extending from the early antecedents of anatomy and physiology, this "brain first" doctrine continues as an active scientific enterprise and approach to address brain disease, developmental disorders, and

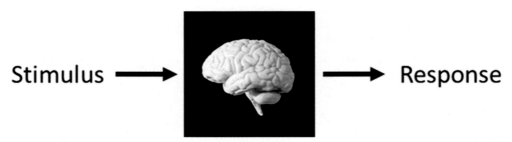

The Many-Neuron Brain Model

Stimulus ⟶ [brain] ⟶ Response

Figure 3.4 The impact of functional neuroimaging has revolutionized conventional neuroscience and introduced neural systems that provide scaffolding for specific behaviors.

psychiatric conditions. The transformative impact of neuroimaging on understanding principles of brain mechanisms and the brain's association with behavior now dominates the landscape of neural investigation and theoretical frameworks. The origin of the neuroimaging branch of brain science stems from the seminal work of Ogawa et al.[7] at AT&T Bell Laboratories in Murray Hill, New Jersey, who discovered that during magnetic resonance imaging (MRI) of rats a blood oxygen level-dependent signal in the occipital lobe was increased during "lights on" and decreased during "lights off." He reasoned, correctly, that the signal represented a proxy for neural activity that was responsive to the external stimulus of light. The effect was replicated in humans using functional MRI a year later by Belliveau et al.[8] at Harvard initiating a scientific revolution in noninvasive high spatial resolution investigations of human brain responding to external stimuli and cognitive challenges (Fig. 3.4).

3.1.7 A comprehensive plan to study brain and behavior

Of all the advances in the life sciences, these findings have opened the master flood gate of curiosity about the underlying neural mechanisms of the human brain and the implications for questions like, "What makes us human?" The focus on the brain with the advantages of neuroimaging has catapulted questions about our brains and behaviors to the top of the list of popular topics.

It is often noted that the brain is the most complex organ of the body with over 100,000,000,000 (billion) neurons along with other cells that make more than 100,000,000,000,000 (trillion) connections. These connections are modulated by a multiplicity of neural chemical factors that span spatial scales starting with molecules, and progressing to cells, circuits, and to systems. The complex operations of how brains are made and the cascade of events that lead to the universe of questions related to the functioning human brain are represented by both Zoltán and Tamas in the preceding chapters. The prevailing view is that this biological "hardware" consisting of interconnecting and cooperating neurons leads to the ever-changing repertoire of human behavior, including cognitive processes, emotions, perceptions, memories, and goal-directed actions. However, the monumental task of sorting these structure—function relationships and the linking hypotheses is not for the faint of heart. The pitfalls of oversimplification are common. Returning to the black box model of the brain, input and output functions are connected with elaborate linking hypotheses intended to include neural operations that are *consistent with* the relationships between neuronal functions and observed behavioral domains.

The vision of understanding the operation of every neuron at every functional level as a plan to conquer the challenges of brain disorders has been focused into an action plan referred to as the BRAIN (Brain Research through Advancing Innovative Neurotechnologies, https://braininitiative.nih.gov/about/overview) Initiative. The initiative was launched on April 2, 2013 by US President Barack Obama who announced a Grand Challenge to "accelerate the development and application of new technologies that will enable researchers to produce dynamic pictures of the brain that show how individual brain cells and complex neural circuits interact at the speed of thought."[9] Subsequently, the National Institutes of Health formulated and introduced a 10-year plan to achieve the primary objective of accelerating the development of technology for acquiring fundamental insights about how the nervous system functions in health and disease. A starting point for the BRAIN Initiative is focused on the neural circuits in the brain, including characterization of the component cells, synaptic connections, and dynamic ensembles of activity associated with behavior. This overarching objective spans multiple scales of investigation ranging from the molecular and cellular processes that govern short-range neural circuits to long-range processes that govern complex behaviors observed by neuroimaging of humans.

3.2 The transparent black box reveals the single brain

Existing technologies for brain mapping, predominantly MRI and electromagnetic techniques such as magnetoencephalography, and electroencephalography (EEG), are foundational for the investigation of the human brain under normal and pathological conditions and have expanded the range of behaviors for investigation far beyond the sensory systems. These technologies have contributed extensively to a major branch of neuroscience focused on the correlation of functional brain activity with cognition and behavior. In particular, the explosive growth in brain imaging technologies has led to an operational understanding of specialized neural processes associated with complex cognitive behaviors such as human language, face processing, memory, decision-making, vision and auditory association processes, emotions, learning, and social interactions.

In general, the underlying neural ensembles associated with the sensory systems and fundamental cognitive tasks are (1) localized to specific brain regions and short-range neural circuits that receive and transmit information and (2) are interconnected by long-range pathways between the participating brain regions. Thus two principles of single brain organization emerge. The first is the principle of segregation where specific regions of brain are thought to be dedicated to specific tasks and processing, and the second is the principle of integration where coactive regions in the brain are thought to be interconnected under specific task demands. For example, in the case of the human language system, a region located in the left superior temporal gyrus (STG), often referred to as Wernicke's area, is thought to be specialized for receptive functions of language (understanding and interpreting spoken words). Additionally, a region located in the left inferior frontal gyrus, often referred to as Broca's area, is thought to be specialized for productive language functions (production of speech). These notions of functional specificity are based on lesion studies and observations where damage to these regions reliably results in alterations in either receptive language or productive language, respectively. These two complexes of specialized brain regions are interconnected by well-known pathways, including the arcuate fasciculus and the arcuate uncinate, that transmit information relevant to the processes of understanding language and producing speech. Face processing is another example of specialized processing. Signals originate in the visual pathways and progresses toward increasing specificity in a hierarchical manner as the encoding progresses to inferior temporal structures, including the fusiform gyrus referred to as the fusiform face area. Memory, decision-making, learning, and sensory–motor functions have also been extensively interrogated for single brains in health and disease, and together constitute an era of progress in understanding the stand-alone brain and behavior.

3.2.1 The dyadic unit: brain to brain

As previously noted, humans are innately social. However, neural mechanisms that mediate dynamic social interactions remain understudied despite their evolutionary significance. Live interactions are a core component of our social natures but these aspects of our socialness are not well studied in the contexts of single brain paradigms. The focus on neural coupling between communicating individuals during natural face-to-face dialog highlights the dyad as a dynamic functional unit within this theoretical framework. For example, in this case, the dyad takes on properties that are not necessarily true of either individual alone. These properties may include shared and reciprocal representations of dynamic information transfers modified by both implicit and explicit information. This approach marks a departure from a conventional single brain investigation designed to study modular and specialized processes such as encoding of static and noninteractive faces and cognitive processes. Theoretical frameworks that include neural coupling are unique to dyadic operations and introduce a novel domain for investigation. Primary social cues consist of eye-to-eye contact, facial expressions, acoustical expressions, spoken language, affective touch, and body gestures. How are these cues sent and received? How do these cues entrain or couple the dyad? What guides our acquisition of the cues and how are they interpreted?

The interactive social brain relies on sensory input acquired from the primary sensory systems discussed earlier as well as cognitive systems related to the theory of mind.[10] In a dynamic social interaction between two people, the sensory inputs and outputs are often reciprocal between the partners and contingent so that one partner influences the other such as in mimicry of facial expressions or convergence of linguistic styles and walking cadence. Speech production in its native form assumes that a second brain receives and understands the speech. During the act of sending and receiving verbal information, the two brains operate as a unity as information is shared. This section introduces some of the emerging principles of the neuroscience of "TWO" and the dyadic model of social interactions. We start with a spoken conversation between two people. Is there any system in the brain that is sensitive to or operational during verbal interactions as in a dialog as opposed to a noninteractive monologue?

Communication based on natural spoken language and interpersonal interaction is a foundational component of social behavior; however, theoretical representations of the neural underpinnings associated with communicating individuals remain in the early stages.[11–13] Investigation of dynamic social interactions between two individuals extends the fundamental unit of behavior from a single brain to a two-brain unit, the dyad. For example, canonical human language models are based on single brains and consist of specialized within-brain units for language functions such as production of speech (Broca's region), and reception/comprehension of auditory signals (Wernicke's region), as well as systems associated with high-level cognitive and linguistic functions.[14–16] However, understanding how these single within-brain subsystems mediate the rapid and dynamic exchanges of information during live verbal interactions between dyads is an emerging area of investigation. Two-brain studies during natural conversation are challenging not only because it is necessary to simultaneously record synchronized and spontaneous neural activity within the two-brain unit, but also because the communicating partners are engaged in different tasks: one talking and the other listening. These joint and nonsymmetrical functions within a dyad include complementary transient and adaptive responses as opposed to mirroring or imitation and thus extend the computational complexity of these functions and investigations.

It has been suggested that coupling between neural responses of the speaker and listener represents mutual information transfer functions,[17] and these functions implement neural adaptations that dynamically optimize information sharing.[12] The complexities of synchronized neural activity have been recognized in previous studies where participants recited and subsequently heard the same story.[18] A hierarchy of common activation for the compound epoch with both talking and listening functions was associated with multiple levels of perceptual and cognitive processes. These various levels of abstraction were assumed to operate in parallel with distinguishable timescales of representation,[19] and it has been proposed that live communication between dyads includes coupling of rapidly exchanged signals between these various levels of representation.[12]

3.2.2 Imaging technologies for dyadic interactions

However, investigation of these putative underlying fine-grained interactive neural processes challenges conventional imaging techniques because one brain is only half of the system. The unit dyad includes the simultaneous activity of two brains and novel computational approaches to relate them. This knowledge gap between static and "solo" processes and dynamic and interactive processes is, in part, a consequence of neuroimaging methods that are generally restricted to single individuals, static tasks, and nonverbal responses. Understanding neural processes that underlie dynamic coupling between individuals engaged in interactive tasks such as talking and listening requires the development of novel experimental paradigms, technology, and computational methods. These challenges are largely addressed using techniques such as fNIRS that enable simultaneous acquisitions of brain activity-related signals (as represented by hemodynamic responses) from two naturally interacting and verbally communicating individuals.

fNIRS is based on changes in spectral absorbance of both oxyhemoglobin and deoxyhemoglobin (deOxyHb) detected by surface-mounted optodes.[20–24] These hemodynamic signals serve as a proxy for neural activity similar to hemodynamic signals acquired by functional magnetic resonance, functional MRI (fMRI).[7,22,27] fNIRS is a developing neuroimaging technique that provides noninvasive, minimal risk, localized measurements of task-related hemodynamic brain activity.[26,27] Because detectors are head mounted and relatively insensitive to motion artifacts, superficial cortical activity can be monitored in interactive two-person settings (Fig. 3.5B).

Conventional functional neuroimaging methods optimized to investigate neural operations in single human brains do not interrogate systems engaged during live, spontaneous social interactions due, primarily, to confinement, isolation of single participants, and intolerance to head movements. Given that it is widely appreciated that human beings are predisposed to spontaneous social behaviors, understanding the neural underpinnings is a high priority, yet currently understudied for the above reasons. Development of imaging techniques, paradigms, and computational approaches for natural, dynamic interactions advances opportunities to investigate these relatively unexplored "online" processes.[28–32] We applied an innovative hyperscanning technique using fNIRS to study the interactive processes of human language.

(A)

The Two-Brain Model

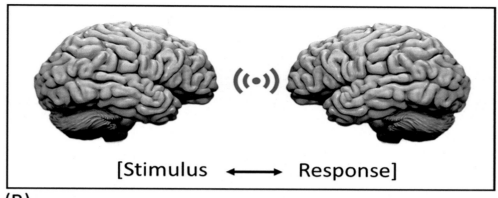

[Stimulus ⟷ Response]

(B)

Figure 3.5 (A) The "two-brain" model. Two interacting brains are conceptualized as a unit during the interaction where stimuli and responses are shared in a reciprocal and dynamic manner. The separation between stimuli and responses becomes vanishingly small and the unified interaction becomes the focus of investigation. (B) Illustration of a dyadic setup to image two brains simultaneously (hyperscanning) during socially interactive and natural tasks such as talking and listening using fNIRS (with permission from the participating individuals). *fNIRS*: Functional near-infrared spectroscopy.

Human language systems are typically investigated by fMRI and employ noncommunicative internal thought processes (covert speech) rather than actual (overt) speaking due to the deleterious effects of head movement in the scanner. Although in some cases actual speaking has been achieved during fMRI,[33,34] these studies do not capture speaking processes as they occur in a live interactive dialog with another person. Neural systems engaged by actual speaking have, however, been validated using fNIRS[35,36] and further extend the technological advantages of fNIRS to the investigation of neural systems that underlie the neurobiology of live verbal communication and social interaction.[11,32,37–39]

3.2.3 The interactive brain hypothesis and spoken language

A recently proposed interactive brain hypothesis provides a general framework suggesting that natural interactions engage neural processes not engaged without interaction.[40,41] This hypothesis has been tested in studies of various communications and interactions. One example is with communicative pointing, a human-specific gesture that is intended to share information about a visual item with another person. In one study, neural correlates of pointing were examined with and without communicative intent using positron emission tomography, and showed that pointing when communicating activated the right posterior superior temporal sulcus and right medial prefrontal cortex in contrast to pointing without communication.[42] Further, in addition to augmented activity during communicative gestures, an EEG study found that resonance between cross-brain signals in the posterior mirror neuron system of an observer continuously followed subtle temporal changes in activity of the same region of the sender.[43] This fine-grained temporal interplay between cross-brain regions is consistent with both coordinated motor planning functions and synchronized mentalizing during interpersonal communications based on gestures. Similarly, neural systems engaged during live speaking and listening with intent to communicate are not conventionally investigated and the extent to which these static conditions represent dynamic neural systems is a current topic of investigation. Dyadic observations enable a direct test of the interactive brain hypothesis during verbal communication by isolating the effects of interaction using both contrast comparisons as well as cross-brain coupling methods (discussed next).

3.2.4 Two brains, one dyadic mechanism for verbal dialog

The interactive brain hypothesis proposes that interactive social cues are processed by brain substrates and provide a general theoretical framework for investigating the underlying neural mechanisms of social interaction. We test the specific case of this hypothesis proposing that canonical language areas are upregulated and dynamically coupled across brains during social interactions based on talking and listening. fNIRS was employed to acquire simultaneous deOxyHb signals of the brain on partners who alternated between speaking and listening while doing an object naming and description task with and without interaction in a natural setting. Comparison of interactive and noninteractive conditions confirmed an increase in neural activity associated with Wernicke's area, including the STG during interaction (Fig. 3.6). However, the hypothesis was not supported for Broca's area. Cross-brain coherence determined by wavelet analyses of signals originating from the STG and the subcentral area was greater during interaction than noninteraction. In support of the interactive brain hypothesis, these findings suggest that interactive tasks such as a conversation between two people engage dynamically coupled cross-brain neural mechanisms that include mechanisms that transmit and receive interpersonal information. A further finding confirms that the within-in brain connectivity between neural signals originating in

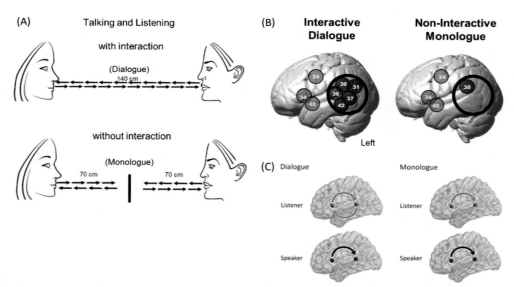

Figure 3.6 Interactive versus noninteractive language. (A) In the dyadic paradigm, speakers either engage in a two-way interactive communication about objects and their descriptions (top) or a one-way (monologue) description (bottom). Both conditions involve talking and listening behaviors; (B) comparisons of the neural responses during the monologue and dialog conditions reveal increased activity in Wernicke's area during the interactive dialog (left) relative to monologue (right); (C) findings of the Granger causality analysis reveal variation in connectivity between Broca's and Wernicke's areas during dialog and monologue conditions. Signal transmission is bidirectional during listening in the dialog conditions whereas the direction of transmission is top-down (Broca to Wernicke) in other conditions. *References: From Hirsch J, Adam Noah J, Zhang X, Dravida S, Ono Y. A cross-brain neural mechanism for human-to-human verbal communication. Social Cognitive and Affective Neuroscience. 2018;9:907-20. https://doi.org/10.1093/scan/nsy070. Ono Y, Zhang X, Noah JA, Dravida S, Hirsch J. Bidirectional connectivity between Broca's area and Wernicke's area during interactive verbal communication. Brain Connectivity. 2021 Jun 15. https://doi.org/10.1089/brain.2020.0790.*

Wernicke's and Broca's during interactive language is bidirectional for the listening task and unidirectional for talking during interaction and during both talking and listening when the spoken language occurs without interaction (monologue).[44]

These findings provide a keystone building-block for two-person neuroscience by documenting that interpersonal interaction, as in a very simple conversation between two people, is associated with brain responses that are specifically aligned with interpersonal interaction rather than the canonical separate functions of speaking and listening. The many conversations with Tamas and Zoltán have jogged my thinking to consider the important questions of whether these interactive systems are innate or learned. Given the dominating property of "socialness," a convincing argument can be made for evolutionary selectivity for social cues and interpretation. However, the alternative model for developmental processes that are modulated by environmental cues can also be made. The ultimate truth could also lie in the middle of these two positions proposing that our social behaviors are both favored by innate properties and enhanced by developmental plasticity. One undeniable fact in all of this is that at birth the human brain is "primed" to develop orderly maps of separate sensory inputs due, primarily, to the "holes in the head" designed to detect signals from the outside world. In the following section, we move on to the visual system to consider the generalizability of these interactive findings and test the hypothesis that systems for interpersonal interaction are also distinguished from the canonical face processing systems.

3.2.5 Face-to-face contact: a fundamental dyadic event

Understanding the neuroscience of dynamic face-to-face human social interactions remains a challenge, despite frequent calls for an increase in second-person neuroscience approaches,[28] and for an interactionist approach[41] to study concepts like the "we-mode."[45] Key to these theories is the idea that communication or mutual engagement between people (i.e., two or more partners jointly sharing information with one another as in eye-to-eye contact) involves additional neural networks and social dynamics as compared to performing the same task alone.[31] Further, it is assumed that the analysis of two brains together can reveal a new dimension of neural activity not observed by studying one brain at a time. However, defining the distinct neurocognitive components of communicative social interactions remains challenging.

Eye contact is a fundamental component of face processing and a diagnostic hallmark in a number of developmental disorders, including autism spectrum disorder (ASD) and psychiatric conditions such as schizophrenia and depression. Intermittent eye-to-eye fixations are widely thought of as particularly potent stimuli. Literary claims that "the eyes are the window to the soul" have been noted for centuries and taken as self-evident. For example, wisdom regarding the social significance of eye gaze is long-standing in classical literature and may have its source in a quote from Marcus

Tullius Cicero's *Orator*: "Ut imago est animi voltus sic indices oculi"—for as the face is the image of the soul, so are the eyes its interpreters.[46]

Eye-to-eye contact might be poetically thought of as a "spark" that ignites a "social connection" between two people. These connections can vary in significance. For example, claims of "love at first sight" are not uncommon and suggestive of a quintessential role for eye-to-eye encounters. I recall a sign in the 14th Street subway station in NY City at one time that read, "Don't look at strangers." Somehow, the assumption was that a mutual glance with a stranger would be an invitation for an unwanted encounter. Once an 8-year-old boy told me that he did not want to look people in the eye because he did not want them to know his secrets. Although there are many circumstantial and cultural variations related to the interpretation of interpersonal eye gaze, eye contacts generally signal salient cues such as levels of engagement, emotional status, intention, judgment, and an array of nuanced exchanges of social information, including an invitation for interaction.

However, regardless of the interpersonal significance, the neural mechanisms underlying real face processing and direct eye-to-eye contact with another person are not well understood and are active areas of research. This research builds upon decades of prior research on face and object processing that reveals localized specializations for faces and specialized objects. Technical developments in fNIRS now enable full-head acquisitions of brain signals acquired simultaneously on two individuals under naturalistic conditions with eye-tracking measures to extend this research to the investigation of faces during live interaction. This combination of hemodynamic signals and eye-tracking enable identification of visual acquisition processes when real and simulated faces are viewed and the eye-contact moments between the two participants. These ongoing and prior studies reveal that eye-to-eye contact is a special case of face processing and processes for real faces are distinguished from simulated faces by activity in localized regions, including in the right temporal parietal junction (TPJ)[47,48]

The appreciation of real faces is special case of face processing. The perception of a dynamic/real face requires many complex factors to be interpreted in real time to facilitate socialization and communication.[49–52] Real face-to-face contact is a dynamic and interactive behavior in which face cues are reciprocally exchanged and activity within neural networks specialized for facial recognition, dynamic motion, emotion, and socialization is expected to play a fundamental role. It is well-known that subliminal presentations (less than 17 ms) of static faces with fearful expressions influences reaction times and decision processes.[53] It is a reasonable inference that many of the "back-channel" cues that signal social interactions are also either subliminal or barely detectable. Although face processing is actively studied, the pathways are not entirely understood. However, it is known that they include the TPJ, fusiform gyrus, ventral occipital temporal cortex regions, and the posterior superior temporal sulcus.[54,55] Additional anterior temporal gyrus and prefrontal lobe structures have also been shown to play a role in face interactions, including the inferior and medial frontal gyri. Although neural activity specific to perception of faces has been observed in the inferior occipital and fusiform gyri, perception of dynamic eye gaze has been associated with higher processing areas in the superior temporal sulci and TPJ.[56–60] These ventral stream regions have been previously associated with the perception of movement.

While these areas have been shown to be involved in static and dynamic facial processing, the mechanism of information exchange and regulation of circuits that upregulate attentional mechanisms related to real and dynamic eye-to-eye contact between partners in social interaction is not well understood. Previous studies have explored the role of eye movement behaviors, including blinking and attention regulation in a social circuit that is more active in joint attention tasks compared to simple eye gaze or during randomized video sequences.[50–52] The significance of these findings related to understanding the exchange of information in live face-to-face interaction is enhanced by the relevance of eye-contact behavior and social interaction difficulties that are characteristic of ASDs, social anxiety, and schizophrenia.[61–63]

Here, the specific neural responses across dyads while making face and eye contact were compared to when each subject alone interacted with a prerecorded dynamic video of the face of a partner (Fig. 3.7A). In the case of the real partner, we hypothesized that detection of dynamic stimuli, such as facial expressions and eye movements known to occur in the real face condition, will elicit neural activity that is not present when subjects perform the same task with a prerecorded video sequence of a dynamic face. Specifically, we observed increased activity in the right TPJ during the perception of real faces as compared to the perception of the dynamic video faces, (Fig. 3.7B) and increased cross-brain coherence of signals originating from areas of the cortex associated with visual and social functions.[48]

Similar findings were observed (Fig. 3.8A and B) when the comparison face was a human-like robot.[64]

3.2.6 Is there a dyadic domain?

What is established by the above is a "drop in the bucket" for the new "neuroscience of two." However, this entry level "proof of principle" points the way to a new wave of future directions. Some of these future directions are anticipated by Fig. 3.1 and emerge from the web of connections illuminated by our 19 conversations. What are the guiding questions for dyadic neuroscience in health and medicine? As in any new branch of science, the fundamental unit of inquiry is foundational. Here the fundamental unit involves components of two brains. These are not well characterized. For example, what unit is established by two visual systems working together to create an eye contact? What is the unit established by two language systems working together in

Eye-Gaze paradigm: "Real" vs. Video

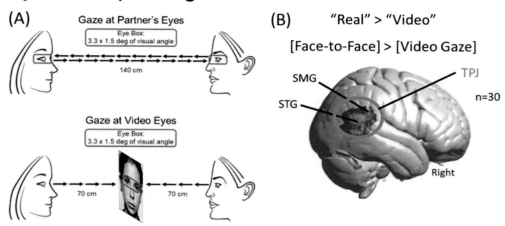

Figure 3.7 Face-to-face versus face-to-video face. (A) Experimental paradigm shows two conditions: gaze at a partner's eyes (top row) and gaze at video eyes (bottom row). (B) The contrast of real eyes greater than video eyes is shown for the sample of 30 participants. Findings are consistent with expectations for social functions with increased activity during the face interaction in regions that include the SMG and the STG. These regions are included in the TPJ, a region that has been previously shown to be sensitive to face processing. *SMG*, Supramarginal gyrus; *STG*, superior temporal gyrus; *TPJ*, temporal parietal junction. *References: From Noah JA, Zhang X, Dravida S, Ono Y, Naples A, McPartland JC, Hirsch J. Real-time eye-to-eye contact is associated with cross-brain neural coupling in angular gyrus. Frontiers in Human Neuroscience. 2020;14:19. https://doi.org/10.3389/ fnhum.2020.00019 Hirsch J, Zhang X, Noah JA, Ono Y. Frontal temporal and parietal systems synchronize within and across brains during live eye-to-eye contact. NeuroImage. 2017;157:314—30. https:// doi.org/10.1016/j.neuroimage.2017.06.018.*

a dialog? What is the fundamental unit established when hands engage in a greeting handshake? These units belong simultaneously to both brains and have some similarities to a phone call that requires a caller and a receiver that are connected by a telecommunication system and a protocol that transmits the interactions. However, the stimuli for these dyadic interactions may be transient and very brief dynamic signals not traditionally investigated in face–object–language paradigms based on known stimulus properties and measured outputs. See Figs. 3.2–3.4. Our brains are designed to extract information and react to these paired events, and yet we know little about these seminal operations, let alone what the stimuli consist of. How do these social systems develop in the brain? What happens when they do not develop?

3.2.6.1 The dyadic approach to understand social differences in ASD

ASD is a pervasive developmental disorder characterized by atypical behavioral responses to face and eye contact with reduced responses to emotional cues conveyed by facial

Eye-Gaze paradigm: "Real" vs. Robot

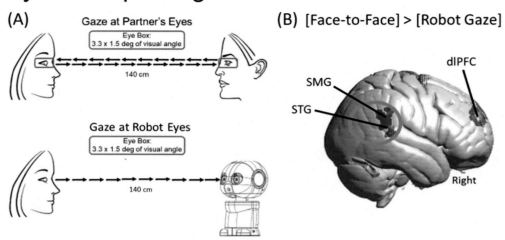

Figure 3.8 Face-to-face versus face-to-robot. (A) Experimental paradigm shows two conditions: gaze at a partner's eyes (top row) and gaze at robot eyes (bottom row). (B) The contrast of real eyes greater than robot eyes is shown for the sample of 15 participants. Findings are consistent with prior findings shown in Fig. 3.7 and expectations for social functions with increased activity during the face interaction in regions that include the SMG and the STG, regions included in the TPJ. *SMG*, Supramarginal gyrus; *STG*, superior temporal gyrus; *TPJ*, temporal parietal junction. *Reference: From Kelley M, Noah JA, Zhang X, Scassellati B, Hirsch J. Comparison of human social brain activity during eye-contact with another human and a humanoid robot. Frontiers in Robotics and AI. 2020;7:209. https://doi.org/ 10.3389/frobt.2020.599581.*

dynamics. Current models of face and eye processing in ASD are based primarily on noninteractive paradigms where data are acquired in single subject situations rather than dyadic paradigms that include live social interactions. However, the behavioral differences are related to live social interactions, and emerging dyadic approaches for investigation of the underlying neural systems associated with specific behavioral characteristics offer an impactful advance. Ongoing two-person hyperscanning studies are currently testing the hypothesis that the neural systems that underlie live face-to-face interactions in typically developed participants are hypoactive in ASD. Preliminary findings are consistent with this hypothesis and further suggest that differences in face processing in ASD may be related to atypical functioning in neural pathways known to be sensitive to visual sensing.

In keeping with the curiosity-driven conversations over the previous year and a half, we have expanded our investigations of dyadic-related properties to address issues with impact beyond conventional brain science. Here we ask how the quality and meaning of interpersonal dialog alters neural systems. In one model the canonical regions associated with language production and reception would be modified (a local systems model), and in another model, the brain would adopt a constructionist strategy and "design" a neural response appropriate for the needed strategy.

3.3 Dyads in agreement and disagreement

Everyday conversations and face-to-face encounters in a real social world are made up of situations in which agreement and disagreement are components of transactions and negotiations communicated by language and facial expressions (Fig. 3.9). While linguists have investigated the behavioral aspects of verbal interactions[65,66], and the neural correlates of spoken language exchanges within dyads have been previously described[67−69]; and face-to-face interactions have also been previously described[38,48] understanding how these multiple neural systems adapt to extended dialectical discussions between partners remains an open and timely research area. The relevance of insight regarding the neurobiology of human dyadic behavior during expressions of

Conversations between dyads in Agreement vs. Disagreement

Figure 3.9 An illustration of gestures associated with communicating hotly contested views. *Photo by Afif Kusuma on Unsplash.*

congruent and incongruent opinions is highlighted in times of extreme political and social division. This includes our current political landscape with deeply divided issues on topics such as COVID-19 vaccines, masks, and social practices; supreme court rulings related to abortions and a woman's right to choose; the validity of the 2021 presidential election; and international affairs such as the withdrawal from Afghanistan. These hotly debated issues and divided opinions bring home the urgent need for better understanding of our biological underpinnings related to conflict and consensus.

A fundamental question in social and interactive neuroscience relates to how dynamically interconnected neural systems between two brains adapt to social conditions such as agreement or not. It has been proposed that in a natural dialog emergent perceptual experiences of dyads are shared to achieve a common level of understanding.[70] How does this occur? How do the canonical neural systems for language, face processing, and social systems adapt when challenged by agreement or disagreement with another person?

To address these questions, we compared patterns of neural activity from extended verbal dialogs that contrasted preassessed mutual agreement or disagreement. In such an interaction "the dyad" functioned as a social unit where both individuals were "linked" together by the exchange of spoken information in the foreground and a simultaneous stream of ongoing social cues such as face and acoustic processing occurring in the background. During normal conversation, these "back-channel" cues are assumed to be used to signal understanding, affect, and willingness to cede or accept the speaking role.[71] Their effect is such that, although subject to cultural variation, human speakers tend to take turns in conversation so that overlap of speech is avoided and interspeaker silence is minimized. For example, across the 10 typologically different languages examined by Stivers et al.,[72] the mean interspeaker gap was approximately 200 Ms. This short latency between turns suggests that talking and listening functions occur nearly simultaneously during typical conversation where listeners necessarily formulate their response before the incoming speech has been completed. Thus talking and listening, as it occurs in a natural dialog, can be modeled as a single two-person (dyadic) function where processes for "sending" and "receiving" implicit and explicit information occur simultaneously and reciprocally. We hypothesize that all of these putative functions are modulated by levels of dyadic "attunement."

3.3.1 Dyadic mechanisms during "attunement" and "misattunement"

We have addressed the questions related to language systems during prosocial and contentious interactions by recording neural responses and speech concurrently during live face-to-face discussions of topics where the interlocutors were either in agreement or disagreement, a common feature characteristic of verbal exchanges between two individuals.[73] The dialectical misattunement hypothesis[29] proposes that disturbances in the reciprocal unfolding of an interaction result in a change of the dyadic state, thus predicting a behavioral and neural difference between the two conditions employed in this study. Findings in this investigation were consistent with this prediction and contribute a theoretical neural framework for the effects. The experimental aims were twofold: first, to test the hypothesis that large-scale interactive language processes during spontaneous verbal face-to-face communication are modified by expressions of agreement and disagreement; and second, to advance understanding of how dynamic and adaptive information processing related to jointly sharing and creating information leads to potential entrainment between interacting humans. See findings in Fig. 3.10.

Given the relevance of conflict and diverse opinions in our current political and cultural conditions, dyadic studies aimed at understanding the neural underpinnings of

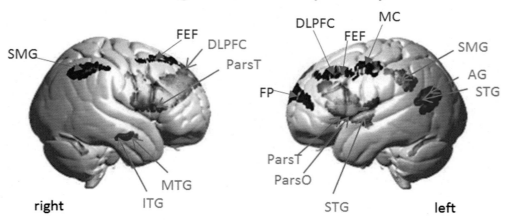

Figure 3.10 Talking during disagreement (red) and agreement (blue). The contrast between the two conditions (disagreement and agreement) reveals widely distributed and unique activity patterns for each condition. Specifically, in disagreement, bilateral frontal lobe activity is prevalent, including the dorsolateral prefrontal cortex and frontal gyri, left hemisphere SMG and Wernicke's area on the left hemisphere, including AG and STG. In contrast, the activity pattern associated with agreement includes FEF and MC. These sensory systems may be associated with face-to-face interaction consistent with the observation that SMG on the right hemisphere is also active. Prior findings illustrated in Figs. 3.7 and 3.8 are consistent with the association between left SMG and interactive face processing. *AG*, Angular gyrus; *FEF*, frontal eye fields; *MC*, motor cortex; *SMG*, supramarginal gyrus; *STG*, superior temporal gyrus. *Reference: From Hirsch J, Tiede M, Zhang X, Noah JA, Salama-Manteau A, Biriotti M. Interpersonal agreement and disagreement during face-to-face dialogue: An fNIRS investigation. Frontiers in Human Neuroscience. 2021;14:601. https://doi.org/10.3389/fnhum.2020.606397.*

conflict and consensus could prioritize novel applications for neuroscience in domains such as world peace, local governments, private and corporate institutions, religious organizations, and interpersonal interactions, for example. A theme throughout the 19 conversations was related to how critical it is to direct scientific inquiry to topics beyond the single brain and body and reach into real-world issues. Evidence of separate neural processes under conditions of agreement and disagreements is a small step in this direction.

3.3.2 Dyadic convergence

Converging patterns of coordinated behavior are frequently observed during interpersonal interactions such as walking in step[74] or synchronized applause.[75] One model for dynamic and cooperative behaviors during dialog, a very specific example of coordinated behavior, proposes a counter-phased pattern of intention to speak that is driven by a common syllable rate between dyads such that oscillating processes within the brains of speaker and listener are mutually entrained.[76] Such a framework predicts neural entrainment between the two interacting brains. Empirical support for this idea of neural coupling by mutual entrainment during a language task has been provided by a functional MRI study[34] in which the brain activity of a speaker was first measured while they told an unrehearsed life story; next the brain activity of a listener was measured while they heard the recorded audio of the story; and finally the listener's comprehension of the story was assessed using a detailed questionnaire. Using intersubject correlation analysis, Stephens et al. found that the speaker's and listener's brains exhibited joint, temporally aligned patterns of neural coupling that were correlated with the extent of comprehension (for example, such patterns were significantly diminished when the story was in a language unknown to the listener). These findings suggest that neural coupling between the speaker and listener represent not only oscillatory patterns indicating speaking and listening turn-taking behaviors, but also the transfer of mutual information.[12,17]

3.3.3 Neural coupling: a feature of the dyadic unit

A hierarchy of neural activations associated with the compound dyadic epochs where both talking and listening functions occur at the same time has been associated with multiple levels of perceptual and cognitive processes assumed to operate in parallel with distinguishable timescales of representation.[19] These conjectures have contributed to the proposal that live communication between dyads includes neural coupling of rapidly exchanged signals between these various levels of representation.[12,77] Empirical support for this theoretical framework is found in the hyperscanning investigation of simultaneous talking and listening during a natural dialog paradigm presented earlier[67] See Fig. 3.6. In that study, increased cross-brain neural coupling was observed between the STG

(BA 42, Wernicke's area) and the adjacent subcentral area (BA 43) during talking and listening with interaction but not during the monologue (noninteractive) condition. This finding is consistent with the within brain connectivity also found during interaction as well (Fig. 3.6C). Together this theoretical framework and these empirical findings support a biological underpinning for dynamic interactive behavior evidenced by neural coupling during language tasks. We test this hypothesis for the agree and disagree conditions predicting that cross-brain coherence will be different during agreement and disagreement conversations but without a prediction of which condition will be the highest.

Cross-brain neural coherence of hemodynamic signals originating from fNIRS is established as an objective indicator of synchrony between two individuals for a wide variety of tasks performed jointly by dyads.[78] Representative examples include coordinated button pressing,[79] coordinated singing and humming,[80] gestural communication,[43] cooperative memory tasks,[81] and face-to-face unstructured dialog.[68] Similarly, neural synchrony across brains has been shown to index levels of interpersonal interaction, including cooperative and competitive game playing,[82–85] imitation,[86] coordination of speech rhythms,[87] leading and following,[88] group creativity,[89] and social connectedness among intimate partners using EEG,[90] and fNIRS.[91] Cross-brain synchrony has also been shown to increase during real eye-to-eye contact as compared to mutual gaze at the eyes of a static face picture.[38] Neural coupling of fNIRS signals across the brains of speakers and listeners who separately recited narratives and subsequently listened to the passages were found to be consistent with comprehension of verbally transferred information.[92] These findings also replicate previous findings using fMRI.[34] All contribute to the advancing theoretical framework for two-person neuroscience,[31,36,82,93] and to the emerging proposition that neural coupling between partners underlies mechanisms for reciprocal interactions that mediate the transfer of verbal and nonverbal information between dyads.[50,60,94]

Findings of cross-brain coherence during agreement and disagreement (Fig. 3.11) are consistent with the conclusion that verbal interactions when the dyad is in agreement include greater neural synchrony than when the dyads are in disagreement. This synchrony is found between brain regions in the ventral and dorsal parietal streams consistent with prior findings for face-to-face interactions and also dialog interactions.

Wow! Cross-brain neural synchrony during agreement is greater than during disagreement. There is so much more to understand here. Does this have relevance to possible interventions that might encourage prosocial discussions and consensus? Is it possible to isolate the principles of neural and body processes that govern these responses? It is a "happy idea" to imagine that investigations of real-world dyadic interactions might have a positive impact on resolutions of disagreements but this is only imagination at work. Nonetheless, one of the positive outcomes of the many conversations with Tamas and Zoltán has been encouragement toward brave thinking beyond the data in hand. The following dyadic experiment similarly hints at relevant social impact.

Cross-brain Coherence: Greater during Agreement

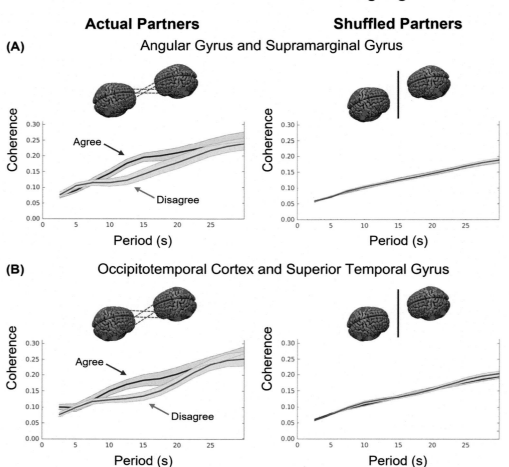

Figure 3.11 Cross-brain coherence. The *Y* axis on all four panels represents the correlation of signal wavelets between the partner brains. The *X* axis on all four panels represents the wavelet period in seconds. The top panels (**A**) are based on the cross-brain correlation between the angular gyrus and the SMG. The bottom panels (**B**) represent the cross-brain correlation between the occipitotemporal cortex and the superior temporal gyrus. In both cases, these regions are associated with facial and verbal interactions. The left columns show data from actual partners whereas the right columns show activations from partners that were computationally combined, shuffled partners, not real partners. The blue functions indicate the observations during the agree condition, whereas the red functions indicate the observations during the disagree condition. In both cases the cross-brain coherence (neural coupling) for the actual partners is greatest for the agree condition. In the case of the shuffled partners the two functions are inseparable consistent with the conclusion that the coherence between actual partners is due to the specific and individual interactions rather than the performance of the task. *SMG*, Supramarginal gyrus. *Reference: From Hirsch J, Tiede M, Zhang X, Noah JA, Salama-Manteau A, Biriotti M. Interpersonal agreement and disagreement during face-to-face dialogue: An fNIRS investigation. Frontiers in Human Neuroscience. 2021;14:601. https://doi.org/10.3389/fnhum.2020.606397.*

3.4 Two brains from different social worlds: high and low disparity dyads

3.4.1 Dyadic interactions vary depending upon who you are talking to

In socially diverse populations, prosocial communication between individuals from different socioeconomic backgrounds is common. For example, imagine that you get on a bus and sit down next to a stranger. In the normal course of events, you might initiate a "friendly" (prosocial) conversation. Here we ask the question, how would the neural systems that underlie this dialog differ if the person sitting next to you was of a different socioeconomic status (SES) than you? Although socioeconomic differences are known to influence complex social behaviors, the neurobiology associated with live interpersonal interactions between humans with socioeconomic disparities is not well understood. In societies with diverse demographics and egalitarian values, ordinary encounters with prosocial and transactional intent commonly require regulation of prejudices, stereotypes, and communication habits. These challenges are related to diverse ethnic, religious, gender, occupational, and socioeconomic identities of individuals and are an ever-present hallmark of societal norms in daily interpersonal interactions (Fig. 3.12).

3.4.2 Socioeconomic status as a dyadic variable

SES is a well-known category of social stratification that impacts attitudes and communication styles[95,96] as well as a known factor associated with health and well-being.[97] In-group/out-group dynamics are associated with tensions resulting from bias favoring people within the same group (in-group) relative to those within a different group (out-group).[98,99] Although ecologically valid investigations of in-group/out-group dynamics are rare, intergroup interactions employing arbitrary groups confirm in-group versus out-group social effects.[100–104] Neural modulation in social systems has been reported when participants interact with out-group members prior to fMRI scans[105] or view faces of in-group versus out-group members during scanning.[106] Although distinctions related to perceived class, similarity, and diversity may be detected in ordinary interpersonal interactions, they are not necessarily expressed. The importance of prosocial interactions in socially diverse environments places a high priority on understanding their neurobiological and psychiatric underpinnings.

In contrast to prior studies of bias and in-group preferences, an overarching goal of this exploratory study was to understand the mechanisms that are naturally engaged during effective communication between individuals with different group identities. Emerging models propose that complex forms of prepotent responses related to prejudice and stereotyping involve frontal neural systems and structures that have a role in detection of biases and self-regulating behaviors to prevent bias expression.[107] This framework predicts that neural responses will be evident in frontal areas during nonconfrontational conversations between high and low disparity dyads.

Conversation between dyads with high socioeconomic disparity vs. low socioeconomic disparity

Figure 3.12 An illustration of an interaction between high disparity dyads.

An emerging theoretical framework suggests that neural functions associated with stereotyping and prejudice are associated with frontal lobe networks. Using fNIRS and hyperscanning techniques during a face-to-face live communication, we explored an extension of this model to include live dynamic interactions. Neural activations were compared for dyads of similar and dissimilar socioeconomic backgrounds. The SES of each participant ($n = 78$) was based on education and income levels (Fig. 3.13).

Post-scan questionnaires confirmed increased anxiety and effort for high disparity dyads (Fig. 3.14).

Consistent with the frontal lobe hypothesis, left dorsolateral prefrontal cortex, fronto-polar area, and pars triangularis were more active during speech dialog in high than in low disparity groups (Fig. 3.15). Further, frontal lobe signals were more synchronous across

Classification of dyadic disparity

Classification of dyadic disparity was based on two factors:

1. Highest level of education
2. Annual parental (household) income

Point System

Education		Parental (household) annual income:		
Completion or some of:		Less than $50,000		10 pts
high school	10 pts	Between	$50,000 and $100,000	20 pts
college	20 pts	Between	$100,000 and $150,000	30 pts
grad school	30 pts	Above	$150,000	40 pts

The dyadic disparity score was the difference between the total points for the two partners:

- Differences greater than 25 were classified as High Disparity
- Differences less than 25 were classified as Low Disparity

Figure 3.13 Classification of dyadic disparity. The classification of dyadic disparity was based on two factors: highest level of education and annual parental or household income. A point system was used to quantify socioeconomic status of each individual participant as indicated in the figure box. Each dyad was given a "disparity score" that was the difference between the total points for the two partners. If the difference was greater than 25 the dyad was classified as high disparity; if the difference was less than 25, the dyad was classified as low disparity. The two groups were matched with respect to gender and race, variables that are known to influence communication styles. Normalization across variables other than education and income is assumed to rule out effects due to these unintended factors. Both groups of dyads engaged in prosocial dialectic discourse during acquisition of hemodynamic signals. *Reference: From Descorbeth O, Zhang X, Noah JA, Hirsch J. Neural processes for live pro-social dialogue between dyads with socioeconomic disparity. Social Cognitive and Affective Neuroscience. 2020;15(8):875—87. https:// doi.org/10.1093/scan/nsaa120.*

brains for high than low disparity dyads. Convergence of these behavioral, neuroimaging, and neural coupling findings associate left frontal lobe processes with natural prosocial dialog under "out-group" conditions.[108]

Given the relevance of prosocial behaviors to current political and cultural conditions, future studies aimed at understanding live regulation of implicit and prepotent stereotypes and prejudices are a priority. Implicit racial biases, for example, are known to be resistant to modification.[107] However, focusing on mechanisms that represent cognitive assessment of self-regulation, as suggested in this dyadic study, may point to innovative strategies for achieving positive social interventions, and thus extending impact beyond the

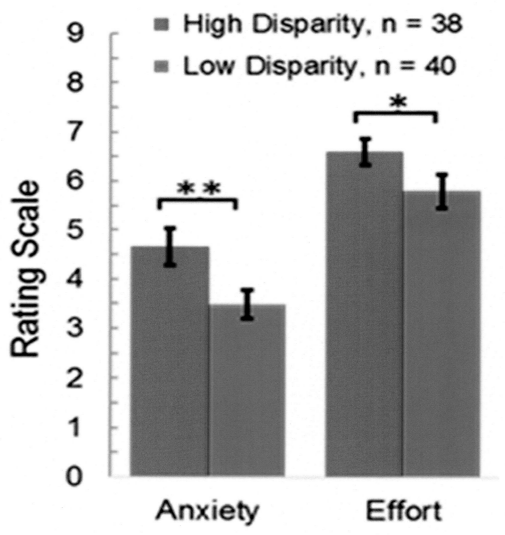

Figure 3.14 Behavioral comparison of dyads. The posttask survey for high disparity (*red*) and low disparity (*blue*) dyads is shown on the *Y* axis for the question related to anxiety and the question related to conversational effort. The high disparity ratings are shown in red whereas the low disparity ratings are shown in blue. In both cases of anxiety and effort, the high disparity dyads were higher. *Reference: From Descorbeth O, Zhang X, Noah JA, Hirsch J. Neural processes for live prosocial dialogue between dyads with socioeconomic disparity. Social Cognitive and Affective Neuroscience. 2020;15(8):875-87. https://doi.org/10.1093/scan/nsaa120.*

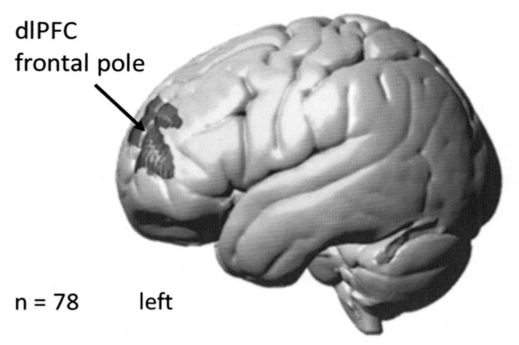

Figure 3.15 Neural comparison of high disparity > low disparity dyads. Findings highlight a neural system called upon for some kind of modulation during prosocial conversations between high disparity dyads. This system includes processes located in the left DLPFC, FP, and PT. *DLPFC*, Dorsolateral prefrontal cortex; *FP*, frontopolar area; *PT*, pars triangularis. *Reference: From Descorbeth O, Zhang X, Noah JA, Hirsch J. Neural processes for live pro-social dialogue between dyads with socio-economic disparity. Social Cognitive and Affective Neuroscience. 2020;15(8):875–87. https://doi.org/10.1093/scan/nsaa120.*

conventional bounds of neural circuitry toward social engineering. It is noteworthy that the effects observed were due to variables that included education and income which are variables that are changeable. It would be interesting to know if the circumstances were altered in either direction would the neural effects be similarly altered?

3.4.3 Dyadic neuroscience and a road map toward big and small questions

The final section of this chapter suggests a pathway for a rigorous and impactful dyadic neuroscience. It starts with the premise that:

3.4.3.1 A single brain is half of a dyadic social unit

Accordingly, the focus is on dyadic interactions and includes those properties that are shared and mutually reciprocal. Questions like what are the stimuli, what are the neural responses, what is the role of neural coupling, and how are they altered in health and disease are active areas of investigation. An advantage of dyadic neuroscience includes the broad range of new questions that can be addressed using coordinated hyperscanning methods and multimodal measurement approaches. These questions bring the advantages of neuroscience in closer proximity to applications that benefit ordinary lives. I list a few examples next, but there are many more and much to learn.

Developmental and degenerative disorders: Are the social differences that characterize ASDs better understood by live dyadic interactions? Could these approaches provide insight into interactive therapies to reduce symptomatology? Can degenerative conditions be improved with social interventions?

Health and well-being throughout the life span: What are the social mechanisms that promote biological health and well-being at all stages of growth and development? What are the effects of social judgments, bullying, and approval versus nonapproval in a dyadic situation? What are the effects of physical well-being on social well-being?

Social bonds between humans and pets: What are the essential elements of a social bond between a human and a pet such a dog or a cat that promote well-being? Are these based on eye-contact mechanisms or some innate match between human and pet social systems? What are the mechanisms of the human dog/cat dyad?

Psychiatric disorders: Can dyadic interactions between structured interviews between patients and psychiatrists be diagnostic or informative for a treatment plan? Can dyadic studies provide mechanistic insight into the underlying neurophysiology of conditions such as depression, schizophrenia, and personality disorders?

Music therapy: Can social interactions be improved by music interventions? What are the basic mechanisms of music therapy? Can dyadic function be improved with music?

Zoom versus "in person" meetings: Why and how are they different? Do live face processing mechanisms differentiate between a virtual or live dyadic mode? If so, how do these differences contribute to conflict resolution, prosocial interactions, and health-care delivery?

Emotional contagion: What are the mechanisms that convey emotion from one person to another in a live dyad? What are the roles and mechanisms of facial expressions and acoustical signals?

Education: Can learning be enhanced by understanding the roles of interpersonal interaction in the transfer of information?

Politics: Can conflict resolution be assisted by the application of principles related to the biology of "attunement" versus "misattunement" in dyads?

Understudied social behaviors: What more can we learn about deception, joint attention, imitation, and reputation management using live (natural) and dyadic structures.

Social touch: Does affective touch engage similar systems as visual and auditory systems?

Development versus innate neural circuitry for social and interactive functions: Are interactive systems supramodal? Are they learned and developed with experience or are they part of the developmental cascade of neural systems?

Social consequences of somatic and endocrine influences: What are they, and how do they work on the neural systems that underly behavior?

Like many advances in science, technology drives the discoveries[109,110]. Here the development of near-infrared spectroscopy combined with multiple peripheral modes of data acquisitions during simultaneous imaging of dyads has been the key to initiating these approaches. However, none would be possible were it not for the prior century of brain science and the ever-changing model of the brain. It is still a mysterious black box, and there is much more to talk about. I hope others are inspired to rely on friendships and spontaneous conversations to infuse creative and novel approaches to this new domain of questions.

References

1. Weber E.H., De Pulsu R. Auditu et tactu. Annotationes Anatomicae et Physiologicae. CF Köhler; 1834.
2. Helmholtz H.V. Handbuch der physiologischen Optik: mit 213 in den Text eingedruckten Holzschnitten und 11 Tafeln (Vol. 9): Voss; 1867.
3. Graham CH, Kemp EH. Brightness discrimination as a function of the duration of the increment in intensity. *J Gen Physiol.* 1938;21(5):635−650.
4. Hodgkin AL, Huxley AF. The dual effect of membrane potential on sodium conductance in the giant axon of Loligo. *J Physiol.* 1952;116(4).
5. Hartline HK. The response of single optic nerve fibers of the vertebrate eye to illumination of the retina. *Am J Physiol Legacy Content.* 1938;121(2):400−415.
6. Hubel DH, Wiesel TN. Receptive fields, binocular interaction and functional architecture in the cat's visual cortex. *J Physiol.* 1962;160(1):106−154.
7. Ogawa S, Lee TM, Kay AR, Tank DW. Brain magnetic resonance imaging with contrast dependent on blood oxygenation. *Proc Natl Acad Sci USA.* 1990;87:9868−9872.
8. Belliveau J, Kwong K, Kennedy D, et al. Magnetic resonance imaging mapping of brain function human visual cortex. *Investig Radiol.* 1992;27(0 2):S59.
9. The White House, Office of the Press Secretary. Fact sheet: brain initiative [Press release]; 2013. Retrieved from http://obamawhitehouse.archives.gov/the-press-office/2013/04/02/fact-sheet-brain-initiative
10. Carter RM, Huettel SA. A nexus model of the temporal−parietal junction. *Trends Cognit Sci.* 2013;17(7):328−336.
11. García AM, Ibáñez A. Two-person neuroscience and naturalistic social communication: the role of language and linguistic variables in brain-coupling research. *Front Psychiatry.* 2014;5:124.
12. Hasson U, Frith CD. Mirroring and beyond: coupled dynamics as a generalized framework for modelling social interactions. *Philos Trans R Soc B: Biol Sci.* 2016;371:20150366.
13. Hasson U, Ghazanfar AA, Galantucci B, Garrod S, Keysers C. Brain-to-brain coupling: a mechanism for creating and sharing a social world. *Trends Cognit Sci.* 2012;16(2):114−121. Available from: https://doi.org/10.1016/j.tics.2011.12.007.
14. Gabrieli JD, Poldrack RA, Desmond JE. The role of left prefrontal cortex in language and memory. *Proc Natl Acad Sci USA.* 1998;95:906−913.
15. Binder JR, Frost JA, Hammeke TA, et al. Human temporal lobe activation by speech and nonspeech sounds. *Cereb Cortex.* 2000;10:512−528.

16. Price CJ. A review and synthesis of the first 20 years of PET and fMRI studies of heard speech, spoken language, and reading. *NeuroImage*. 2012;62:816−847.

17. Dumas G, Nadel J, Soussignan R, Martinerie J, Garnero L. Inter-brain synchronization during social interaction. *PLoS One*. 2010;5:e12166. Available from: https://doi.org/10.1371/journal.pone.0012166.

18. Hasson U, Nir Y, Levy I, Fuhrmann G, Malach R. Intersubject synchronization of cortical activity during natural vision. *Science*. 2004;303:1634−1640.

19. Hasson U, Yang E, Vallines I, Heeger DJ, Rubin N. A hierarchy of temporal receptive windows in human cortex. *J Neurosci*. 2008;28(10):2539−2550. Available from: https://doi.org/10.1523/JNEUROSCI.5487-07.2008.

20. Villringer A, Chance B. Non-invasive optical spectroscopy and imaging of human brain function. *Trends Neurosci*. 1997;20(10):435−442.

21. Cui X, Bray S, Bryant DM, Glover GH, Reiss AL. A quantitative comparison of NIRS and fMRI across multiple cognitive tasks. *NeuroImage*. 2011;54:2808−2821.

22. Ferrari M, Quaresima V. A brief review on the history of human functional near-infrared spectroscopy (fNIRS) development and fields of application. *NeuroImage*. 2012;63:921−935.

23. Scholkmann F, Holper L, Wolf U, Wolf M. A new methodical approach in neuroscience: assessing inter-personal brain coupling using functional near-infrared imaging (fNIRI) hyperscanning. *Front Hum Neurosci*. 2013;7:813.

24. Strangman G, Culver JP, Thompson JH, Boas DA. A quantitative comparison of simultaneous BOLD fMRI and NIRS recordings during functional brain activation. *NeuroImage*. 2002;17:719−731.

25. Boas DA, Dale AM, Franceschini MA. Diffuse optical imaging of brain activation: Approaches to optimizing image sensitivity, resolution, and accuracy. NeuroImage. 2004;23:S275−88. Available from https://doi.org/10.1016/j.neuroimage.2004.07.011.

26. Scholkmann F, Kleiser S, Metz AJ, et al. A review on continuous wave functional near-infrared spectroscopy and imaging instrumentation and methodology. NeuroImage. 2014;85(1):6−27. Available from https://doi.org/10.1016/j.neuroimage.2013.05.004.

27. Boas DA, Elwell CE, Ferrari M, Taga G. Twenty years of functional near-infrared spectroscopy: introduction for the special issue. *NeuroImage*. 2014;85(Pt 1):1−5.

28. Redcay E, Schilbach L. Using second-person neuroscience to elucidate the mechanisms of social interaction. *Nat Rev Neurosci*. 2019;1.

29. Bolis D, Balsters J, Wenderoth N, Becchio C, Schilbach L. Beyond autism: Introducing the dialectical misattunement hypothesis and a Bayesian account of intersubjectivity. *Psychopathology*. 2017;50(6):355−372.

30. Schilbach L. A second-person approach to other minds. *Nat Rev Neurosci*. 2010;11:449.

31. Schilbach L, Timmermans B, Reddy V, et al. Toward a second-person neuroscience. *Behav Brain Sci*. 2013;36(04):393−414. Available from: https://doi.org/10.1017/S0140525X12000660.

32. Schilbach L. On the relationship of online and offline social cognition. *Front Hum Neurosci*. 2014;8:278.

33. Gracco VL, Tremblay P, Pike B. Imaging speech production using fMRI. *NeuroImage*. 2005;26(1):294−301.

34. Stephens GJ, Silbert LJ, Hasson U. Speaker−listener neural coupling underlies successful communication. *Proc Natl Acad Sci USA*. 2010;107:14425−14430.

35. Zhang X, Noah JA, Dravida S, Hirsch J. Signal processing of functional NIRS data acquired during overt speaking. *Neurophotonics*. 2017;4(4):041409.

36. Scholkmann F, Gerber U, Wolf M, Wolf U. End-tidal CO_2: an important parameter for a correct interpretation in functional brain studies using speech tasks. *NeuroImage*. 2013;66:71−79.

37. Babiloni F, Astolfi L. Social neuroscience and hyperscanning techniques: past, present and future. *Neurosci Biobehav Rev*. 2014;44:76−93.

38. Hirsch J, Zhang X, Noah JA, Ono Y. Frontal temporal and parietal systems synchronize within and across brains during live eye-to-eye contact. *NeuroImage*. 2017;157:314−330.

39. Pinti P, Aichelburg C, Lind F, et al. Using fiberless, wearable fNIRS to monitor brain activity in real-world cognitive tasks. *JoVE (J Visualized Exp)*. 2015;(106)e53336.

40. De Jaegher H, Di Paolo E, Adolphs R. What does the interactive brain hypothesis mean for social neuroscience? A dialogue. *Philos Trans R Soc Lond B Biol Sci.* 2016;371(1693):20150379.
41. Di Paolo E, De, Jaegher H. The interactive brain hypothesis. *Front Hum Neurosci.* 2012;6 (163):1—16.
42. de Langavant LC, Remy P, Trinkler I, et al. Behavioral and neural correlates of communication via pointing. *PLoS One.* 2011;6(3):e17719.
43. Schippers MB, Roebroeck A, Renken R, Nanetti L, Keysers C. Mapping the information flow from one brain to another during gestural communication. *Proc Natl Acad Sci USA.* 2010;107 (20):9388—9393. Available from: https://doi.org/10.1073/pnas.1001791107.
44. Ono Y, Zhang X, Noah JA, Dravida S, Hirsch J. Bidirectional connectivity between Broca's area and Wernicke's area during interactive verbal communication. *Brain Connectivity.* 2021;. Available from: https://doi.org/10.1089/brain.2020.0790. Online ahead of print.
45. Gallotti M, Frith CD. Social cognition in the we-mode. *Trends Cognit Sci.* 2013;17(4):160—165.
46. Cicero 46 BC Orator, Chapter 18, Section 60. In: Wilkins AS, editor. Oxford Classical Texts: M Tulli Ciceronis, Rhetorica. Oxford, England: Oxford University Press; 46 BC.
47. Kelley M, Noah JA, Zhang X, Scassellati B, Hirsch J. Comparison of human social brain activity during eye-contact with another human and a humanoid robot. *Front Robot AI.* 2021;7:599581.
48. Noah JA, Zhang X, Dravida S, et al. Real-time eye-to-eye contact is associated with cross-brain neural coupling in angular gyrus. *Front Hum Neurosci.* 2020;14:19. Available from: https://doi.org/10.3389/fnhum.2020.00019.
49. Chang L, Tsao DY. The code for facial identity in the primate brain. *Cell.* 2017;169(6):1013—1028.
50. Koike T, Tanabe HC, Okazaki S, et al. Neural substrates of shared attention as social memory: A hyperscanning functional magnetic resonance imaging study. *NeuroImage.* 2016;125:401—412. Available from: https://doi.org/10.1016/j.neuroimage.2015.09.076.
51. Koike T, Sumiya M, Nakagawa E, Okazaki S, Sadato N. What makes eye contact special? Neural substrates of on-line mutual eye-gaze: a hyperscanning fMRI study. *Eneuro.* 2019;6(1). Unsp Eneuro.0284-18.201910.1523/Eneuro.0284-18.2019.
52. Lachat F, Hugueville L, Lemarechal JD, Conty L, George N. Oscillatory brain correlates of live joint attention: a dual-EEG study. *Front Hum Neurosci.* 2012;6. ARTN 156 10.3389/fnhum.2012.00156.
53. Etkin A, Egner T, Peraza DM, Kandel ER, Hirsch J. Resolving emotional conflict: a role for the rostral anterior cingulate cortex in modulating activity in the amygdala. *Neuron.* 2006;51(6):871—882.
54. Haxby JV, Hoffman EA, Gobbini MI. The distributed human neural system for face perception. *Trends Cognit Sci.* 2000;4(6):223—233.
55. Kanwisher N, McDermott J, Chun MM. The fusiform face area: a module in human extrastriate cortex specialized for face perception. *J Neurosci.* 1997;17(11):4302—4311.
56. Cavallo O, Lungu C, Becchio C, Ansuini A, Rustichini, Fadiga L. When gaze opens the channel for communication: Integrative role of IFG and MPFC. *NeuroImage.* 2015;119:63—69.
57. George N, Driver J, Dolan RJ. Seen gaze-direction modulates fusiform activity and its coupling with other brain areas during face processing. *NeuroImage.* 2001;13(6):1102—1112.
58. Hooker CI, Paller KA, Gitelman DR, Parrish TB, Mesulam M-M, Reber PJ. Brain networks for analyzing eye gaze. *Cognit Brain Res.* 2003;17(2):406—418.
59. Mosconi MW, Mack PB, McCarthy G, Pelphrey KA. Taking an "intentional stance" on eye-gaze shifts: a functional neuroimaging study of social perception in children. *NeuroImage.* 2005;27(1):247—252.
60. Saito DN, Tanabe HC, Izuma K, et al. "Stay tuned": inter-individual neural synchronization during mutual gaze and joint attention. *Front Integr Neurosci.* 2010;4:127. Available from: https://doi.org/10.3389/fnint.2010.00127.
61. Schneier FR, Kent JM, Star A, Hirsch J. Neural circuitry of submissive behavior in social anxiety disorder: a preliminary study of response to direct eye gaze. *Psychiatry Res Neuroimaging.* 2009;173 (3):248—250.
62. Senju A, Johnson MH. Atypical eye contact in autism: models, mechanisms and development. *Neurosci Biobehav Rev.* 2009;33(8):1204—1214.
63. Tso IF, Mui ML, Taylor SF, Deldin PJ. Eye-contact perception in schizophrenia: Relationship with symptoms and socioemotional functioning. *J Abnorm Psychol.* 2012;121(3):616.

64. Kelley M, Noah JA, Zhang X, Scassellati B, Hirsch J. Comparison of human social brain activity during eye-contact with another human and a humanoid robot. *Frontiers in Robotics and AI*. 2020;7:209. Available from https://doi.org/10.3389/frobt.2020.599581

65. Pickering MJ, Garrod S. Toward a mechanistic psychology of dialogue. *Behav Brain Sci*. 2004;27 (2):169−190.

66. Babel M. Evidence for phonetic and social selectivity in spontaneous phonetic imitation. *J Phonetics*. 2012;40(1):177−189.

67. Hirsch J, Noah JA, Zhang X, Dravida S, Ono Y. A cross-brain neural mechanism for human-to-human verbal communication. *Soc Cogn Affect Neurosci*. 2018;13:907−920.

68. Jiang J, Dai B, Peng D, Zhu C, Liu L, Lu C. Neural synchronization during face-to-face communication. *J Neurosci*. 2012;32:16064−16069.

69. Jiang J, Borowiak K, Tudge L, Otto C, von Kriegstein K. Neural mechanisms of eye contact when listening to another person talking. *Soc Cognit Affect Neurosci*. 2017;12(2):319−328.

70. Garnier M, Lamalle L, Sato M. Neural correlates of phonetic convergence and speech imitation. *Front Psychol*. 2013;4:600.

71. Schegloff E, Jefferson G, Sacks H. A simplest systematics for the organization of turn-taking for conversation. *Language*. 1974;50(4):696−735.

72. Stivers T, Enfield NJ, Brown P, et al. Universals and cultural variation in turn-taking in conversation. *Proc Natl Acad Sci USA*. 2009;106(26):10587−10592.

73 Hirsch J, Tiede M, Zhang X, Noah JA, Salama-Manteau A, Biriotti M. Interpersonal agreement and disagreement during face-to-face dialogue: An fNIRS investigation. *Frontiers in Human Neuroscience*. 2021;14:601. Available from https://doi.org/10.3389/fnhum.2020.606397

74. Zivotofsky AZ, Hausdorff JM. The sensory feedback mechanisms enabling couples to walk synchronously: an initial investigation. *J Neuroeng Rehabil*. 2007;4(1):1−5.

75. Néda Z, Ravasz E, Brechet Y, Vicsek T, Barabási AL. The sound of many hands clapping. *Nature*. 2000;403(6772):849−850.

76. Wilson M, Wilson TP. An oscillator model of the timing of turn-taking. *Psychonomic Bull Rev*. 2005;12(6):957−968.

77. Gregory SW, Hoyt BR. Conversation partner mutual adaptation as demonstrated by Fourier series analysis. *J Psycholinguist Res*. 1982;11(1):35−46.

78. Zhang X, Noah JA, Dravida S, Hirsch J. Optimization of wavelet coherence analysis as a measure of neural synchrony during hyperscanning using functional near-infrared spectroscopy. *Neurophotonics*. 2020;7(01):1. Available from: https://doi.org/10.1117/1.nph.7.1.015010.

79. Funane T, Kiguchi M, Atsumori H, Sato H, Kubota K, Koizumi H. Synchronous activity of two people's prefrontal cortices during a cooperative task measured by simultaneous near-infrared spectroscopy. *J Biomed Opt*. 2011;16:077011.

80. Osaka N, Minamoto T, Yaoi K, Azuma M, Shimada YM, Osaka M. How two brains make one synchronized mind in the inferior frontal cortex: fNIRS-based hyperscanning during cooperative singing. *Front Psychol*. 2015;6:1−11.

81. Dommer L, Jäger N, Scholkmann F, Wolf M, Holper L. Between-brain coherence during joint n–back task performance: a two-person functional near-infrared spectroscopy study. *Behav Brain Res*. 2012;234:212−222.

82. Cui X, Bryant DM, Reiss AL. NIRS-based hyperscanning reveals increased interpersonal coherence in superior frontal cortex during cooperation. *NeuroImage*. 2012;59:2430−2437.

83. Liu N, Mok C, Witt EE, Pradhan AH, Chen JE, Reiss AL. NIRS-based hyperscanning reveals inter-brain neural synchronization during cooperative Jenga game with face-to-face communication. *Front Hum Neurosci*. 2016;10:11.

84. Piva M, Zhang X, Noah A, Chang SW, Hirsch J. Distributed neural activity patterns during human-to-human competition. *Front Hum Neurosci*. 2017;11:571.

85. Tang H, Mai X, Wang S, Zhu C, Krueger F, Liu C. Interpersonal brain synchronization in the right temporo-parietal junction during face-to-face economic exchange. *Soc Cognit Affect Neurosci*. 2016;11:23−32.

86. Holper L, Scholkmann F, Wolf M. Between-brain connectivity during imitation measured by fNIRS. *NeuroImage*. 2012;63:212−222.

87. Kawasaki M, Yamada Y, Ushiku Y, Miyauchi E, Yamaguchi Y. Inter-brain synchronization during coordination of speech rhythm in human-to-human social interaction. *Sci Rep.* 2013;3:1692.

88. Jiang J, Chen C, Dai B, et al. Leader emergence through interpersonal neural synchronization. *Proc Natl Acad Sci USA.* 2015;112:4274−4279.

89. Xue H, Lu K, Hao N. Cooperation makes two less-creative individuals turn into a highly-creative pair. *NeuroImage.* 2018;. Available from: https://doi.org/10.1016/j.neuroimage.2018.02.007.

90. Kinreich S, Djalovski A, Kraus L, Louzoun Y, Feldman R. Brain-to-brain synchrony during naturalistic social interactions. *Sci Rep.* 2017;7(1):17060. Available from: https://doi.org/10.1038/s41598-017-17339-5.

91. Pan Y, Cheng X, Zhang Z, Li X, Hu Y. Cooperation in lovers: an fNIRS-based hyperscanning study. *Hum Brain Mapp.* 2017;38(2):831−841.

92. Liu Y, Piazza EA, Simony E, et al. Measuring speaker−listener neural coupling with functional near infrared spectroscopy. *Sci Rep.* 2017;7:43293.

93. Konvalinka I, Roepstorff A. The two-brain approach: how can mutually interacting brains teach us something about social interaction? *Front Hum Neurosci.* 2012;6:215.

94. Tanabe HC, Kosaka H, Saito DN, et al. Hard to "tune in": neural mechanisms of live face-to-face interaction with high-functioning autistic spectrum disorder. *Front Hum Neurosci.* 2012;6:268. Available from: https://doi.org/10.3389/fnhum.2012.00268.

95. Vaughan E. The significance of socioeconomic and ethnic diversity for the risk communication process. *Risk Anal.* 1995;15:169−180.

96. McLeod JD, Owens TJ. Psychological well-being in the early life course: variations by socioeconomic status, gender, and race/ethnicity. *Soc Psychol Q.* 2004;67:257−278.

97. Sapolsky R. Sick of poverty. *Sci Am.* 2005;293(6):92−99.

98. Levin S, Van Laar C, Sidanius J. The effects of ingroup and outgroup friendships on ethnic attitudes in college: a longitudinal study. *Group Proc Intergroup Relat.* 2003;6:76−92.

99. Tajfel H. Social psychology of intergroup relations. *Annu Rev Psychol.* 1982;33:1−39.

100. Dunham Y, Baron AS, Carey S. Consequences of "minimal" group affiliations in children. *Child Dev.* 2011;82:793−811.

101. Lemyre L, Smith PM. Intergroup discrimination and self-esteem in the minimal group paradigm. *J Person Soc Psychol.* 1985;49:660.

102. Otten S, Wentura D. About the impact of automaticity in the Minimal Group Paradigm: evidence from affective priming tasks. *Eur J Soc Psychol.* 1999;29:1049−1071.

103. Brewer MB. In-group bias in the minimal intergroup situation: a cognitive-motivational analysis. *Psychol Bull.* 1979;86:307.

104. Turner JC. Social categorization and social discrimination in the minimal group paradigm. *Differentiation between social groups: Studies in the social psychology of intergroup relations.* London: Academic Press; 1978:101−140.

105. Richeson JA, Baird AA, Gordon HL, et al. An fMRI investigation of the impact of interracial contact on executive function. *Nat Neurosci.* 2003;6:1323−1328.

106. Van Bavel JJ, Packer DJ, Cunningham WA. The neural substrates of in-group bias: a functional magnetic resonance imaging investigation. *Psychol Sci.* 2008;19:1131−1139.

107. Amodio DM. The neuroscience of prejudice and stereotyping. *Nat Rev Neurosci.* 2014;15:670−682.

108. Descorbeth O, Zhang X, Noah JA, Hirsch J. Neural processes for live pro-social dialogue between dyads with socioeconomic disparity. *Soc Cognit Affect Neurosci.* 2020;15:875−887.

109. Pinti P, Aichelburg C, Gilbert S, Hamilton A, Hirsch J, Burgess P, Tachtsidis I. A review on the use of wearable functional near-infrared spectroscopy in naturalistic environments. Japanese Psychological Research. 2018;60(4):347−73. Available from https://doi.org/10.1016/10.1111/jpr.12206.

110. Pinti P, Tachtsidis I, Hamilton A, Hirsch J, Aichelburg C, Gilbert S, Burgess PW. The present and future use of functional near-infrared spectroscopy (fNIRS) for cognitive neuroscience. Annals of the New York Academy of Sciences. 2020;1464(1):5. https://doi.org/10.1111/nyas.13948.

Discussion 1—20th November 2020

1 How to transcribe the Zoom discussions?

Tamas: I'll tell you one big thing for me about recording. When I talk to people—for example, when I speak with journalists about one of my papers on a recorded interview—it's very uncomfortable for me when they ask me to repeat what I have just said. I cannot repeat the same exact sentence or paragraph. Because if you said, "Oh, I forgot to record, let's start again," that thing that I just told you is out of my head.

Zoltán: Once you have the recording, how do you get the transcript?

Joy: Okay, so let me tell you, I have a very clever research assistant. I sent the recording to him, and I said, "Figure this out. We're sure that there are automatic transcription programs out there." So he did, and I'm using a transcription platform called Otter. What we do is send the audio file to otter.ai, and it comes back within an hour transcribed. Now the transcription, as you can see, is not all that good, but it's good enough to get an idea of what was said.

Joy: Also, this is saved, so as soon as we leave the meeting, it will start making the file. That usually takes it about 15 min, and then I get the files, I send them to Ray, and then I get them back.

Zoltán: And besides the actual text, does otter also recognize that different people speak?

Joy: Sometimes it does, and sometimes it doesn't. It seems to vary in that.

2 The brain as a "black box"

Joy: So, Tamas started his chapter with the concept that the nervous system is made up of individual cells that are connected through synapses, the neuron doctrine. Just before the neuron doctrine, there were these really influential people like Helmholtz and Weber who thought about the brain as just a black box, but they were interested in what the black box did. They developed this whole school that's called psychophysics, in which you look at the intensity of a stimulus, and then you measure what comes out of the brain. So you have psychometric functions for the intensity of light on different backgrounds, and you transform functions of how detectable that is. It's the same with sound; stuff goes in and stuff comes out, but it's been transformed. Then you can make inferences about how the brain is processing sensory stimuli by the metric that relates the input and output. All of this was done around the early 1800s.

3 Buildings to bricks as brains to ion channels: what does the ion channel tell us about the brain?

Tamas: If I may say something as a naive person who is ignorant about many issues, for me, the brain remains a black box. This is in spite of all the older and recent discoveries. Let me give you an example. There was a great talk recently by an ion channel person here at Yale. I was listening to the talk, and he was very precisely focusing on specific ion channels, and then ion channels in general. Then he discusses that in depth in a very amazing and intuitive direction. Nevertheless, at the end of the talk, I understood nothing more about how the brain works, and neither did it give me a new direction to conceptualize possibilities. So for me, in a way (not undermining his outstanding and beaconing work), I learned nothing about the brain. This is where I come back to my usual argument: contemporary neuroscience reminds me of an approach you might take when you want to try to understand how a building is built. I know it's a wrong analogy. Nevertheless, you want to understand how a building is put together, and you start by examining a brick. Then you go deep inside the brick, exploring the smaller constituents. Once you go deep inside of the brick, you will understand everything about the brick, but you will understand nothing about how the building is put together. This is where I feel neuroscience has been heading in the last 100-some years.

Joy: Well, that's why I think that it's useful to start from the black box approach.

Tamas: Yes, right. Yeah, that's why I'm saying that it speaks to me.

Joy: My humble beginnings in science started with that model, because the legacy of these old psychophysicists was alive and well at that time. It was replaced, of course, when the Nobel Prize of Hartline came on board. Then there was Hubel and Wiesel, and people began probing the brain, futzing with the neurons, and looking at the machinery—looking at the bricks in the black box. But the black box still existed. It's a transformer machine; stuff goes in and stuff comes out. What comes out is different from what goes in, and therefore you have a transformer. I think that part of our idea here is to just as you say: we must go back to the bigger picture. That's why I wanted to kind of start my thinking, my historical thinking, and I think it really adds to yours.

Zoltán: What was, going through your narrative in your chapter Joy, is that if you start with the discussion of the eye-to-eye contact, it becomes easy to grasp your story that the brain is another side of a dyad. I thought that if you actually start with that, and then you move to these ideas of black boxes, it might be best. During eye contacts we connect our brain with others.

Joy: That's an interesting idea.

4 Eye-to-eye contact and social consequences

Joy: You know, in the 14th Street Subway Station in New York, there is a sign that says, "Don't smile at strangers. Don't look at strangers."

Tamas: So let me tell you one remarkable experience of mine in Naples, Italy. I have been spending a significant amount of time there over the past 25 years, because my wife is from there. If you go to Naples as a foreigner who doesn't understand their culture, the way people drive cars and the whole traffic situation is absolute chaos for an outsider. If you are a pedestrian, and you want to cross the street in Naples, you will never going to be able to cross the street if you wait for cars to stop at the cross-walk. Specifically, if you look into the eyes of the driver, they definitely won't stop for you. If you establish the eye-to-eye connection, you have already lost the battle of trying to cross the street. On the other hand, if you just step down and start to walk across the street, everybody stops without having any emotional outbursts. It's an amazing thing, how eye contact, in that given circumstance, diminishes your ability to actually move forward.

Joy: Well, eye contact is one of the most salient of all social cues, and it's a call to action. Coupled with body gestures, facial expressions, and even words, it really is a compelling human tool that connects us and connects our brains via the eye contact. So it's very, very important. I kind of agree—one could start with the eye contact story, you know, how poets, fiction writers, and our grandmothers understood the value of eye contact long before neuroscientists ever got on board.

Tamas: Or you could even talk about contemporary events. If you look at Desert Storm, all of the allied soldiers were wearing sunglasses. The idea was that you should not make eye contact with the locals, because that could trigger a negative reaction.

Joy: Exactly. So yeah, the salience of eye contact and its meaning in social situations is well known. We're the first people that I know of, and most people agree with this, we are the first people to actually study this together in real time.

Zoltán: A tangentially related issue that occurred to me about this is disability. When you are blind, whether you're born blind or you become blind as you move forward in your life, how do those impairments come into play in these kinds of contexts?

5 Eye-to-eye contact, blindness, and other deficits

Joy: The brain is very, very resourceful. I've just completed a study with one of my graduate students from Gallaudet University (which specializes in education for deaf and hard of hearing people) who went out to Seattle, Washington, and did an imaging study on individuals who were born both deaf and blind. We had eight Deaf-Blind participants in four pairs. They communicate with each other by touch, and they use a form of sign language called pro-tactile ASL. It's similar to ASL, like a touch version of the visual sign language for deaf people. Often, they touch each other's legs—they use this sign language by moving and touching each other's hands, arms, and legs, and drawing signs and symbols on their bodies. So, my student went there with our para-digm that we've published—talking and listening, talking and listening with interaction and without interaction, my exact paradigm—and amazingly enough, she gets good data. I mean, I said all along, "How can you ever do this?" Again, there are only eight people, you know? Little did I know, she got beautiful data. We're still processing and getting it, but the bottom line is that the brain repurposes the sensory system. It alters the somatosensory areas and language regions, Wernicke's Area and Broca's Area (the front part of the brain that normally responds to speech production), and they are responding through the sensory system. So what the brain does is that if you can't use one system to communicate, it hijacks another one.

6 Brain development and adaptations to sensory deficits

Zoltán: And that's fascinating about development! You might have some differences between the front and the back of the brain in general, and perhaps some cortical regions are more suited to establishing certain core cortical areas, but I don't think that the deaf and blind cortex knew what it would become at the time of neurogenesis during embryonic stages. So it adopted these functions, and probably all these subcortical switchboards were also adapting to these functions. So it's a thalamocortical system and also striatal system as they are rearranged. This is one of the biggest questions in developmental neuroscience: how do we specialize for certain functions? And then exploring the workings of the brain after these early sensory lesions are just incredible, your studies on these deaf and blind volunteers are fascinating. I actually never heard of anybody who was studying deaf and blind people.

Joy: I think it's a first. I think we have a major article on our hands. The student is also deaf, and she's at Gallaudet, so I'm dealing with her and her committee. She's now finishing her thesis, but we all agree that this is a potentially high impact level publication.

Zoltán: You said that these people were born both deaf and blind?

Joy: Yes.

Zoltán: Do you know how they started to even communicate, and at what point in their lives?

Joy: No, you know, I don't have the individual stories of the participants, but the demographics are really important, and that'll be a part of this. We need their ages, what their life experiences have been, and so on. There don't seem to be any comorbidities. You would think that if one is both deaf and blind, what else they may be dealing with? They kind of all live together in a community, so that's why we were able to recruit them. That's why my student went there. She's a student that I share with the Gallaudet faculty, and she was a graduate student at Gallaudet, but she came to me and asked me to be her thesis advisor because she wanted to do this kind of work. There were not many options for doing this kind of work. We sent her off to Seattle with two portable fNIRS units. The project just blows my mind. She came back, and we analyzed the data, so I know that the results are solid.

Zoltán: Let me understand this better. Do these individuals live with their families, do they go to an institution for the day, or are they there continuously?

Joy: As I understand it, they're adults who are living together in a kind of assisted living community. They're not with their families. I don't know the details of that.

Zoltán: It would be interesting to understand how they developed during childhood and so forth.

Joy: I need to know a lot more about it. One of the reasons that I don't is because it's quite difficult communicating with my student, who is deaf, so we communicate by writing and with an interpreter. I've communicated initially about the science and the work. Now that she's done, I should get her thesis this week, actually. There'll be ways to ask questions about the demographics and the histories of each of the eight people in this study. I think this example truly puts the whole idea about the brain, I don't want to say upside down, but in a unique place—that it's not as simple as people might think. I mean, I know the idea has been around for a long time that these structures and functions, you know, step into each other's roles under certain circumstances, but these are very intriguing aspects of brain function.

7 Atypical development and brain

Zoltán: If you do a very early intervention to alter sensory input to the brain in an animal model, the kinds of changes you see are just enormous. The intervention experiments of Pasko Rakic are very famous. He did them at around embryonic days 50–60 in the macaque. Then you see a complete change of the occipital pole. Normally, the macaque cortex is a flat surface at the back of the brain. When Pasko removed the retina very early in the embryonic stages and these animals did not develop retinal input to the thalamus, the cerebral cortex developed unusual foldings, so-called sulci and gyri right in the occipital pole, and there was a significant reduction in the extent of the primary visual cortex that you can identify on histological sections because of the highly prominent layer 4. His interpretation of these data was that in spite of having no eye input to the occipital cortex via the thalamus, you still have a visual cortex with prominent layer 4 that still developed from a protomap for that particular cortical area in the occipital pole. However, these manipulations probably did not influence the autonomous development of the thalamocortical projections. They develop with or without retinal input. One could interpret these results as, you still have a visual cortex because the thalamocortical system developed normally initially, but since there was no retinal input, it started to degenerate and receive less influence to produce a visual cortex. That's why you still have a visual cortex, but it is smaller. If we could ablate the thalamocortical connections from the very beginning, you probably wouldn't have any type of visual cortex. Also, for such rearrangements in the cortex we have to consider not just the thalamocortical connections, but also the corticothalamic connections. In fact more projections are going back to the thalamus than the thalamus projects to the cortex. This ratio is about 1:10. After sensory manipulations they are also rearranged. If a thalamic nucleus that normally receives sensory input from the eye or ear does not receive this input during development, it starts to receive certain types of cortical projections. It starts to have an entirely different loop system that is processing the cortical information from other areas that are active. I just can't imagine how a deaf and blind thalamocortical connectivity emerges and how it is changing throughout life in the blind and deaf your volunteers.

Joy: Yeah, it can be incredible.

Zoltán: While I was listening to Joy about her ideas that brain is another side of a dyad, I had an idea. I always wondered whether it would be possible to image the brains of conjoined twins. Conjoined twins are born physically connected to each other because the early embryo only partially separates to form two individuals. These are extremely rare events, around 1 in 200,000 live births. As surgical methods develop more and more of them get separated early, but there are some who remain physically linked together for the rest of their lives. There are many types, some are joined facing each other at the chest, but otherwise have two independent heads and bodies; others are joined near the navel (belly button), and may share abdominal organs such as liver. And you can have craniopagus twins who are joined at some area of the skull, and some of these twins might even partially share parts of their brains. My assumption is that there are fewer and fewer of these pairs because they are having operations early in life, but they could tell a lot about how the body is represented in the brain and how the brain controls the body in situations where these conditions develop from very early stages. What would be the representation of a shared body in two different brains? How would those two brains interact to coordinate movements within a differentially innervated shared body? How would the metabolic or endocrine state of one body influence the two brains through shared or separated circulation? Or, if there are two bodies how would those bodies be controlled from a partially fused brain? These ideas are drifting away from our original discussions; nevertheless, it is likely remarkable how communication occurs within and between the brains and the periphery in these circumstances, and it may be similar in the case of the deaf and blind individuals.

Discussion 2—27th November, 2020

1 Many points of view: the "elephant" and what is a brain?

Zoltán: I was thinking how we could maybe represent how we each think about the brain in a graphical format. Joy likes this black box diagram. This is our starting point now. We have a brain, and we don't care how the brain actually looks or what kind of cellular elements it contains, how it is constructed, whether it's a pigeon brain, a monkey brain or a human brain. We put in some stimulus, it has an effect and produces some kind of a behavior, it predicts what will happen, and then it will adjust the behavior.

Tamas is thinking about these questions in a different way. He is extending it to metabolism and gender, and looking at how that's changing the behavior of the brain. It's probably the same brain, but it's in a different status. For Tamas the autonomous nervous system is part of the brain and he would even argue the liver is part of the brain. I would go even further, because I'm looking at the brain through the eyes of a developmental evolutionary biologist. I'm looking at how the genome (and then the ultimate readout of the genome) and development are changing the structure, and how those are affecting the behavior of that brain. I'm more interested in how that "black box" was put together and what factors had a detrimental impact on the development of this black box that will result in altered behavior.

Joy is looking at the brain not in isolation, but with other brains, and the way interactions between brains and individuals affects how the black box reacts to a stimulus or input and changes or reinforces its predictive abilities. For Joy, brain is another side of a dyad. All of us are right, but we are all wrong at the same time, as well.

This is why John Godfrey Saxe's (1872) "The Blind Men and the Elephant" poem that I put into my chapter is so relevant. We each study different aspects of the functions of the nervous system. We can't even agree on what constitutes the brain or the nervous system. Tamas likes to include the entire body in this definition, while Joy advocates the study of more than one brain at a time. I consider the brain as the product of a developmental program that evolved during millennia. The life experience of an individual and through the genome the entire evolutionary history are embedded into the brain during development when the structure of the brain is taking shape as the result of the interactions between an unfolding genetic program and the environment. It does not mean that we are wrong, but we have different opinions based on the specialised area we each study. The three of us are probably also right in our field, but we look at the brain in very, very different ways. We are "The Blind Men and the Brain."

2 The whole body is part of the brain

Tamas: Okay, I agree. Can I make a point? Can you go back to your schematic that you are proposing for the cover illustration of our book? The way I look at myself is that if you simplified that schematic, I'm actually within this box, within the central box, within the brain box. My main point is that there is no brain on its own; that actually, the whole body is part of the brain. The truth is that if you simplify the diagram, then I am within the brain.

Zoltán: So this is just a very primitive representation.

Joy: Exactly. That's what I like about it.

Zoltán: Even better: we could expand these diagrams, and instead of just looking at the blue, you would put something around it and say "body" or "not body."

Tamas: Yes. From this primitive perspective, as you're depicting Joy's take on it, whatever I think about is in that blue box. It's within the green and blue central boxes, and I hate to say that you too. The stimulus is, in my case, whatever comes from outside of the body that eventually gets converted. It can be visual or auditory, but in my case, it's mainly fuel and some other things.

Joy: That's kind of a point where you and I really intersect, because you add something to the conventional brain that lives in the cranium.

Tamas: Yes, yeah.

Joy: I think that the brain is not a single unit until it is connected to other brains. Yes. So you and I add something to that box, and Zoltan is still trying to figure out how the thing works, or how it comes about.

Zoltán: But if you look at the development, it is not like this.

Joy: Okay, okay, but I wanted to go back and say we need graphics, we need pictures just like what you have done here. We need to make our points as graphically as possible.

Zoltán: We evolved those figures a lot, and now we can think about how to extend them. You asked me to talk about development a bit more today. The single point I want to make is that there is a very big difference between how this computer was put together and how our brains were put together.

Joy: 100% agree.

3 How was the brain put together?

Zoltán: The computer was put together with a particular pattern of fixed wiring, and it's rigid. One switches it on and runs the programs on it. Our brain is put together by itself, and it's dynamic. It's switched on while it's booting *and* while you are putting it together. That's a very, very big difference. What programs you run on that self-assembling circuit will actually determine the connections it will have, it will influence the assembly of its own circuits. That's why metabolism, nutrition, drugs, alcohol, genetic influences, and influences of society will all affect how your black box is put together. In a way, while I'm inside the black box, I'm trying to figure out how that black box is wired up, how the cells interact and interconnect, and what can go wrong, but you cannot understand that without actually looking at the bigger picture.

4 The brain needs a body

Tamas: So, as I said before, if you look at Nenad Sestan's dead pig's brain being brought back to life, now the entire thing is different. It's gone. That's what he patented, it's how he perfuses the brain. As soon as it hinges on that, by definition, he's basically talking about my take on the whole brain: that without the rest of the body, you have no brain. Every aspect, everything you think about, and even what you know, what Joy's thinks about, what I think about, it's definitely gone. What's coming into the brain from the rest of the body predicts everything that the brain eventually will do with any other stimuli that hits the brain. I mean, it's fantasy, but if you chop the head off, you die, and the brain is dysfunctional.

Zoltán: At least for a short period, that brain will preserve the life experience, the developmental program, or the things you went through your life, what kind of sport you did, or what musical instruments you played on or what languages you learned. All these are preserved in those circuits that are still there in that isolated brain, all that is preserved from you and your life.

5 One brain—two bodies versus two brains—one body

Tamas: I understand, and I agree with that. So one interesting thing. I talked to Nenad last week, and I told him about our conversations. I told him specifically about Zoltán's ideas about what to examine on the conjoined twins. He's interested in all of this, and he would go even further. Rather than examining conjoined twins with one body and two heads, he suggested to examine conjoined twins with one head and two bodies who have one brain, but four limbs. There is one such pair of twins that he has been following. I don't know how many single-brain, two-body twins exist in the world. Do you know about that, Zoltán?

Zoltán: The prevalence of conjoined twins is fortunately very low, around 1 in 200,000, and the separation procedures are also much more advanced nowadays. Examining such twins would be incredible.

Tamas: Exactly, because it's the opposite of yours. It's one brain and two completely separate bodies.

Zoltán: There are probably two body representations in the same brain, perhaps in similar areas. It would be incredible to compare the sensory periphery representations in such brains.

Tamás: I was thinking about it.

Zoltán: Abby and Brittany Hensel are perhaps the most famous conjointed twins and I can send you the links to their videos (https://en.wikipedia.org/wiki/Abby_%26_Brittany). They do not want any additional publicity, but to study them would provide an incredible opportunity to learn more about how brains interact. To what extent can you interact through the joined nervous system, blood, and skin? Imagine that you interview them separately, so the other one can't hear what you are talking about, but the other can feel stress, anxiety, or happiness signaling through the body. Some of the endocrine signaling generated in one of them might influence the other.

Joy: Oh, that would be a remarkable thing. Well, we are now at a level where we've developed methods to do mapping simultaneously on two brains using functional near-infrared spectroscopy. That would be an amazing study to do. For the documenting, though, what we already know is that the brain can reorganize itself. I like what you said—"our brain was put together by itself." The brain can do this. There are many, many examples out there. For example, I had a patient when I was doing functional imaging at Columbia University, and we had an incidental finding where somebody came in, and his head was largely water, yet he appeared perfectly normal. What he had, of course, was untreated hydrocephalus. His brain was this thin little band around the circumference of his inner cranium on one hemisphere, essentially, and yet, he appeared perfectly normal. He had a college degree, a job, a wife and family.

Zoltán: And this was an "incidental finding," that person or the family did not realize this huge amount of damage that already happened to the cortex from this hydrocephalus. I completely agree that some of the cortical functions are overrated. Basically, we think that the cortex is doing a lot of things, but perhaps many of these functions are actually done by lower centers, perhaps the striatum, thalamus, or the brainstem.

Joy: I think that this was all cortical. The conformation of the cortical substrate could work perfectly well in different configurations other than the wired up system that we see in a brain that looks normal.

6 Lot of ways to make a brain

Zoltán: Probably this person had the condition of normal pressure hydrocephalus. In these patients the cerebrospinal fluid's pressure is mostly normal, but during the night when they sleep there are some temporary elevations in their cerebrospinal fluid pressures, so called plateau waves. As the fluid builds up temporarily, it causes the ventricles to enlarge and the pressure inside the head to increase temporarily, but during these times it is still compressing brain tissue, especially the cerebral cortex which eventually will lead to severe symptoms, such as deviation during walk, dementia, and urinary incontinence. There is a chronic process of cortical loss in this condition, initially nothing is noticed, it is all compensated by other brain areas, but eventually, the symptoms appear and from one day to another, the person decompensates. In some cases of normal pressure hydrocephalus, the cortex becomes like a paper-thin sheet, but initially, you don't even notice that there is a problem, then later there are dramatic consequences. So some of these patients can go to university, they even have excellent scores on their exams, but then they suddenly lose higher functional capabilities and drastically deteriorate. These cases question the dominant role of the cerebral cortex in various functions.

7 Cortex may be overrated

Zoltán: If you look at some movies on YouTube, you can find some recordings from a cat that has no cortical connections with the brainstem and spinal cord (https://www.youtube.com/watch?v=wPiLLplofYw). This cat's brain is not connected to the rest of the brainstem and the spinal cord, but it can adjust its gate pattern according to the speed of the conveyor belt. Yes, the gate is absolutely fine. So, I usually show this video to medical students to illustrate that for these functions, you do not need cerebral cortical function. Also, there is a mutant mouse that is extremely interesting, showing a similar point. This transgenic mouse with 'cerveau isole was developed by André Goffinet in Brussels. Without going into the details, in this conditional KO mouse, there are no projections from the cortex to the thalamus and nothing entering the cortex from the thalamus because of genetic manipulations. So, André knocked out the expression of a cell polarity gene in the region of the developing internal capsule. The thalamocortical projections don't make it to the cortex, and the cortical fibers don't reach the thalamus. They are completely interrupted; this is why it is also called the "cerveau isole" mouse. If you look at this mouse compared to a typical mouse, you will see that it is difficult to distinguish which has no connection with the cerebral cortex. They are walking and swimming fine, and they can even walk on the grid at the top of the cage, just like the normal control, although perhaps there is a little bit of ataxia. They do all this without the connections with the cerebral cortex. I use these cases to show that we overrate the functions of the cerebral cortex. I always say that when we talk about higher cognitive functions we should really mean higher functions, such as planning, emotion, judgment, production and understanding of language, reading, and mathematics. Walking around, swimming, and basic perception and reflexes are not higher cognitive functions.

Joy: Built on this animal model, the mouse can swim, but would this mouse be able to find and learn the location of a platform if you were to place it into a water maze?

Zoltán: These experiments were not done, because the mice die at the age of three weeks, but my prediction would be that they would have difficulties in learning these tasks.

Tamas: Do they die because they can't continue after suckling?

Zoltán: The suckling is finished around postnatal day 21, so they are three weeks of age.

Tamas: So can they find their own way to survive? What is the cause of the death at that age when they die?

Zoltán: We don't know whether their death is connected to the brain function or some extraneuronal abnormality. It sounds like if it's connected to three weeks, that's exactly the end of the nursing by the mother.

Tamas: Yeah, that's the weaning. That's it.

Joy: Yeah. So can they forage for food, for example?

Zoltan: I would be interested in exploring how independent they are at that stage. I think we still have the colonies at Oxford with Simon Butt, so we could explore these questions.

Tamas: Are the animals smaller than the controls?

Zoltán: Yes, they are a little bit smaller. I think the body weight was probably published by Andre, so we can have a look. That's a very good example where Joy talked about the paper-thin cerebral cortex in hydrocephalus, and even with that damaged cortex, or rather, without the cerebral cortex, you can function for a while.

Joy: But you imply that you are not sure whether the cortex has delegated all these jobs. Can the cortex delegate its function to other parts?

8 Cortex gathers information from outside

Zoltán: There are two issues here. Can the cortex delegate function to subcortical structures, such as striatum, cerebellum, or thalamus? And the other question is can we transfer one cortical function into another? There are some functions that the cortex can delegate to other cortical areas, but there are some functions that can't be delegated to other parts. As far as we understand, it has to do with the primary entry systems—the optic nerve transmits photo responses from the retina to a particular part of the thalamus, for example, and it's pretty hard to get that part of the thalamus to respond to sound, instead. So there are some parts in which the input is defined from early stages, and this input is relayed to the cortex. Some of the plasticity after sensory areas are lesioned occurs on subcortical levels. There is, of course, a great deal of cortical plasticity, but we should not forget all of the changes before and within the thalamus. A lot of cortical functional plasticity is apparent after a brain lesion due to tumor or stroke, and some rearrangement occurs that can lead to functional recovery or rehabilitation. Another part of the cortex will take over. Some functions can be quite flexible.

9 Cortex organization remains flexible across the life span

Joy: And you could take somebody in their eighth decade of life, teach them how to use the piano, and watch the cortical regions associated with their hands expand with fMRI, for example. That's evidence that the brain remains flexible and adaptable throughout one's lifespan.

Zoltán: There's also a very big difference whether you have a chronic, protracted ablation of that region, or whether it happens suddenly.

Joy: I'm sure that's true.

Zoltán: Andre Goffinet's mouse never really developed the cortical connections, so it had to deal with all this input at the level of the spinal cord, brainstem, thalamus and striatum—whatever was available for them on the subcortical level. That is a very big difference from an acute lesion. But just imagine that you are missing just a tiny fraction of a particular subtype of your GABAergic interneurons. Most likely, you would have a much bigger phenotype than this "cerveau isole" mouse has, so it's better not to have a cerebral cortex than to have a scrambled up, developmentally abnormal cerebral cortex.

Joy: So even subtle alterations of cell composition in the cortex can have dramatic effects.

10 Cortical cells are relatively slow to turn over

Zoltán: I would like to justify why I look at this "elephant," the brain, the "black box" through development and evolution, and I shall urge you both to also consider development. If you look at your hardware, the brain you have in your skull is very different from any other body parts. If you look at your various body parts, we really change these tissues quickly. The tissue turnover is staggering in some parts of your body! You replace some of your intestinal lining in three days, taste buds on your tongue in ten days, your lungs in a couple of weeks, red blood cells in four months, liver in six months. That's why we have "dry January" when we stop drinking in January so the liver can recover a bit after Christmas and New Year. You also replace your nails, hair, and bones. Even your heart is replaced a bit, but if you look at the neural retina and the neurons of your brain, there is hardly any replacement.

Tamas: Are you sure about the heart?

11 Cell replacement in other body parts

Zoltán: Yes, there is some replacement even in the adult. You replace 10% in about 10 years, but I am not an expert in the heart. There are some stem cells that you can induce to replace some heart tissue in the developing mouse. This is much more prevalent than in the adult, but it is still present.

Tamas: I was going to bet on no replacement in the heart. Actually, my understanding of the heart was that its cells are quite stable, but maybe not.

Zoltán: What I know is from the seminars by my colleagues in my department from Paul Riley's group. He described that the pericardium has some dormant stem cells, and you can reactivate them in the adult. They have neural crest origins, and you can produce cardiac muscle from them. During development they have this ability very broadly, but in the adult much less. Nevertheless, there is still a bit of replacement. In the blind cavefish, this regenerative capacity and mechanisms for replacing heart tissue is present throughout life, so the cavefish is used to understand why it keeps the ability to regenerate in the adult.

Tamas: Can you regenerate cardiomyocytes?

Zoltán: Paul Riley had some papers on this 10—20 years ago, but I do not follow this literature closely, although I am interested in the possibility that the meninges of the brain, which also have crest origin, could have such function. We have a study with my colleague and friend Tomohiro Matsuyama where we showed that after perinatal stroke some meningeal progenitors can produce some neurons.

Tamas: OK. So what is happening in the brain? In the cerebral cortex?

12 Two systems of replacement and regeneration

Zoltán: There are two issues. One is normal replacement, and the second is replacement after injury. Most of the generation of the neurons is done in utero, and there are very few regions where there is still some neurogenesis. There is no mitosis of mature neurons. In the second case, you can activate these dormant progenitors, and they do the job; however, this second case is still very much debated.

Tamas: I thought the first part was also still debated.

Zoltán: Yes, this goes through phases. Initially, the consensus was that there is no neurogenesis in the adult. Later, it was discovered by Bayer and Altman that there are limited sites of adult neurogenesis in the mouse and rat, and these sites kept expanding. There was even some suggestion that there was adult neurogenesis in the cerebral cortex; however, the most recent results in humans question whether even the dentate gyrus or the subependymal zone can have continued neurogenesis beyond childhood.

Tamas: When it's not maturing anymore? Can you update us on these recent developments?

Zoltán: Arturo Alvarez Buyalla from UCSF, San Francisco, had a string of papers followed by others that suggested that there is no continuous neurogenesis in the human dentate gyrus of the hippocampus or in the subependymal or subventricular zone beyond the first few years of life. In mice, these regions keep going, but not in humans.

Tamas: In humans—

Zoltán: No, not in humans. That's a very good point, because I was showing you mouse brains here, and remember this big debate between Pasko Rakic and Elizabeth Grove around twenty years ago? Elizabeth suggested that there was BrdU incorporation through progenitors into newly produced cortical neurons when you inject this S-phase marker into an adult monkey. It turned out that they were mostly looking at satellite neuroglia and satellite glia rather than neurons. Basically, there is no adult neurogenesis in the monkey or human cortex, and the definitive paper came from Scandinavia, where they looked at the C14 levels.

13 C14 and evolution

As you know, a human is just a "crazy ape," as Albert Szent-Györgyi wrote. The United States and USSR were doing surface nuclear testing during the Cold War, and this caused a huge elevation of C14 in the biotope. If you look at the bark of a tree, you can see the huge elevation of the C14 in the rings associated with 1963. After '63, it went down because they signed the limited Nuclear Test Ban Treaty. I was born in '64, one year before Balázs, Tamas' brother.

Tamas: Yes, I know.

Zoltán: My Mum was pregnant with me at the peak of that Carbon 14 elevation. To see whether my neurons were born in '64 or later, we could test the C14 levels of my neurons. My brain would be the best tissue to use to test that. What is the level of carbon 14 in the cerebral cortical neurons in my brain? If the level is the same as in '63–'64, then it means that there was very little replacement of my neurons, right? So in my black box, there's a definitive set of neuron I'm still operating with that were produced in the city of Nagykörös in Hungary. Basically, these experiments were done on donated brains by a Scandinavian group, and this is exactly the case. If you look at human C14 levels in these neurons, it's exactly the same as when you were born, so they suggested that there is a very little replacement of your cortical neurons. However, the glial cells were 10 years younger, on average.

Joy: Now, I'm really confused by something. So where is this C14?

Zoltán: After the surface testing, the C14 was disseminated around the globe by dust, clouds, and air. It reached its peak around 1963–64. C14 ended up in the food that my mother who was pregnant with me consumed (in fact, in every pregnant mother in those years). C14 is also in the DNA of the neurons. If neurons do not divide, then the C14 levels remain high. If they divide, they change nucleotides, and these new nucleotides will have a different level of C14 dependent on the levels in the environment.

Joy: So C14 is in the DNA in the nucleus of the cell?

Zoltán: Yes.

Joy: So they had to image the carbon 14 levels on a single cell level?

Zoltán: Exactly. This was the technical breakthrough to monitor this "experiment in nature."

Tamas: So you're talking about the DNA?

Zoltán: Yeah, the DNA in the nucleus of a neuron.

Tamas: So we're not talking about the rest of the cell? Because those parts are probably replaced, but the DNA is probably not. You have DNA repair.

Zoltán: You are right, that could also change the C14 levels, but I doubt that there is too much of this in nondividing, postmitotic neurons, but I am not an expert in this area.

Tamas: So, what I'm talking about is that the membrane composition is likely continuously changing for many reasons, and that has multiple implications. One would assume, wrongfully or rightfully, that the changing membrane—just simply the membrane constitution—will have an impact on signaling event, which signals come through and which not. So, even without changing this actual cell, they are continuously changing.

Zoltán: Very good point. Even if the nucleus is untouched, you still have all this replacement. Maybe you replaced the rest of the cell over the years.

Tamas: Then the DNA has repair mechanisms and all those kind of things, so there could be a dynamic turnover of the cell with unpredictable status at any time, changing membranes and other cell constituents.

Zoltán: Yes, but on neurobiological terms, it's still the same cell.

Tamas: Okay. Including your liver, which is the same despite the fact that it's being replaced. So what I'm trying to say (arrogantly) is that this is semantics, basically, where you draw the line of what is special, what is not special? So what do we know about function? Is the cell any different if it harbors carbon 14 or not?

Zoltán: We tend to think that it doesn't matter for the life of the individual whether your nucleotides in your DNA, which was produced 56 years ago, in my case, have higher or lower C14 levels. We are now using this peak of the C14 levels to calculate the timing of the generation of the DNA in my neurons. The other components of the cell, such as water and proteins can be replaced, but since there is no division of neurons with new synthesis of DNA the C14 levels will continue to reflect the levels that were around when that DNA was synthesized at the time of the S-phase of that division that produced that neuron. Since there are no additional neurons with lower C14 levels that were incorporated to my brain after the C14 levels fell, we can say with certainty that my neurons were born in 1964 with me. Obviously, C14 can have damaging effects on the DNA. I don't think that these levels cause that much damage on the life of a single individual, but I just don't know.

Tamas: I agree with you. As just one example, my grandfather was a pulmonologist in the '40s, '50s, and '60s. He performed an enormous amount of X-ray imaging, for example, without really protecting himself.

Zoltán: Mm hmm.

Tamas: And he lived to 99 years of age.

Zoltán: Probably, you had higher exposure to radiation with your constant intercontinental flights over the years than your grandfather.

14 Neurogenesis and injury

Joy: It's difficult to predict the effect of any of these things.

Zoltán: I was just using the C14 to birthdate the cells in the human brain of someone like me, born in 1964, but what is clear from all these studies is that the brain has very little turnover of the neurons. However, the brain is comprised of many other cellular elements. In fact neurons are the minority compared to glia. Well, you might change the microglia, and probably there's a huge turnover of your astrocytes, but at least the neurons do not change, they are not replaced, they do not just dissolve. You could use that argument to emphasize the importance of development. Once you have a set of neurons in your brain, you will have to live with them for the rest of your life.

Tamas: OK.

Zoltán: There are recent papers that Pasko likes very much showing that even in the newborn brain, you have a little continued neurogenesis in the dentate gyrus of the hippocampus and in the subependymal zone, but even in these selected regions, the neurogenesis fades out very fast. Basically, these are the principles we now have. There is no mitosis of mature neurons, no or very limited neurogenesis in the cerebral cortex, and you have some limited amount of so-called postnatal (I wouldn't really call it adult) neurogenesis.

Joy: How about the injured brain?

Zoltán: Great question. We don't know yet whether the injured brain has a higher capacity for neurogenesis. Perhaps you can activate some kind of mechanism with special conditions, just like you can do in neonatal mouse stroke models. My point was that your in utero life is the most important part for your nervous system because this is when you get this awesome machinery, and you have to generate these cells in a particular sequence. These cells have strict fate determination, perhaps as early as the last S-phase of their final division, and then they have to migrate and stop in the right place. When these cells leave the germinal zone, they have an address, they have a destination, but if the migration machinery is not there, or there are issues with the molecular or environmental factors, then they scramble up. They produce a faulty circuit, which is not as good at performing the intended functions. That's why development is very important. This is my point for the box, that we cannot ignore development.

Joy: But does it matter how you get those neurons into the black box?

Zoltán: As a developmental biologist, from the genome and the upbringing, I can tell you what kind of black box you will have. You will have to cope with the tools you have in your black box during your entire life.

15 Fine-tuning the "black-box"

Zoltán: Let me show you a few examples. All this brain development is starting as an unfolding genetic program that interacts with the environment. I don't want to bore you with all these details; I spent my scientific life studying this. Your neurons are generated, and they migrate closer to their destination when you are born. You start your life, and start your extrauterine interactions with your environment immediately using that awesome, but not yet finished, "computer" that is still assembling, but you already have to respond to the environmental cues. In fact, the environmental cues will optimize the wiring so that you can cope with life much better as an individual. You fine-tune your black box according to your environment! That's why these early years are so important.

Joy: How early do you assemble these circuits? In utero or postnatal life? Or both?

Zoltán: The maturity of the circuits at birth depends on the species. In some species, this is pretty much done when they are born or hatch. For instance, when a turtle is hatching, it has a very mature visual system, and when they are leaving the egg, this visual system has already been assembled. The retinal projections in the brain are wired up using spontaneous activity patterns, so when they hatch, and they see the rising sun in the morning, they can actually go straight to the ocean. That's why in certain cities, people switch off the lights when the turtles are hatching so that they do not get confused, and they know where to go. Other mammals are also more mature than humans. Imagine horses. When the little horses are born, they can already run and follow their mothers within a few hours. Even newborn whales can follow their mothers immediately after they are born. For us, it takes much longer.

Tamas: We are hopeless when we are born.

16 Humans have a very long period of postnatal development

Zoltán: Yes, in utero MRI movies showed that babies are very active. There are all sorts of very interesting interactions between the baby and the mother in utero, but when we are born, we are completely hopeless. We have to spend a lot of time to learn posture, taste, motor programs, thermoregulation, somatosensory awareness. That is the biggest difference between us and all the other species. Our development occurs over a very, very protracted length of time. If we compare our brain development with the chimpanzee, the chimpanzee brains grow rapidly before birth, and then the growth of the brain will level off very soon after birth.

Joy: Yes, there is a big difference.

Zoltán: Whereas in humans, the brain grows rapidly before birth and in the first couple of years into childhood. We peak around postnatal year 11. This is when your cerebral cortex is the thickest. This has to be considered when you learn languages, piano, tennis, skiing, etc. After that, there is a steady decline of the thickness of the cerebral cortex for many reasons.

Tamas: Now let me ask you this. You may know the answer about the elephant and another person. Another animal I'm interested is the crow. What about those? Do they also show similar developmental kinetics to the monkey or to us?

17 Humans versus crows

Zoltán: It is very difficult to compare developmental kinetics between birds and mammals. I was just talking about similarities and differences between primates, chimpanzees, and humans.

Tamas: I am interested in crows because they have the largest brains relative to their body size. My understanding is they have one of the most sophisticated brains in birds.

Zoltán: If we consider the brain as a black box, then the crows could outperform some mammals in their performance in spite of having very different arrangements in their brain, and they follow a very different developmental program.

Zoltán: So let me show you this video from BBC (https://www.youtube.com/watch?v=cbSu2PXOTOc). It was filmed in Oxford.

Sound from the film: "Alex studies wild birds, which he releases after three months of research. This one is nicknamed 007, and it's about to attempt one of the most complex tests of the animal mind ever constructed (at least, Alex believes it is). The bird is familiar with the individual objects, but this is the first time he's seen them arranged like this. Eight separate stages must be completed in a specific order if the puzzle is to be solved. This video shows that the bird succeeds. The bird has to get the short stick first, then he is using the short stick to collect the three stones to get the longer stick from the box. Then the eighth and final stage is to use the long stick to get the reward. This video is remarkable."

Joy: So, the bird was shown this whole thing before, and then it was mixed up?

Zoltán: The bird had seen the individual components, but not in this particular long sequence of tasks.

Tamas: It was not a first try to use, though, I think.

Zoltán: My understanding is that it was in a sequence that the bird has never seen before, and this whole thing comes back to it. So, you asked me about the timing of crows' brain development. I don't know about that. I do not know how long it is, but what is clear is that they have a completely different brain organization. We are used to the mammalian type of brain organization with a six-layered huge isocortex. Birds and reptiles have a very different organization.

Joy: How can we compare them?

18 Alternative brains

Zoltán: We chop up the "black box," and we look at how neurons are organized and how they are interconnected. We can identify the cells that the afferents contact from the sensory organs, and we can also identify what are the output cells from these different brains. If you section a "black box" of a mammal, then it looks like this, but if you start chopping up other brains (which I also did over the last thirty years), such as brains of kangaroos, turtles, iguanas, or a crocodile, then things are very different. They have a very different brain organization. They have this ball of cells protruding into the lateral ventricle, and they have a pathetic looking very small dorsal cortex. The same is true for the iguana and the crocodile. This ball-like structure is called the dorsal ventricular ridge or wulst, depending on the species, and it's not at all like the cortex.

Joy: So, where would be the equivalent part of that "ball" in the mammalian brain, and where is the cortex in the reptiles or in the birds?

Zoltán: These are the fundamental questions of comparative neuroanatomy. Some of these questions are still not resolved. Everybody agrees which part of the bird brain is homologous to the mammalian striatum and to the hippocampus, but there's still a huge debate about what is homologous to what when comparing this ball of cells in birds and reptiles to brain areas in mammals.

Joy: Okay, but for us, it doesn't matter where the cells were coming from; what is important is the function of those cells.

19 Function versus structure and form

Zoltán: Yes, but just having similar functions does not satisfy developmental and evolutionary biologists. We want the lineage, clonal relations, and indeed, if you record from this ball of cells of the dorsal ventricular ridge (DVR), then there is a very sophisticated representation of the visual world. For instance, does the ball of cells of an iguana have very similar receptive fields to a ferret visual cortex? Moreover, they have multiple representation, suggesting that they have at least three different visual representations, so they have multiple areas to represent vision, just like in the mammalian visual cortical areas.

Joy: Interesting.

Zoltán: It doesn't matter whether it's a ball or a layered structure; you can have very sophisticated functions. That's why I usually show that movie of the crow to students in my lecture, so they appreciate that crows are smart even without the cerebral cortex, and they can do certain things extremely well.

20 The big question for evolutionary neurobiology

Joy: What do you, as a developmental and evolutionary biologist, want to understand about this ball?

Zoltán: The biggest question for us developmental and evolutionary biologists is to understand where the components of the black box arrived from. Do they have a common origin or not? We are just looking at where these individual elements of the box were coming from. For us, the big question is whether we have similar bricks or building blocks in the avian brain and the mammalian brain. You probably know very well how Harvey Karten, the pioneer of this field, spent so much time exploring these questions. He suggested that the two brains use very similar principles and very similar elements and have similar connectivity between these similar elements, and indeed, we completely agree with this. The connectivity of the avian and mammalian brains between these elements is very similar, but where I disagree with Harvey is about where similar groups of neurons are coming from. In a bird brain, these circuits are generated from a completely different sector of the neuroepithelium than in mammals. They are derived from different sectors of the neuroepithelium; therefore, they are not homologous. It is not just having similar bricks that are arranged in different manner. The mammalian brain developed new bricks that did not evolve in reptiles or birds.

Joy: Analogous?

Zoltán: Yes. So, in my opinion, based on our own lineage studies, they are derived from very different sectors of neuroepithelium. For instance, the thalamic recipient neurons in birds and mammals might have similar gene expression networks that contain RORbeta and other genes, but they still come from different sectors of neuroepithelium. Nevertheless, it's surprising how similar the gene expression and connectivity are. There is a homoplasty between these structures. There is convergent evolution in which different neuronal groups evolved analogous traits.

21 Principles of connectivity

Joy: But if you are just looking at the black box, you don't care about this, correct? I think one of the things that we do care about universally, and it's sort of clear for all of us, is connectivity. Tamas cares about connectivity of this neurocircuitry to the other parts of the body. Zoltán cares about the connectivity within the system that makes it work. I care a lot about the connectivity between two brains, and I'm asking the question do the principles of that connectivity across two brains recapitulate the principles of connectivity within brains and occur across brains and other body parts? Are there principles of connectivity that bind us together in some way?

Zoltán: I completely agree that this is a question for developmental neurobiologists and evolutionary biologists, because basically, you have different constraints on the system that—I almost said that we decided to evolve this way, but that is wrong. We did not decide anything. We, I mean ancestors of our species as a population, were tinkering with our development, and that produced different outcomes in different individuals, and the best of these for that particular environment at the time was selected out. The developmental program produced an adult structure with a possible differential behavior, some more advantageous, which was selected out or not. So, we ended up with these different developmental programs to produce a mammalian brain.

Joy: What was the biggest change that produced the mammalian cerebral cortex?

Zoltán: The biggest difference between the reptilian, avian, and the mammalian brain is how the different sectors changed their neurogenesis by changing their progenitors and the way the newly born neurons migrate. By changing the type of cells you produce and the way you position them, you can generate brains with very different organizations. Basically, in the ball of cells in reptiles and birds, in the dorsal ventricular ridge, you have a lot of neurons generated, but they pile up, form the dorsal ventricular ridge, and they can form circuits that are very sophisticated. In mammals, these cells do not protrude into the lateral ventricle; they migrate out, form the lateral amygdala, endopyriform nucleus, and the claustrum, and they perhaps also contribute some cells to the lateral cortex. It's still debated. In mammals, all these cells moved out ventrolaterally. The interneurons, the GABAergic interneurons, are generated both in mammals and sauropsids (birds and reptiles) at the same site. The biggest change was in the neuroepithelium in the dorsal cortex in mammals, which contains various progenitors that produce a huge variety of cortical neurons that are arranged in a laminar fashion. Our brain exploded in the dorsal aspects in mammals, and in particular, in primates. So, this germinal zone is producing more and more and more cells, and that's why we have a different sector of the neuroepithelium, which is generating all of these cells.

Joy: That is fine, but why is it important?

Zoltán: For a neuron-at-work modeler, it might not be an important question, but for us evo-devo people, it is *very* important. It is a key question because the two systems evolved differentially, they rely on different developmental algorithms, and they have different developmental abnormalities. If you have altered development, then you will have a different structure in that black box, and your behavioral repertoire will be limited according to what you have in the black box and how they are assembled. So, I can show you examples where certain aspects of brain development are damaged, and then it has a huge effect on the outcome.

22 Abnormal brain development: consequences

Tamas: I can see that. Brain developmental abnormalities are quite common in the population, and there is quite a big impact on society.

Zoltán: I would like to go even further than that and become a bit more philosophical, and perhaps talk about the responsibility of society, how to deal with that situation. We have brain with different capacities. We have different ways of getting through life with what we have now. We are born with a set of neurons and we use those neurons for the rest of our lives.

Tamas: That's true. No, I agree with you on that concept completely.

Zoltán: But imagine when you start life with a major handicap, and I want to show you an example, which is shocking and really surprising. You know, that Zika virus was going through the whole world a couple of years ago, and you know that in Latin America, even if you knew you had Zika infection in the first trimester and your child is affected, abortion is out of the question. Basically, we have now thousands of kids with congenital Zika syndrome, with very small brains, and they have to get through life with that brain and have limitations on their entire life.

Tamas: Yeah. Very sad.

Zoltán: I would like to go further with this. If you have one single inflammatory response during pregnancy, this can influence neurogenesis of the fetus. Helen Stolp (who was in my laboratory many years ago) demonstrated this in mouse experiments with a single lipopolysaccharide injection to the pregnant mother to experimentally mimic the immune response to a maternal bacterial infection during pregnancy, then you can influence neurogenesis of fetus. Helen detected changes within eight hours, you have detectable reduced neurogenesis in those embryos, and there is a damage in the structure of the neuroepithelium. Then you have less cells generated that were supposed to be generated at that time. There is no turning back, you missed that opportunity to generate those cells, and the brain will be altered and the behavior can change.

Joy: And this could have an effect on the rest of the life of that individual.

23 Fetal alcohol syndrome

Zoltán: Exactly. I wanted to show that if you have a similar intervention, perhaps with maternal alcohol consumption, you would have a long lasting effect on the life of that individual!

Tamas: And this comes with effects on the whole family, and it has an impact on society.

Zoltán: I was shocked when I read the statistics on these issues. If you have some genetic problems with the migratory mechanism during brain development, there is not much one can currently do to influence this genetic susceptibility. But there is also self-harm and harm to the unborn child. Alcohol consumption during pregnancy is much more common than we think, and it can have a devastating effect for the rest of the life of that individual.

Tamas: Actually, alcohol probably has a very pronounced impact, but even if you were to have a very drastic diet, as a mother, with your child inside of you can have a tremendous impact on your brain development, really, which we published with Marcello a few years ago. Then overnutrition, undernutrition, or acting on this hypothalamic circuit can completely rewire the dopamine system, which then eventually will have a major impact on ambulatory locomotion in the offspring, how they respond to cocaine, and how they deal with that kind of stuff.

Zoltán: Of course, you can have many things that can impact brain development during pregnancy and they can all act in combination with diet, infection, drug abuse—cannabis or cocaine—but I looked into alcohol, and how it was discovered at University of Washington. Basically, people talk about fetal alcohol syndrome.

Joy: Can you recover from this?

Zoltán: Yes and no. It has a lifelong disability; it is not curable as such; the child is not just growing out of it. However, if you diagnose it early, and then you have intensive therapy, it can make an enormous difference in the prognosis of the child. There is a small window of opportunity, up to the age of about 10–11, when you have a better chance to recover functions. You remember, we just talked about this, this is the peak of your thickness of the cortex. You can achieve some recovery if there is early intervention, and the recovery depends on the severity of the fetal alcohol syndrome.

Joy: Basically, your life is severely affected and destroyed.

Zoltán: These children will have a different life and it is nothing to do with them why they will have that life. One in a hundred babies born in the United States and Europe have fetal alcohol syndrome (FAS). How mild or severe the effect is depend on other factors, such as diet, and other things. 30% of the pregnant women in the US and Europe, are drinking alcohol during pregnancy. 95% of FAS will have mental health problems. 60% will have disturbed school experience.

Tamas: And probably they will disturb all the others as well.

Zoltán: 60% will have trouble with the law, and 55% will be confined in prison, drug or alcohol treatment centers, or mental institutions. 52% will exhibit inappropriate sexual behavior, and more than 50% of the males and 70% of the females will have alcohol and drug problems themselves. 82% of them will not be able to live independently. It's a predictable outcome on the structure of the brain. I don't know whether there is a structure or correlation, or the severity of the fetal alcohol syndrome is actually measured by looking at these parameters in the face such as the low nasal bridge, small eye openings, short nose, but I don't know how well they correlate.

24 Induced changes to brain development during pregnancy

Tamas: I want to say something that is absolutely not inappropriate in my view but is politically not correct, which relates to your earlier statement that if you induce inflammation with the lipopolysaccharide injection or whatever you did to those mice, then you have a long-term impact on the offspring's life. So, vaccination does similar things. You know, this has been my view from discussing the book, but in my view, it has been pushed to the fringes, because if you argue that vaccination can cause altered brain development, they label you as some sort of a bigot. On the other hand, you do have the very fundamental biology behind where inflammation can lead in these mouse experiments.

Zoltán: I understand that. I completely agree it should be studied in much more detail. If the animal experiments of Helen apply to inflammatory responses induced by vaccination in human, then I would suggest that, if it is possible, do all the immunization way before pregnancy so that you do not have this inflammatory response during pregnancy especially during early stages of neurogenesis. However, we do not have any concrete evidence that vaccination has such effect and we do not have any evidence to suggest that it has an impact on brain development. We need to understand whether the cytokines after such immunization can act on progenitors and change the program, and whether this change of program, would result in some changes in the brain that was produced.

Tamas: I think this should be studied further and if it is not good, then immunization should be pushed to other stages. Maybe later stages when you will have less neurogenesis, maybe it's less of an issue.

Zoltan: Apparently, COVID-19 vaccination is safe at any stages of pregnancy, but I am not familiar with all the clinical study that stated this.

25 Potential effects of early interventions during periods of plasticity

Joy: You know, is there any thinking in your models about the impact of perhaps social intervention early on, or at a critical time? We all know that when a baby first arrives on Earth, they're so dependent upon social contact information between parents or whoever the caregiver is. Is it to the extent that this social interaction or those social influences could actually help mitigate some of these inborn neurological circumstances? It's a question. I mean, we're not going to answer that now, but this is a really important question, I think. And it sort of is a point where the kind of things that we think about the two-brain part of it could actually make a difference. I see opportunities to think about the role of social interaction in terms of a biological phenomenon: when I want understand how the brain does these fast "Send" and "Receive" operations, they're also influencing the whole brain. To what extent does that benefit us developmentally?

Zoltán: These are fantastic ideas! So, we have these elements of the nervous system produced very early on, and then we have these elements at our disposal to generate the circuits to cope with the environment and produce the behaviors so that we are okay, but these can be achieved in many different ways. Eve Marder is looking at the central pattern generating circuits in the crustacean stomatogastric nervous system in crabs and lobsters.

Tamas: Interesting.

26 Social interventions to achieve social adaptations

Zoltán: She's looking at how crabs and lobsters can generate the same behavior with many different circuits. The circuits adapt to produce that oscillation and behavior, and even by changing the connections, changing the channel expression. These creatures live happily, but they produce their behavior by using many different circuits.

Joy: That makes perfect sense. I like that.

Zoltán: We have very different brains, and we get through life with these different brains by adapting very differently, and if you are brought up in a caring family, you read, you have stability, even if you have some problems, like dyslexia, or some other issues, you will able to cope. Now, if you have this support, you can adapt, but that's why the social interactions are so important, especially in the child.

Joy: Yeah. That's great. I think I think it fits perfectly with everything. So, I see these three legs are coming together right now.

Tamas: I'm sure you know that's the clearest example of the diet inducing a massive downturn in your well-being. I think these issues are interplaying with each other to have an overall impact on how the brain functions and what the brain does.

Zoltán: I mentioned diet for different reasons, because if you have an abnormally put together circuit, schizophrenia or some other conditions, which anatomically can be very mild, can have a considerable impact on behavior. If you put that person on a ketogenic diet, you might mask some of these symptoms.

Tamas: You know, the medications that are effective, at least on the positive symptom side, are the psychotropic drugs that cause massive obesity and diabetes. We argue that those must be targeted in all those circuits to make them function in a different way. Definitely, all these things will have an impact on both development as well as on adult functions. One thing that we looked at years ago was the effect of C-section on brain development in mice, and how profound the impact has on its own versus natural birth. There are many, many different ways to be able to look at this.

Discussion 3—22nd January, 2021

1 A silver lining from hardship to wisdom

Joy: We're talking about how deprivation can induce positive things in society. Look at how the new government in the United States is becoming the most diverse ever. I don't think it would have happened if not for the past 4 years.

Tamas: No, I don't, either. I think that we're benefiting from the really horrific time that we've been through now, as we reconnect.

Zoltán: In Hungary, you remember when we had the first free elections in 1990, we were swinging from one extreme to the other. We Hungarians, when we jump at the top of a horse, we can never end up at the top, we either jump too little or too much. This is reflected in the fluctuations in the free elections.

Tamas: At the end of the Second World War, there was a profound impact that synchronized people. Negative attitudes diminished, and people were focusing on the positives. Good things happened out of those horrific times.

Zoltán: I wonder whether COVID will have such an impact.

Tamas: I am not so sure. Here is an example of my youngest son. We have been trying to keep him from getting COVID from the beginning. He is 19, and he is a first-year college student whose senior year in high school was basically ruined. He has been taking classes online from his dorm room. He became restless, and could not keep himself away from his friends at one point. No matter how much we told him that there is a great likelihood that he will get it and bring home the virus, my son was and is oblivious to these kinds of discussions. I think this is the case for many, many people. I mean, it's true for society at large, and not only in the United States I was saying to Joy that in certain people, there is a genuine, inert, genetic, understanding of these things, so they don't need to go through horrible things to understand the implications of behavior, versus the larger part of society does not get it up to the moment they, themselves, go through hardship. In a way, Zoltán and I are "lucky," because we were born into and experienced a society with intolerance and hardship.

Joy: Yeah, but maybe it's a frontal lobe function. What does it mean for the future? So if you look at teenagers, they have to develop the frontal lobe, you know, the gambling, the risk taking, everything is different for a teenager than for adults. I completely buy that for this particular age group, but it's also true for so many adults. Educated, intelligent people in their 30s, 40s, 50s, 60s, they just do not have the ability to comprehend the impact of what was going on.

Tamas: Basically, I don't want to advocate for war and stuff like what's going on right now, but it seems to me, and that's my personal view, is that hardship is absolutely necessary for the larger group to move forward in a way that is benefiting the whole.

Zoltán: Somebody in the US posted on Twitter, "Why should we sign up for the Paris agreement so that France is benefiting?" They don't understand that we are in it together. It's like if somebody were to say that the Geneva agreement is benefiting the Swiss.

Tamas: Exactly.

2 What is wrong with the typical human brain?

Joy: I see what you mean. Just because it's called the Paris Agreement or Geneva Convention, it's the whole world that is benefiting, not just Paris or Geneva. I think we're seeing examples of the limitations of the human brain. So many of us, as humans, have difficulty understanding probabilities, understanding our role in the bigger picture. Some people are very limited in their ability to see their own little space in relation to the big space. I think it has to do with weaknesses of our species with respect to our sociability.

Zoltán: Why do you think this is, for most people? Is it a developmental thing that you evolve this way versus that way, or that social interactions shape it?

Joy: I think that that's a very important question. Of course, I don't have an answer. I think that our brains are designed to learn based on coincidence and events. For example, if you do something and something else happens, then you learn to do or not do that thing. That is not probabilistic thinking. That is a biologically-driven coincidence gathering technique. Isn't it the same one that we use when we make inferences from a small sample to something much bigger? My point is that I don't think we're designed for big thinking by default.

Tamas: I see. I completely agree and buy into that. My interest in feeding is actually speaking to that, because if you focus on your daily survival, then big thinking is definitely not going to be part of it. At the same time, we've had these discussions about making decisions about eating. A person came to us few years ago talking about food and fMRI of the brain. He was looking at the brain's response to visual cues of chocolate cake versus broccoli. Then he made the argument that if you were intelligent, if you understand the health consequences, then you would choose broccoli. Actually, if you were intelligent with regard to evolution, you definitely would not take the broccoli. You would definitely take the chocolate cake. With the chocolate cake, you assure you will put on weigh in the form of fat, so that if tomorrow and the day after you have nothing to eat, you are fine. That is long term planning! The guy who showed the broccoli and the chocolate cake also showed a picture of six-pack abdominal muscles. If you think about a six-pack, it is definitely not a positive evolutionary thing to have, because again, you're not going to make it through winter with a six-pack and little food.

3 Coincidence detection and the developing brain

Zoltán: How do you explain what Joy was talking about, coincidence detection and learning? That's absolutely true for the developing brain. At that time, the kinetics of the developing brain is very different from the adult brain. The stimulus and activity patterns linger for much longer in the developing brain than in the adult. I think there is a reason for that: because then your coincidence detection will be better if things last a bit longer. For instance, in mouse development, in an embryonic mouse at days 16−17 (so just before birth), if you zap the thalamus with a stimulus delivered from an electrode, you will elicit a slow depolarization in the cortex for over 300 ms. If you do the same at the end of the first postnatal week, then you have about 150 ms, or even less, and then in the adult, it's much faster. There is a reason why it's very advantageous for that young animal to have this different timeframe to establish this coincidence. What I don't understand is that you can have a single event when you develop an aversion for a food or something. So how do you explain that?

Joy: You're specifically talking about food, or anything at all, like remembering details around you when 9/11 happened?

4 Single-event food aversions

Zoltán: I'm asking more about foods in particular that you have eaten only once, and you disliked. How is it that you have that memory?

Tamas: Well, we have this strong ascending pathway coming from the vagus nerve, through the pons and brainstem to the amygdala, central amygdala. So there, you've activated a large set of subcortical networks, which are going to feed up to the cortex. It's a very broad arousal of those areas in response to a threat, which I think makes sense, because if it's some sort of a poisonous thing, you definitely don't want to eat it again, it's reasonable that you would be so responsive to that. Although I must tell you, when I was about six years old, I loved tripe, which is stewed prestomach of the cow. I don't know, Joy, do you know tripe?

Zoltán: Have you ever had tripe? It is an acquired taste.

Joy: Probably not.

Tamas: I loved tripe back then as a kid, prepared as a Hungarian stew with paprika. In front of our apartment, on the other side of the street, there was this bistro. Every now and then on Thursday, and only Thursdays, they had tripe on the dinner menu. I loved it so much that once I ate a huge amount, but then I really got sick, and I couldn't eat it for years. Later, I recovered completely, so I have been loving it again for the last 40 years. My taste aversion went away. Some people have the same experience with tequila. In any event, aversion can also go through extinction.

Joy: Interesting. It's really interesting. I had nothing so severe, but I think the coincidence of the food and sickness is more salient than the fact that the food made you sick. Once, as a child, I think I was getting the flu. My mother served a normal dinner that night, and she served squash as a vegetable. And to this day, I really can't eat squash, but it wasn't that the squash made me sick; I was getting sick when that association was formed.

Tamas: Let me give you another example that speaks to this point of yours. We used to go to St. Kitts to work with nonhuman primates. In the early spring of 1999, my wife was a few weeks pregnant with our first child. She was in the stage of pregnancy when she would get nauseated easily. In the hotel where we were staying, they were grilling hamburgers. There was also a band playing Caribbean music dominated by steel drums. She had nausea attacks right after dinner two days in a row. For the next 10 years, whenever she heard Caribbean music with loud steel drums, she would get nauseated.

Joy: Very interesting. A single event associated with a negative outcome can stay with you for a very, very long time. It must somehow modify the brain.

5 Cerebellum and single-event learning

Zoltán: I usually think that a single event that stays with you for a long time, is having to do with the cerebellum and the motor system. For example, when you're little and you learn to ride a bicycle in a single event, you get it and it stays with you for the rest of your life. You may not pick up a bicycle for 50 years, but you can still ride it. And that's almost a single event.

Joy: I have these experiences over and over again in dancing. I mean, it's not the kind of thing that's a common experience, but when training for a new dance movement, once I get it, it is there, and it's there forever. It only takes one moment.

Tamas: Like swimming, swimming is the same, at least in principle. If they throw you in the water, you will swim. Although swimming is innate.

Joy: So these single trial events are very much a part of our life and our survival. Again, because I experienced this in the motor world, I think about it as the cerebellum converting it to almost a reflexive response. You know, when we talk about motor memory, the cerebellum can hold on to these patterns of activity.

6 The vestibular system and single-trial motor learning

Tamas: What is the role of the vestibular system in that?

Joy: Well, I'm sure that is known, but my knowledge is limited.

Tamas: The reason I am asking is that in the past 10 years, it became a "revolutionary" approach in behavioral neuroscience to head-fix an animal, place it on a rolling ball, and record from cells while the animal is in a virtual environment "freely" behaving. A guy came to Yale a few years ago for a job interview, and he was one of the pioneers of this technique. In his talk, he made the point that this revolutionary approach allowed him to prove that the original works of Hubel and Wiesel were inaccurate. They were inaccurate, he said, because in those studies, animals were anesthetized. In his set up, animals are awake and "freely" moving. I completely bought into the notion about anesthesia, but then I suggested to him that these studies he did in mice could also be done in humans. You can head-fix human subjects and put them on a treadmill or a ball, for that matter. He said that it would be stupid because humans would not tolerate such a setup, as it is really not comfortable. While with that answer he simply manifested his gross ignorance about animals and their "feelings," my point to him was that actually the biggest problem, in my view, of head-fixing is that you completely eliminate the vestibular system from behavioral control and perception. In sum, what I told him was that he replaced the pitfall of anesthesia with the pitfall of lack of vestibular control of behavior and neurons involved in behavior. He just would not get it. I tell you, this is my biggest issue with contemporary neuroscience: in the name of new technology, some people feel that they are now in the position to pontificate, not recognizing that there are always caveats and pitfalls.

Joy: Well, our work on the same idea is at the interpersonal interaction level. The discovery is that interaction engages a neural system that we haven't seen before. It's been overlooked because we haven't done experiments that allow it to be observed.

For example, when people talk, they also talk with their hands, they gesticulate. When you're thinking about how the brain is organized, the motor and sensory strip. . ..

7 Languages and the roles of supplementary hand gestures

Zoltán: I don't think it's that simple as that. My wife is fluent in three languages, and can switch from one to the other according to which family member she is addressing, and she gesticulates much more when she's speaking Italian, which is one of her mother tongues. She is completely bilingual with French, but she's not using her hands the same way when she's speaking French, though.

Tamas: Very interesting.

Zoltán: She's almost never using her hand when speaking English.

8 The brain and acquired versus native languages

Tamas: Basically, you're saying that the brain process is completely different for different languages. I knew an individual, my late ex-grandfather-in-law, back in Hungary in 1987. He had a stroke. He spent 10 years between 1945 and 1955 in Siberia as a prisoner of war. He learned Russian while there, but never spoke Russian after he came back to Hungary. But after he had the stroke (in 1987), he was lying in the bed in the hospital, and he started to speak Russian. He couldn't, he would not even speak Hungarian. From that point on, I was thinking that it might be possible, especially if you are multilingual from early on, that you might develop Alzheimer's disease using one language but not have it when you use another. The idea behind this being that different neuronal pathways might be used in significantly different languages.

Joy: I was the one who published that paper in Nature that originally showed that if you learn a second or third language late in life, the production of that language occupies separate spaces. Conceptually, it's all the same in the back of the brain, Wernicke's Area, but as you learn separate languages late in life, it takes up different spaces from your native language in the front of the brain, Broca's Area. When I first made this discovery, and before it was published, I gave grand rounds in Neurology. The Department Chair of Neurology came up to me afterwards and said, "Hirsch, I could have told you that. We see this in our patients all the time; people will have a stroke, and damage an area that affects one language more than the other. It's very common in practice." The difference between him and me is that I got it published in Nature, and I documented it with an understanding of the neural circuitry.

Zoltán: One could envision that under any neurological situation, there could be a differential impact on a person depending on the language they speak or how late they learned a language. I think my wife learned the motor automatisms of hand gestures with Italian; it is somehow bound together, whereas no similar gesticulation programs are associated to her English. Or, think about your colleagues who have Japanese origin, but lived in a different country for a long time. Their body language is changing when they switch from one language to another.

9 Different languages/different brains?

Tamas: I have a different question. Zoltan and I have a structurally very different language, Hungarian. It is fundamentally different from Indo-European languages, English, German, Italian, Russian, etc. My question is if you learned entirely different languages, even if you are raised with them from childhood, would you be using different parts of the brain to process those two languages?

Zoltán: Joy, in your paper, you compare different combinations.

Joy: No, not very well. I tried to, but I didn't have a large enough sample size. In truth, the question that you ask is one that has sort of baffled linguists for a long time because it's not really clear exactly how patterns of thinking differentiate between different linguistic systems. This has been studied a lot with Chinese and Asian languages relative to the Eastern European languages, and also Hungarian and Finnish.

Tamas: In fact, I tell you Finnish, which has very different words, very few words, maybe two or three, are similar to Hungarian and Estonian. But I remember once I was flying from Madrid to Helsinki with Finnair. I had a couple of drinks before I would take the plane and I was about to go to sleep, then when I was not really comprehending what's going on in the background, than in this semi-sleeping state, Finnish sounded like Hungarian.

Zoltán: In the sixth year of medical school, I did my pediatric anesthesiology practice in Meilahti Helsinki University Central Hospital in Finland with other students from Szeged. We could not understand Finnish. However, some words and the grammar seemed very similar. Let me give you a few examples comparing words in English, Finnish and Hungarian where the only difference is the ending wowel: blood-veri-vér, hand-käsi-kéz, water-vesi-víz, horn-sarvi-szarv. Finnish kept the word ending vowels, while Hungarians did not pronounce them and ultimately they disappeared. There are many other words that are less similar, but the roots must have been similar: we-me-mi, you-te-ti, what kind of-millainen-milyen, fish-kala-hal, head-pää-fő/fej.

Can it be generalized that if you learn a language late in life, then you have separate areas in Broca's Area. If you learn it early, it's almost like an amalgam right?

Joy: At least one doesn't separate it with the resolution of imaging tools that we have now.

Zoltán: These discoveries on language representations and ideas on functional recoveries in multilinguals have to be emphasized, for example, in language education. Actually, there is a good reason why you should be learning multiple languages. It is good for your brain, for your cognitive reserves later in life. Among other benefits, it's benefiting your health later. In protection against Alzheimer's disease, currently the strongest effect is to be a multilingual, but also, and Joy you might like this very much, dancing. Some studies in Germany showed that dancing had a very positive effect.

10 Does dancing protect against Alzheimer's disease?

Joy: It has been reported that dancers don't get Alzheimer's. Of course, one can make up all kinds of stories about why that's true, if it is. I think it might have to do with how much of the brain is used all the time. I mean, when you are dancing, you've got the music coming in, so you're synchronized with the music and another person. There is no part of your body that isn't working in relation to some other part of your body. It's like, you get this huge, big brain, and it becomes one thing, this whole brain is devoted to doing one thing.

Tamas: I would say that not only your brain, but in order for you to utilize properly your muscle, you obviously need to have the proper fuels for your muscles. So you need to have your liver, pancreas, and all of those tissues lined up to fine-tune the ability to move.

Joy: Interesting.

Tamas: I mean, this is one of those things I've always been thinking about when you talk about behavior. You cannot, in my view, from my very limited angle, you cannot dismiss the liver, the pancreas, the kidney, the heart, and so on, because they are the ones that actually make sure that if you intend to manifest the behavior, you can actually manifest it.

11 The food axis connects brain to body to mind

Zoltán: I was just talking to one of our common friends. We had a couple of grants rejected on both sides, and we are now thinking how to team up. Then we were talking about the mouse model she has, which has tagged leptin receptor-expression. I told her that it's not surprising that the hypothalamus is full of leptin receptors. What is very surprising is that the hippocampus is also full, as well as some layers in the cerebral cortex, like layer 6b, which is absolutely fascinating. What is surprising for me is that these receptors—leptin, orexin, and ghrelin—have representations straight in the cerebral cortical circuits in circuits that we consider to be the anatomical substrates of our highest cognitive functions.

Tamas: Those are, as much it's surprising to you, it's not surprising to me at all.

Zoltán: It's suggesting that you have a direct interaction between your body and your brain state; that something that is produced in your fat cells has a direct effect on the neuronal circuits that are responsible for you your high level cognitive functions on your thinking.

Tamas: Fifteen years ago, we reported that ghrelin, a hormone from the gut, goes directly to the hippocampus and controls synapse transmission and related behavior. And again, for me, it just makes sense. As we have been discussing, the brain is not compartmentalized, and it needs to be informed by the body.

12 What is this book about?

Joy: This is the topic of our book. Isn't this why we're here?

Zoltán: It took me two months to realize. I remember that, Tamas, you invited me to Porto Allegre in Brazil to a meeting that wasn't in my field of expertise. The whole meeting was on metabolism, regulation of hunger and appetite. You invited some of the best people in the field and me. The meeting was an eye opener for me. I never learned so much on a meeting than on that one. Also, all the world experts in your field were giving some fantastic provocative presentations. Now I realize just how good that meeting was. Honestly, for me, going to meetings outside of my field has always been much more stimulating.

Joy: I think we have a Big Idea on our hands here, and it might relate to the book. I'm thinking about what we're doing here, having a dialog among unexpected friends. I mean, yeah, we would never get together in a meeting, and we are scientists that really, we don't naturally speak the same language. And here we are, we're putting together this book based on dialogs across disciplines. We cover behavior, we cover neuroscience, we cover basic physiology, and we cover development. Our idea is to find the common ground and enrich that common ground by our differences and the new questions.

Tamas: Just to add to the silo aspect of neuroscience, here is another of my experiences. It's an individual who is very smart, greatly accomplished, tremendously accomplished, and very specifically focused on one particular thing. I've known him for a long time. This person looks at me, looks at my work in the hypothalamus, and basically sees me as a monkey in the circus who can ride a bicycle. That sort of says: you're cute, you're silly, but really, what you're doing is not truly *science*. I think that's a fundamental problem in probably all disciplines. If you cannot talk to somebody who works on ion channels about anything that isn't an ion channel, or as Zoltán pointed out, his field is preoccupied with only transcriptional factors and nothing else, then it's very unfortunate.

13 Synergy to creativity

Joy: So in a way, we're advocating a binding together of disciplines in order to create something new. The idea is that we are seeking to relate what each of us does to the other's work—I mean, I would love it if my two-brain work could fit in and be enriched by your work.

Tamas: I always love to synergize. For me, synergy is the thing, but I think synergy in this case is coming from a combination of each of our diverse backgrounds. In a way, you're making the point that diversity is a very helpful thing to have when you want to solve complex problems. If you have very similarly trained, similarly raised individuals, it's very difficult to see how they're going to be able to make transformational solutions.

Discussion 4—10th February 2021

1 Individual variations in tolerance of food shortages

Tamas: We had this paper published a couple of weeks ago on anorexia nervosa, in which individuals—predominantly adolescent girls—start to engage in diet and exercise, and they start to starve themselves and lose weight. A large number of these individuals can do this for a prolonged period of time. Luis Glück, who got the Nobel Prize for Literature in 2020, struggled with anorexia nervosa for four decades. The intriguing thing about this disorder is that people can live long lives. But there is a subset of people with anorexia nervosa who die. In fact, other than suicide, it is the psychiatric disease with the highest mortality rate. In this paper, we interrogated how neurons that control hunger in the hypothalamus may be relevant for manifestation of anorexia nervosa phenotypes in an animal model. Key aspects of this paradigm are food restriction, period of peri-adolescence, and access to a running wheel. When mice are food-deprived and provided a running wheel, they just love to run. When hunger-promoting neurons were impaired in these mice, they jumped on the wheel and started to run, but they all died. In contrast, when in the same condition, hunger-promoting neurons were activated, they ran and would never stop and never die of starvation or disease related to food restriction. The key finding was that the way these neurons controlled the survival and endurance of these food-restricted mice was by controlling fat in the blood. The hungrier you are, the more fat you mobilize from your fat stores. The less hungry you feel because of the impaired functioning of these hypothalamic neurons, the less fat you mobilize from your fat stores. It is really fat that you use when you have no incoming food. Also, long distance runners are running on fat and not carbohydrates. So, these mice with anorexia nervosa died on the running wheel because they could not mobilize their fat as alternative fuel. But if they were provided the limited amount of food in the form of fat, they all survived. In our view, a key aspect to become anorexic and survive is to have these neurons that confer a very specific ability to overreact to the cue of low food availability. It is a rare case.

For example, Zoltán and I, who have engaged repeatedly in diets, we never ended up in in the spectrum of anorexia nervosa, as most people do not. But those, we claim based on this recent paper, who have hypersensitive hypothalamic neurons that trigger other brain structures to become compulsive about continuous movement might be more prone to developing anorexia. While the overall phenotype of people with anorexia nervosa is clearly dangerous and a potential threat to health and survival, these same phenotypes could be tremendously advantageous for migration. Because these are the individuals who, when food is scarce or not available, are triggered to move and have the ability for long-distance, endurance movement. They are capable of surviving for prolonged periods of time with no available food. Thus, intriguingly, these hypothalamic hunger-promoting neurons that drive endurance exercise may be in the center of migration. So what I'm going to make the point for and trying to see how I could explore it in an experimental way is to argue, or at least raise the hypothesis, that these neurons are in the center of human population migrations. As an example from the animal kingdom, hummingbirds migrate from Mexico through the Gulf of Mexico to the southern part of the United States. They fly nonstop from Mexico to Texas, and they use 1.3 g of fat to make this very significant trip. I'm trying to come up with a set of experiments to actually test whether these hypothalamic neurons are fundamental for population migration, then make the leap and see if it is possible to interrogate whether the migration of humans out of Africa could have been driven by such traits. There are genetic correlates of anorexia nervosa; therefore, I believe that the possibility, in our view, a key aspect to become anorexic may be to have these neurons—a very specific ability to overreact to the cue of low food intake. It is a rare case.

2 Puberty and energy metabolism

Joy: You know, just a sidebar, let me just inject this, and then we'll go back to your topic. As I'm listening to you, I'm remembering what some of the anthropologists at the American Museum of Natural History talked about when I was curating the brain exhibit there. And that is, there's something very special about human puberty in that it's the time when teenagers start to show conspicuous, often unexpected behavior changes. We break all the rules, we tear down our family's traditions, we start something new, we move out of town—we move on. Something these anthropologists have thought is that this period in our development has partially enabled our migrations because that's when we become restless—kids move on, they get out of town—that it's during those teenage years when they're able to break the bonds, break the traditions, and move on. I was just wondering if that has any relationship to this notion that you have of metabolism?

Tamás: It is, I believe it is. We have another unpublished paper that we have been trying to publish over the last couple years, in which we actually show that these neurons, when you interfere with them in the adolescence period, animals have long-lasting cortical alterations and behavior alterations. If you step outside of these periods, then we didn't see that impact.

Joy: Isn't that interesting? It is that the two ideas might actually have some synergy? Of course, we're just making it up. I think you're absolutely correct. So when I thought about puberty, I always thought about, yes, the child has to leave the house because they cannot tolerate their parents anymore. That brings back the other thing, that is, the migration is what we've seen by these individuals who are in that mode of moving on.

Tamas: Pasko had some fundamental works on puberty and cortical organization.

Zoltán: Once I listened to a lecture, when Pasko was elected Fellow of the Royal Society, given by physicist who was modeling the geological conditions around the time Homo Sapiens was migrating out of Africa. Apparently, it was a miracle that humans managed to migrate. If I understood the lecture, his message was that there was a very small window of opportunity, because that route is normally closed. There was a very, very narrow opportunity to migrate. Listening to you, it might be the case that those people who had these genetic predispositions could migrate, and now a large fraction of humans have them. We could look at the prevalence of this ability across the globe.

3 Genes for anorexia

Tamas: Actually, there is a GWAS study that was published regarding anorexia nervosa. It is indeed a great idea to look at how those genes might be distributed in the population.

Zoltán: For instance, if you look at a map, about the distribution of how the ability to digest milk and the evolution of that ability, it's surprising how well it coincides with the spread of dairy farming. Where you have very heavy production of milk, for example in Britain, most people can digest milk. But some people in the Far East, they can't tolerate milk.

4 Metabolism and the migration out of Africa

Tamas: So this was one of my eureka moments in the recent past (while working on this book): Hypothalamic AgRP neurons drove the migration of humans out of Africa. This week, I got very excited, you know, I've been sort of getting in a bit of an intellectual plateau over the last couple of years. But it's exciting me because if you think about, it's very obvious. And again, the underlying idea is not mine.

Zoltán: There was a symposium hosted by the Allen Brain Institute on Evolution, where the migration within and out of Africa was discussed. One of the talks looked at how we differ from each other genetically in twin studies (human/human), human/chimp, human/mouse, human/broccoli etc. Basically, this talk ended up looking at the milk tolerance. Where can people digest milk? Did diversity of milk intolerance exist in different locations in Africa? In some of the African populations it is actually very frequently observed. Where is the milk tolerance in other regions, what enzyme is involved, and how this developed and was distributed in the population? It's increasingly evident that in Europe, milk tolerance is high because of agriculture and milk production. But in areas where this wasn't introduced, it never really developed. The distribution among pastoralist populations and how it evolved is actually correlated with where the gene appeared to cope with milk, which has been localized to East Africa. It started around 3000—7000 years ago. It is correlated with when we started keeping animals as livestock. So everything had to be just right. The right time to migrate out of Africa, and the dry season to be able to walk out. You had to have these genes you are talking about in relation to anorexia nervosa to be able to walk and survive.

Tamas: Did the speaker talk about that?

Zoltán: No, he did not. I'm just trying to put it together again, not because they put this forward, but because I am exploiting it for arguing for these extra signals you are talking about. The GWAS data and some other information that we have will be analyzed.

Tamas: We can evoke many of these phenotypes by altering these hypothalamic neurons. The great uniqueness of these cells is that they are sitting outside of the blood brain barrier. They are capable of sensing a lot of signals coming from the periphery, which makes them a great way of conveying information from the periphery to the central nervous system and then they have many, many, diverse pathways to ascend to cortical regions, including neuronal signals and humoral signals from the periphery, controlled by the endocrine hypothalamus-pituitary and the autonomic nervous system.

Joy: But it brings up more about the philosophical questions as well, you know, it's very interesting. I'm just trying to think if there are ways to actually sort of make predictions and test your hypothesis, given that we have an unknown group of people that have and have not, right, but you can distinguish people based on the representation of these neurons, right? Correct. And their behaviors are different. They're distinguished behaviorally as well. I'm wondering if one could design behavioral experiments that would make predictions based on decisions, lifestyle, habits, capabilities, physical things? I don't know. So it seems that they have a better way of putting together what energy is available and how much that should be. Or they could be on it, computing the availability of energy for the time that they could run? Yeah, you know, I was listening to you, I sort of think about my own physiology, but I'm somebody who can go long periods of time and not eat, I just eat when I'm hungry. And I certainly have never been anything close to anorexic. I just have what I consider efficient metabolism. Now, as a runner, I've never been a sprinter. But boy, can I run long distances, I can literally almost go to sleep and keep running. I've always been able to do that.

5 Metabolism for long distance running

Tamas: And for long distance running, you really need to be able to switch to lipids. You cannot use glycogen for long distances, some of your muscles really need to switch to fat. If you look at long distance runners, they have the best capacity to run almost immediately on fat.

Joy: I don't do that much running anymore, because I do competitive ballroom dancing instead. But when I was running, I would run with friends, who were like sprinters, they were so much faster than me. I mean, for me, a seven-minute mile was a major accomplishment. . . .. so was an eight-minute mile. My running partner could do much better than that. Most of the time, I would run an eight-to-nine-minute mile and be happy with that, but I could run so many of those. Once I started running, however, I never wanted to stop, I could just run and run and run and run, but relatively slowly.

Tamas: These hypothalamic neurons organize fuel availability for the muscle, but also fuel availability for the brain, during development as well. What I was thinking of when I thought about fuel efficiency and brain development is that differential outcome of cortical development in different species may be related to how those cells in the brain that are capable of utilizing and transforming fuels for their differentiation, proliferation, migration and so on.

6 Oxygen and metabolism

Zoltán: Nobody knows how the progenitors that generate our brain utilized fuels. Once, I was really surprised that somebody used some kind of a metabolic marker to identify when the neuronal progenitor was switching from normal epithelial proliferation to neural production, there was some kind of a metabolic shift. I think I have heard it on a meeting from Magdalena Goetz or Wieladn Huettner that glycogen granules increase with the transition of neuroepithelial cells to apical radial glia. This is the stage when the epithelial proliferation to neuronal production occurs. Nobody ever linked metabolic changes to cortical neurogenesis although some clinical conditions, such as susceptibility to Zika infection could depend on these issues. There are lots of studies on hypoxia—ischemia and everyone agrees that oxygen is an important thing in neuronal development, but not much work is done on other important aspects of metabolism.

Tamas: But when you talk about oxygen, you're talking about metabolism. I mean, oxygen is not some sort of a miraculous thing that enters cells and does something independent of metabolism.

Zoltán: Yes, oxygen consumption is just one indicator of a certain metabolic pathway. There are many subtleties which are somewhat underexplored areas of research. Metabolic changes could have contributed to evolution of the brain, it could be one of the reasons why the brain differentially develops in one species vs another.

7 Why are there 6 layers of cells in the cortex?

Joy: Another question: Why are neurons arranged in that number of layers in the cortex?

Zoltán: Why are there six layers? I think because we are anthropocentric and anthropomorphic: as researchers, we like to give numbers, and six is a really good number to give. We could have come up with different numbers than six. Of course, there are not only six layers—there are many more layers in the cortex. But Brodmann, Cajal, Economo, and others, started molding everything to six layers. If you look at the primate brain, there are many more layers than six. We kind of squeezed this hexalaminar scheme to all mammals so that we know what is what. Now when you revisit these ideas, you could relocate some of the layers to others. For example, there is a boundary between layers three and four, and Brodmann had some inconsistencies. Now, based on the transcriptome analysis, we are relocating these layers. Your original questions was why it is good to have these similar cells in the same lamina in the same region. There are many arguments that suggest it's good to have cells with similar properties in the same area because then you save on the wiring. Wiring is expensive, in evolutionary terms, in metabolic terms, so it's good to put them all together. Producing and arranging the same type of neurons together might be also more efficient when you separate the production of different neuronal subtypes in space and time. Also, if you put similar neurons in the same region, then you can produce maps within those sheets of neurons. You can process information in a different way, using very similar algorithms, but separated from one another. That's also why it's really good to have roughly the same building principle across the cortex, because then individuals can allocate these circuits that are (to some extent) already prepared based on what you need in life. Whether you want to become a dancer, runner, or violin player or tennis player, individuals can allocate these areas to different functions during life, with a little bit of modification. When we get a bit older, it's a more difficult to transfer functions and change relative proportions, but for the developing brain having a layered universal large blank sheet of nervous tissue with a prototypical circuit that we can adopt for certain functions is ideal. Perhaps this is why cerebral cortex is such a success story.

8 Neural rehabilitation

Joy: We published a neural rehabilitation paper once, it was a single subject paper, but it was really cool. It was a United Airlines pilot, and he had endured so much turbulence that he had actually lost the ability to move the ring finger of his hand. So what the surgeon did was to connect the control tendons in his thumb to his ring finger, so it could control his thumb and then control the finger. Eventually, he was able to independently control the fingers again. The question was: did the old neural area associated with that finger light up again and start working, or did the thumb area somehow adapt to control the finger? There's an answer. Which do you think it is?

Zoltán: Because it sounds cool: the thumb area adapted to control the finger.

Joy: Yes, that's right. That's right. The thumb area controlled the finger. You want to move the thumb and finger separately? One would imagine that they are located in slightly different areas. So you don't want one to be so close to the other. Well, what happened is that the functional brain area around the thumb, the active area, took on a new responsibility. The old area that wasn't actively controlling the transferred finger anymore didn't light up again. As the neurons differentiated, they must have developed a new recruitment system for blood to the transplanted area for that finger. That is my reasoning there. This speaks to Zoltán's point. You were suggesting that it makes sense if you organize it that way, it's more efficient for the reasons you all mentioned. Also, this arrangement makes it more efficient to bring in the energy to support that function, because if you look at the blood vessels in the cerebral cortex, everything is distributed radially.

Zoltán: Maybe they can open certain regions and vascularize it a bit more. Roy and Sherrington published the first paper on this at the turn of the previous century. They suggested that the color of the brain is changing where you have activity because of the changes in the vascular flow. Maybe the whole cerebral cortex is built like that.

Joy: In fact, I was able to use functional MRI at a high enough resolution to tease that out, but the mechanism for that is still left to be discovered. The truth was that we could see with functional MRI that the centroids of activity for the transplanted and adjacent fingers came together. The functional area wasn't just the transplanted finger, and it wasn't just the adjacent finger.

Zoltán: How do we rearrange the map?

9 A role for neuroplasticity

Joy: How do you rearrange the control system? You know, neural plasticity is one of the most interesting questions. The brain is adaptable. The interesting thing about this man, he was about 85 when this adaptation occurred. This was not a young pilot. This was an old pilot, who had endured years and years and years of trauma to his hands, and this neural adaptation still occurred.

Zoltán: Michael Merzenich, John Kaas and Mriganka Sur and others did some experiment where they were stimulating three digits of the monkey, maybe just one hour per day for a couple of weeks. Then the cortical representation of these stimulated figures in area 3b was substantially larger and this increase was present 3 months later.

They did it in humans with fMRI, as well. They trained human subjects to do rapid sequence of finger movements for 3 weeks, only about 10−20 minutes/day and they demonstrated much larger region of activation with fMRI during this task. Interestingly the change persisted for several months. You probably know this literature very well. There is a huge functional expansion in the cortex if we perform some tasks repeatedly. But you also have lots of subcortical components involved in reallocating these areas—for example, the spinal cord and the thalamus—before the cortex is involved. So every level is capable of plasticity.

Joy: Again, going back to my sidebar, back to my time as the curator at the American Museum of Natural History, these guys had so much wisdom, these anthropologists had this word *pita morphism*. And it meant Forever Young. That was something that described us humans is that we, as senior citizens, know how to play, that we play our entire lives. I mean, look at us, we were, in a way, unlike other animals that become mature, and lose that playfulness. That doesn't describe many of us, you know.

10 Longevity and neuroplasticity

Tamas: Very good point! For example, my grandfather was 98 years old, and I was having a conversation with him. And he told me how he was really concerned about his 10-year investment plan.

Joy: The question I have is, is this the attitude that keeps you young? Or you have this trait that comes together, are they coupled? Can you adopt a behavior that never feeds back to your past? Or this or that behavior is a static trait? You know, it's a chicken and egg problem. But it's, by default that you are like that, or that's why you are doing so? Well, I think it's a question that it's worthwhile to entertain for everybody. I think part of our adaptable brains, there's something very unique about our brains. I *think*, I don't *know* this, because the work hasn't been done. But I would imagine the human brain is very special, in that you could take some people in their 80th year of life and teach them to play the piano, you could do that. Or you could take somebody like me at the age of five, and you could not. The fact is, we retain that plasticity, learn throughout our entire lives, and we remain playful in many ways. What is the neurobiological basis of being human?

Zoltan: I have one more story on plasticity. My brother Béla came to visit us from Basel where he is an ear-throat-nose surgeon. He came to my college room at St John's and noticed that I had some glasses with these prism lenses that bend or refract light such that objects viewed through them appear displaced to the side. I inherited these glasses from Mitch Glickstein who left some of his old books for me when he died and with John Parnavelas we collected the books from his UCL office. I use these glasses for a demonstration of visuo-motor plasticity by prism adaptation on my tutorials for the second-year medical students. With these glasses the visual input can be easily altered by horizontally-displacing prism goggles. When you put these glasses onto the students and then you throw the ball to them, initially they cannot catch the ball. However, after throwing the ball a number of times, their coordination gradually adapts and improves. When they are doing well, you remove the glasses and immediately after removing the goggles the students will have problems catching the ball before they recover again. Measuring the performance one can quantify the adaptation that has taken place. That's an example of a very fast form of learning. I had two of these and Béla, immediately asked me if I could give him one of the glasses. And I said: you are an ear-throat-nose surgeon who is actually curing patients, why do you need the glasses with the strong prisms for your clinical practice? I was really surprised by his answer. He said that he has a lot of patients for tumor surgery. These patients face operation and sometimes nerves or muscles have to be removed to remove the tumor and that can produce functional losses. Patients are worried about losing motor functions. They usually ask him: doc, will I ever recover? How can I retain function or relearn function? And Béla wanted to use my extra set of glasses to demonstrate re-learning to patients who will ask this question in the future. Some of the problems cannot be articulated properly to patients; nevertheless, with the glasses you can demonstrate that there is a chance for re-learning and they can eventually restore and relearn certain functions. He took this to his patients to demonstrate that something is possible to learn or adapt to.

Zoltán: Roger Sperry said that the sum of our two brain hemispheres is more than the individual. If you only look at the characteristics that are associated with the right hemisphere or the left hemisphere, they are largely different.

Joy: People like to make things up when it comes to the brain. It's easy to say, "Oh, these are right brain qualities, those are left brain qualities." I mean, I think that's for poets. I mean, what evidence do we have for it?

Zoltán: Well, after a stroke, you may lose specific functions. You lose function in your right hand because you have a left hand motor cortex lesion.

Joy: That's understandable. I mean, so yeah, anything that's crossed, the visual system is crossed the motor systems across those, those kinds of things, you can track the pathways, and that is just following the roadmap. That makes sense, but ascribing an artistic personality to the right brain, and an engineering personality to the left brain, that's, you know, give me a break.

Zoltán: These are very old ideas. Sperry formulated these ideas, and said that society as a whole ignores the right hemisphere; you are only rewarded for abilities you have in your left hemisphere. So literacy, language, organization, problem solving, those are rewarded, but enthusiasm, artistry, to look at things in a holistic approach, are somehow not rewarding. That's what he said 70 years ago.

Joy: Maybe so, but I'm not sure that the evidence for associating functions to the left hemisphere or the right hemisphere is correct. A personality is a whole brain concept. I'm assuming without reading into it or reading about it, that not only humans are capable of having a sense of humor, for example.

Zoltan: I think there's data on monkeys and dogs.

11 Human humor

Joy: What is the actual underlying premise of humor? Is it known? What are the underlying neurobiological drivers of humor? What is the adaptive value of it? I mean, I kind of understand the value, but I don't know if that has been pursued in depth. Do you know anything about development of humor?

Zoltán: There are a few papers that have been published on it, but they just scratched the surface. I don't think there's a mechanistic understanding of what it is. It's one of these complicated whole brain things that involves multiple systems and we're not good at studying multiple systems, long range networks. For me, for example, love is a much more primitive manifestation of potential brain function, than humor.

Joy: Yeah, obviously, chemical imbalance, and then you develop this sort of addiction to an individual, The point is that you are excited. But there are not many, as far as I'm concerned, biologically complicated aspects, because you need to have that in order to bring people together. But humor is something more abstract, seemingly, but it must serve some very fundamental purpose. I listened to the radio one morning, and they were talking about smiling, what was the origin of smiling? Basically, what they said was that we primates are pretty aggressive. So when you see somebody new, whom you don't know, then immediately you are aggressive, and you want to show your canine teeth. When you see somebody, instead of these canine teeth show offs, now you smile, but I don't buy that. I know that if you're misleading others, it makes sense. Because to get an advantage, you need to sort of mislead the person. **And** there's another factoid to add to this, I read a number of papers on this recently, on how much larger and more diverse the musculature and innervation of the human face is relative to our closest comparable primate face. I've been interested in facial expressions as a part of our language, a back channel language, a social language, for which you don't need words at all. You look at somebody and there's so much infor-mation on their face, particularly if you let dynamics happen, like a facial expression, you get a frown, a smile, a scowl, a gesture of some kind. This information in that gesture, we as humans use that form of signaling to a much greater extent than our closest primates. I think that the smile is a sentence in a language of who I am, who you are. There's something about a smile that's also very contagious, that if I smile at you, you're very likely to smile at me. This forms a synchrony between our brains that sets up a social connection. I think it makes it harder for us to violate a connection that has been set up between us, and this connection prevents aggression that might happen without that expression.

Tamas: That sounds very reasonable.

12 Effects of social reciprocity

Zoltán: Also, how you look at a baby from the very beginning of its life is highly important because of oral facial synergy. A surgeon told us that cosmetically, it may be better to perform a facial correction surgery in cleft palate a bit later, but they operate earlier because the life of that baby with a serious dysmorphism may be tremendously impacted by how people look at them. Imagine that baby is constantly looked at by faces that reflect distress and anxiety, rather than smile and calmness. Looking at happy faces around you early in life is important. Surgeons suggest to resolve this very early on, because of that.

Joy: The reciprocity! And it's our brains that are tuned to other faces. That tuning actually makes this reaction reflex. You don't decide to smile, you just smile, because there's a reflex almost, to smile back at someone who's smiling at you. So I agree, I think this is very important. But how important our faces are in socialization and jumpstarting our connections to each other is an extraordinarily important topic that I think we're just beginning to tap.

Tamas: I just walked the dog downstairs just a couple minutes ago. You walk the dog, and this is a very cute dog. People are making comments and they look at you, but now that we have this mask, you cannot really express yourself. If I think that somebody is looking at me trying to gesture about the dog, and I keep on looking at that person, it can also be taken as some sort of intrusive staring if you don't see the entire face.

Joy: It really truncates our primary social communication, without words at all. You approach somebody and you have a sense of whether you trust them, whether you can engage with them, whether you know who they are, you make a judgment almost instantly, and it's all from the face mostly.

Discussion 5—17th February 2021

1 "Gut" and brain interface

Joy: Zoltán, you sent a paper about "gut feelings," a mechanistic explanation of what a gut feeling is, didn't you? Do you remember what it was about? There's a capsule that's swallowed, which then travels through the intestinal tract and records all sorts of information.

Zoltán: Basically, they wanted to learn more about the gut—brain connection, to explore what this "gut feeling" is. They enhanced humans using the minimally invasive vibrating mechanosensory probe, which combined electroencephalography and electrogastrography with signal detection and emissions quantifying "perceptions" of the brain, stomach, and body.

Joy: The responses following the ingestion of the vibrating capsule are remarkable. With this probe, they mechanically elicited hormonal and neuronal responses from the gut.

2 Ascending signals of the autonomic nervous system and brain functions

Tamás: In my view, this is a quasi-remake of vagal stimulation studies. In those, vagal stimulation was shown to have a multitude of effects in the brain, including impacts on higher brain functions, epilepsy, and other neurodegenerative processes. The gut-brain axis has been implicated even in the etiology of Parkinson's disease, in which gut pathologies precede those of the nigrostriatal dopamine system. This entire field is in its renaissance. The complexities of the autonomic system are being rediscovered and interrogated with new tools in a manner that was previously impossible to accomplish. The pendulum is now moving to explain the control of brain functions via ascending signals of the autonomic nervous system. There is clearly much to be discovered and learned here that will benefit brain research and medicine in general.

Joy: I really like this thesis. It's a good example of the kind of motivation we have for writing this book: the notion that the brain doesn't just talk with itself to control behavior, or to put it differently, it is not sufficient to understand the brain from within.

3 "Gut" control of eating behavior: a paradigm shift

Tamás: Let me come back for a moment to the "pendulum" issue and the appreciation of the autonomic nervous system, with an initial emphasis on control of eating. There is an enormous body of literature on how the gut controls eating via the autonomic nervous system. It was a very sensible idea as, for example, the stomach expands in response to food intake. Chemicals, such as cholecystokinin, the hormone secreted by the upper small intestine in response to fat and protein, were identified to affect those ascending neuronal pathways that were assumed to elicit adaptive behavioral changes to, for example, explain obesity. While the studies were conceptually great, at that time, the tools were not there to become more granular about these questions. Then came the nuclear explosion of the field: the discovery of leptin, a fat-derived circulating "stuff" in 1994/5. Bam! Vagus out, circulation in!

Zoltán: This reminds me of stomach ulcer. 40—50 years ago, they treated ulcer with vagotomy. Then there was the discovery of the relationship between Helicobacter pylori and ulcers. . .

Tamas: Exactly! Same paradigm shift with a broader implication. First, it provided an immediate remedy for a rare, but debilitating form of obesity. Second, it transformed the field of neuroscience of the hypothalamus (considered "less worthy and not really neuroscience" by mainstream neuroscience) to a cutting-edge experimental brain research area. Third, the past 25 years provided a remarkable blueprint by which the periphery communicates (through the circulation) to the brain. But the vagus was out. . . One of the best examples of this "de-vagusing" period relates to another metabolic hormone, ghrelin, which was discovered at the end of the 90s. This hormone is generated in cells of the gut and has a major impact in propagation of hunger when food is not available. The mechanism of action of ghrelin was found to be via circulation, a conclusion that was heavily relying, in part, on experiments employing vagotomy.

Zoltán: When there are paradigm-shifting discoveries, people tend to throw out the baby with the bathwater. Vagotomy must have had an effect on ulcers as well as on ghrelin's action despite the primary mode of action. With vagotomy, you change systemic balance of intertissue communications, which predicts that processes associated with Helicobacter or ghrelin must be modified. Even subtle modifications of biological processes can be fundamentally important for individual variability, which is so crucial in functional brain disease. Unfortunately, what is considered "high impact" in any field today deals with extreme scenarios showing robust influence. Leading journals and pharmaceutical companies want to have "breaking news" or a "blockbuster" rather a substantive contribution to a problem.

Tamas: Ghrelin is a great example for that, as well. Soon after we and others showed that ghrelin affects hunger via the hypothalamus using overlapping pathways that leptin uses, we started to interrogate whether this hunger signal may also be relevant to other brain functions, such as learning, memory, neurodegeneration, and reward. This was back in 2005—2006. There was a clear and robust impact of ghrelin on the hippocampus and the ventral and dorsal striatum with experimental proof for its involvement in learning and memory, reward, and neurodegeneration associated with

Parkinson's and Alzheimer's Disease. Other than one of these studies being picked by the New York Times as one of the inventions of the year, there was no interest in these findings by the pharmaceutical industry. I went to those companies who were pioneers in ghrelin discovery and also had very robust neuroscience programs. They had zero interest in these ideas! Zero! One reason was that in both of these big pharma companies, ghrelin discovery was occurring in their metabolism divisions, which were almost entirely segregated from their neuroscience programs. It was like two people speaking completely different languages. Clearly, the financial incentives of the subdivisions did not foster meaningful crosstalk. That combined with the conceptual rigidity of contemporary neuroscience and metabolism research made it a futile effort to try to convince them that this is of great value for them and potentially for the patients. Because of this, I find it reaffirming, amusing, and grotesque that one of the recently founded biotech companies focusing on gut-brain communications to address neurodegenerative disorders is headed by the very person who declined to pursue such ideas back in the beginning of the 21st century.

4 Where are the models to connect brain function and body signals?

Joy: What you just said corroborates the idea that the gut controls complex brain functions. My experience of the neuroscience community is that these factors outside the brain are not fully appreciated, but one of the reasons is that there aren't models for connecting the ideas. I mean, your notion that people might learn better if they are in a state of hunger. Why is that?

Tamas: What is the model? What is the predictor that could be testable in this regard? Those types of questions haven't been fully explored or investigated. Eventually, when I was thinking about experiments in behavioral and circuits neuroscience using awake animal models, almost all of them, I wouldn't say 100%, but the majority of the paradigms hinge on food restriction and providing food rewards for animals. So the issues of hunger and its role in complex behaviors have been inherent from early experimental approaches in neuroscience. The quantified and qualified behavior wouldn't happen without food being involved. It's a fact, on the other hand, that human studies that you do, Joy, do not consider the metabolic state of the subjects when you are studying them.

5 Developing models and findings to connect brain function (learning) with real faces

Joy: That's absolutely right. Although let me tell you an interesting factoid here. I'm on a thesis committee for one of the students at UCL, and she's studying learning in the presence of faces. The preliminary data she has in her thesis is that when students are able to see the real faces of their teachers, that actually promotes learning. If they are given the information with just a slide or without the face present, then learning is diminished. The paradigm that she used for teaching and learning was a really good one. She had the participants actually learning things. She was teaching them about unfamiliar objects, rare instruments, ancient artifacts of one kind or another, miscellaneous things that most of us, no matter how educated we are, would know nothing about. So she, as the "teacher," showed the "students" pictures of objects, explained what they were, and taught them some basic information about the items. Then the students were tested on this information. When the information about each object was explained by the teacher with her face in the student's view, learning was far superior than when they were given information with just a slide without the teacher's face. Also, the teacher's face could be seen over a virtual connection or in real life, face to face with the student. There was a difference between this virtual face speaking to the student, like we're doing now, as opposed to the face to face, real life instruction, which was found to be superior, as well.

Zoltán: This year, when most of our teaching was online, the students specifically requested that when we record our lectures, we should also record our face while we lecture. We have to click on the right settings on Panopto or Zoom for the recordings, it seems that it is important for the students. So, I always record my lectures in that mode.

Tamas: Back when I was in vet school and I attended physiology or whatever seminars or talks by a Professor, it was definitely more impactful than when I studied from a book or only listened to a lecture. When I had to prepare for the exams from books, I understood it as I was reading it, but I completely forgot about it by the end of the page. I used different colored pens to underline different content, but still it was not even close to as efficient as it would have been if I could connect it to a lecture I attended.

Zoltán: Same for me. I still remember some of the lectures I heard from Bertalan Csillik, Gyuri Benedek, Mihaly Bodosi, and many others, whereas I do not really remember the books.

Joy: When I was teaching medical students, I had to stay really close to the curriculum, because it was already written, they had my slides ahead of time, they wanted to know what was going to be on the tests, and so on. But when I was teaching undergraduates at Columbia, professors were encouraged to kind of let it all hang out and share with them, you know, who we are, why we think the way we do, what evidence we have, and that was really quite fun. I think that the students in that environment probably learned more easily than the medical students, but I don't know. Anyway, I'm just getting back to the learning idea. Your notion that it is better when somebody is hungry, I say, it's also better when somebody is face-to-face, and maybe there's some common ground here, learning under arousal of some kind.

6 Is learning better when hungry?

Tamas: Obviously, you need to eat to survive and you need to take resources from the environment. Thus, when you are hungry, you are most alert to everything in the environment, so you can get information most efficiently and move quickly, as well. Otherwise, you're dead. I think that is the conceptual reason why you learn better when you are hungry. In this regard, the lateral hypothalamus and its orexin/hypocretin system is crucial. These are small sets of neurons that send projections diffusely in the brain, including to the cortex, and they play a fundamental role in arousal. Even their input organization is unique for a long projecting excitatory system: they are predominantly innervated by stimulatory inputs, even on their perikarya. For these neurons, noise is the crucial signal, because it is the totality of the information from the outside world, they need to propagate nonfocused attention, so that when the lion comes, you can be ready to move.

7 Beyond the hypothalamus and receptors for orexin

Zoltán: I am taking over these concepts and move them further from where you stop. I think you figured out what's going on in the lateral hypothalamus, but what is so interesting, if you look at the connections and receptors for these peptides in the brain, they are not everywhere, they are in a very specific positions. If you look at the orexin receptors in the thalamus, which is a conglomeration of several nuclei, they are in the so called higher order thalamic nuclei. As you know, Murray Sherman and Ray Guillery and Ted Jones classified the thalamic nuclei to first order and higher order. The first order thalamic nuclei received direct sensory input from the periphery, like the retina and the skin. The higher order nuclei mainly get input from the cortex; they receive strong powerful synapses from the layer five projections from cortex, and then they send their projections back to other cortical areas. So, there is a cortico-thalamo-cortical communication via these pathways. Now, if you look at the receptors for orexin, they are in the higher order thalamic nuclei. These higher order nuclei are thought to be involved in higher levels of perception such as context. When you open up these trans-thalamic cortico-cortical connections and start looking around, you're thinking, "okay, what is important or unusual in my environment, what should I pay attention to more, where should I adjust the thresholds of my perception?" You are now suddenly associating more things with your current perception and linking up all sorts of other cortical areas through these connections, and you also start to have cross-modal interactions and learning. So the distribution of these orexin receptors could be very important to understand brain state control. We found that the lowermost layers of the cerebral cortex have these receptors and these are the cells that project back to the higher order thalamic nuclei in mouse. What we discovered in the mouse, and it looks like this is similar in humans, is that the receptors for orexin in the cerebral cortex are in the cell populations that are important for brain development. They're in the so-called subplate cells. It's really interesting to see that these cells, which have such an important developmental role, exclusively have this receptor to respond to orexin in the cerebral cortex. If you have a direct input from the lateral hypothalamus, then the cortex will get into a different state. Subplate cells are dominant during development, but then they undergo programmed cell death. It is important to regulate what cells are left behind and where. These remnant cells are not just shadows of the subplate, they have their own important function. This is why I am fascinated with these cells, since it has been demonstrated that their distribution and number is different in schizophrenia or in autism.

8 Bottom-up influences on top-down processes

Tamas: I so much respect people like Zoltán and Joy and others who are into the higher functioning of the brain, and realize that there are important decisions to execute from below (hypothalamus).

Zoltán: I completely agree because it is different neuronal circuitry, when you activate the higher order thalamic nuclei. Then those circuits look at the context. They open up such circuits that normally wouldn't be open, or you just have the first order thalamic nuclei doing the "auto pilot" of perception they don't change too much in these states. Even when you are sleeping, you can elicit responses in your auditory thalamus. But you can't do that in the higher order thalamic nuclei. So it looks like when we shut off the brain; we shut it off in a certain sequence. Certain parts are shut off first, and then others; and it looks like this orexin system is actually working on a system which you want to open up when you want to see more details around that stimulus; when you want to see the context, when you want to link different things together.

Tamas: Yeah, to learn. That's why I was fascinated by the term, which I'm not sure if you've heard of, either of you, "nonfocused attention." Is there such a term that is used in psychology, Joy?

9 Attention and arousal

Joy: Is there such a thing? Not that I know of in that form. It doesn't mean it's not a term, but you might simply call that "alertness," when you are nonspecifically taking in all the information that your arousal will allow. For me, "arousal" sounds like a dynamic or a "go," an event that is going from point A to point B vs the nonfocused attention in my own head, which is in steady state, when I'm alert to all sorts of information coming in. There are many theoretical models of arousal vs valence, for example. Valence can be positive or negative and arousal is high or low. I don't know about other terms, but I think that most people would say arousal encompasses what you're suggesting.

Tamas: You mentioned valence, which became very popular in the past five, six years in my own field. One of the biggest problems I have with behavioral neuroscience is that it is predominantly driven by nomenclature and not data. Terms drive conversations, and there are even debates about the meaning of those terms. Now you mix it up with the multilingual component of the research community, you end up with an abstract, arbitrary entity of behavioral neuroscience. If those who are devotees and connoisseurs of this field cannot agree on single terms (which I personally experienced in NIH behavioral neuroscience study sections as an ad hoc reviewer), then how will the data stand the scrutiny of time? On the other hand—and again, I am being arrogant for the sake of argument—if I say I weigh 220 pounds (100 kg), or the mouse eats 3 g of food, there is very little debate that can be made about those parameters.

10 Linking hypotheses between brain function and behavior

Joy: What do you quantify? Behavior can be measured in terms of bimodal responses—you either detect it or you don't, you decided something yesterday or you decided it now. You can measure behavior like the old psychophysicists did, when you ask about, say, how much light does it take for you to have a detection. You have a continuously increasing level of light, then you plot the probability of a detection, and you get a psychometric function, that sort of curvilinear thing that looks like a bell curve tipped on its side, that's a quantification of behavior. But when it comes to linking those behavioral measures to the neurophysiology, that's when it gets challenging to describe.

11 The role peripheral tissues (such as liver), arousal, and high level perception

Tamas: Again, it's going to sound simple, but when you talk about any behavior, or perception or whatever you are looking at from a higher brain activity perspective, are there efforts to figure out what occurs concomitantly in your peripheral tissues? In your liver, pancreas, kidney, etc.? My take on this is that behavior cannot emerge in the absence of a coordinated activity of the executive branch (brain) and peripheral tissues because there is no way you can move a muscle properly to accomplish a behavior without the proper alignment of the internal activity of the muscle, which will rely on the liver, pancreas, kidney, etc. In a way, that is where I see the future of neuroscience. My prediction is that asking those questions and answering them will shed light on issues of the brain that we have been relentlessly seeking. These approaches may also lead you to define and refine the potential correlation between quantified behavior and the propensity of a tissue, let's say the liver, to have vulnerability to chronic disease development.

Joy: So is there a correlate, for example, of you recognizing a face in not only those regions of the brain that you are recording, but also in various parts of your body? Well, it's an interesting question. We also measure arousal, in terms of, say, skin conductance. Pupil diameter is another one that's used to measure arousal, as is heart rate variability. We try to quantify as many physiological responses as we can to support the validity of the brain imaging measures that we take. If I can show that something that I claim is a neural response might be due to arousal, I'm obligated to show that it's correlated with some measure that is understood to be arousal. I don't know that it comes from kidney or stomach, but I do know that heart rate variability, for example, probably has something to do with the vagal nerve. I can make that assumption without too much controversy.

Zoltán: Skin conductance may be another matter that is actually skipped. There's a very good measure of arousal in addition to measuring pupil diameter. Joy, I wanted to ask whether you did any experiments when you present sequences of repetitive stimuli that are occasionally randomly interrupted by an unusual and unexpected stimulus and the reaction of the participant to this "oddball" stimulus is recorded.

Joy: Oh, yes. I have done oddball paradigms. That has a very interesting signature that comes in a particular timeframe, at about 300 ms. This is the P300. So what's that? What it means is there's a positive EEG signal that is generated at 300 ms after onset, and it's associated with a surprise event. So, suppose you're listening to a series of words, or you're seeing a series of pictures of say, houses. You see house, house, house, house, and all of a sudden, there's a monkey! The monkey image gets you the P300. Then house, house, house, and next it's the unexpected thing, and we get the P300. What can be happening at 300 ms? Where does that signal come from? Well, it's interesting. I mean, 300 ms interval is very, very short. Let's just consider it takes about 20 ms for a signal to get from the retina to the visual cortex. So the P300 means that there's something that's happening 300 ms after onset. It's early processing, but the thing about the P300 is that it's not localized. It is a signal that seems to be more or less distributed throughout the brain. So I really think—and this was my theory for many years—that the higher order thalamic nuclei are involved in this, and when you change the context suddenly, they send these signals to activate many other circuits, like "Come on, guys, listen up! There's something interesting going on." It's a readiness signal. It says, "Get ready, something's going to happen!."

Tamas: The hypothalamus, in general, is a very important system in going from sleep to wakefulness, to bring you out of sleep and paralysis, to function in an awake situation appropriately in line with the changing environment. I mean, that's basically what the system does. One thing that I've been very much fascinated by, besides this learning with the hunger cues, are the arousal components. When you go from sleep to wakefulness, the transition takes some time, and it is also unresolved (in my understanding) how your higher brain functions perform, and at what level they perform depending on how far you are from the sleep cycle. I'll give you an example of something that I've always wondered about. Of course, we are different people and individuals may have different traits and behaviors in this regard. But if you wake me up at 6 o'clock in the morning or 5 o'clock, for me, then if you were to tell me that I need to perform a very highly demanding cognitive task, I most likely wouldn't be able to perform at the level that perhaps I would perform at 10 o'clock in the morning or 11 or whenever my peak is. In this regard, these hypothalamic systems are very important. For example, when you are having surgery, I never understood why surgeons need to start their surgeries at 7 o'clock in the morning. I am not certain, based on individual variations in circadian rhythms, that a surgeon might be the most diligent or ready to do things at 7 o'clock in the morning.

Zoltán: Most of the studies on mouse behavior are done at the wrong time. Mice sleep during the day, and are active during the night. However, we are testing their behavior when it suits us, during the day. It is like performing cognitive tests on Tamas at 5 in the morning. Joy, it will be interesting to see what will happen in your setups if you were to shake people up at 2 o'clock in the morning and run those studies.

Joy: Exactly, one could imagine that all of the social senses would be quite dulled.

12 Variations of social behaviors based on arousal and context

Tamas: I tell you my personal experience regarding this. I was invited to be a finalist at the NIH Pioneer Award competition back in 2010. The Pioneer Award presentation is similar to an HHMI presentation: you have to do it in 15 min. I never really practice a talk. I'm somehow constructed in a way that for me, diligent preparation is detrimental.

But for this one, for this particular presentation, I knew I had to practice because I had no choice, I had 15 min to summarize a lot of work that lead to my hypotheses. I needed to prepare myself to give that talk in a very predictable way, in the same manner, even if—and I'm not joking—I were drunk. I did various things to test myself. For example, I woke myself up at 2 o'clock in the morning after drinking a bottle of wine with dinner. I did these things until I was able to give that presentation the same exact way in 15 min. Clearly, it was not a creative process. But under normal circumstances, executing anything that we have to execute is less predictable. You don't know when you're doing them whether you are at the highest level of your abilities or not.

Joy: You know, this is a sidebar, but a few times in my life, I had to give lectures where I had to use the word "I." I had to talk about myself. That was horrifying, it was terrible. You know, if I have my slides up there, and I talk about my ideas, my models, my data, then I'm comfortable, but if I have to say something about myself as a scientist or give an anecdote about my life and explain how it relates to what I do, that's a whole different dimension, and it is very difficult. I think that's part of what we're doing in this book, is sort of integrating people, you know, our lives *within* the life sciences *with* the science. I think you hit on a really important point that these biographical things impact our science, and they have a lot to tell us about who we are as humans, and how we work as scientists.

13 Beyond hypothalamus to cortex or is it the other way around?

Zoltán: Let me use a horrible analogy. I really look at the cortex as the event coordinator or the brain. For example, if you go skiing, but you are up there without any skis on your feet, for you to come down that mountain gracefully is a virtual impossibility. If you put on skis, then depending on your skills, depending on other factors, you can come down easily, smoothly, successfully, and so forth. So for me, the cortex is the key to performance once your skis are on. The more advanced the skill, the more advanced the cortex, the better you're going to succeed. But really, what matters is that a person puts the skis on. But my eye opening moment was, and I think it happened a couple of years ago, when I started looking at where the receptors are for these hypothalamic peptides in the cortex, and where you can see that they have a very specific strategic distribution. They are not just talking to everybody, but they can switch things on and off. When I could see that, for instance, the orexin receptors are in a cell population, cortical layer 6b, that is specifically and selectively projecting to the higher order thalamic nuclei, then I suddenly started to think, "Wow, these 10,000 cells in the lateral hypothalamus can control how these billions of cells in the cortex communicate with each other!." One of my colleagues gave me a few sections of a mouse brain that has reporter gene expression according to leptin receptor expression, so I looked at leptin receptors, and do you know where the leptin receptors are? They are in the cerebral cortex, they are in the cells that project to the higher order thalamus, just like orexin receptors. Layer five and layer 6B, not everywhere, not all of the cortical cells have leptin receptors. They manipulate a very complex circuit on the thalamocortical level, which we always implicated in higher cognitive functions. If these receptors are on these cortical cells, then surely, the hypothalamus is extremely important in regulating higher cognitive function.

Tamas: One of our problems with publishing our work is the "optogenetic catechism." It demands that you must provide the circuit blueprint for your cause-and-effect findings. It is a daunting task to make people understand that these neurons from the hypothalamus communicate to the cortex via a multitude of pathways, as well as involving the periphery by the means of hormones, nutrients, and many other blood and autonomic nervous system-borne signals. It is a problem for us to make the point more forcefully. But in the end, this redundancy in communications made me originally wonder whether the neuronal doctrine, as it stands, is sufficient and necessary to better learn more about the brain.

Zoltán: This whole system is the result of millions and millions of years of evolution. We inherited this system, and we evolved it further. It's not a primitive system, because the small rodent probably didn't have a big pulvinar that was orexin insensitive, but the blueprint was there. Then it was evolving to use that system to its advantage. So let's turn it around, and let's think about the system that Tamas is working on as the most sophisticated system that controls the cerebral cortical functions.

Tamas: I started to work on reproduction from the angle of reproductive neuroendocrinology 31 years ago. What eventually intrigued me was that the same pathway that I was studying for reproduction is also involved in regulation of the thyroid axis, stress, growth, and feeding. So if you have the same hardware, and all of a sudden, this hardware can predictably adjust itself to control and impact multiple functions that need to be coordinated and are crucial for survival, they must have some level of sophistication that we don't yet understand. It most certainly is not explainable by the catechism of the neuronal doctrine.

14 Many working parts become one brain

Zoltán: You don't have only one cortex; you have hundreds of different states, which will adjust your actual cortex that you have at that moment. So, you may have a circuit with specific neuronal elements that are interconnected with synapses, but depending on what part you activate, you have a completely different brain. These hypothalamic peptides can have such a huge effect on this very sophisticated machine, which part is active, which is not, depending on how you adjust things. As Joy put it, one could argue that the most sophisticated aspect of the brain is not going to be the cortex, because the cortex has these subspecialized, simplified areas, which then convey the complex flow of information that comes from beneath.

Joy: Well, I think that it's an interesting perspective to think about how the more fundamental aspects of the brain relate to the cognitive functions we tend to explore in our world, how they give greater credibility to the higher-level cognitive aspects. But if you look at the brain, you know, I'm always impressed with how efficient the hypothalamus or lateral hypothalamus is with the few neurons that do so much. I think if you start to conceptualize things in this way—I'm not saying it's the right way—but perhaps it's going to open up new ways of looking at the whole brain;

therefore, maybe you get answers to questions in a more effective or successful way. I really like the idea of seeing how far we can integrate our understanding of the basic old brain, that evolutionarily has withstood the test of time on the planet, and how it's connected to and integrated with the superficial brain that's been added on.

15 Hypothalamus as a radio station

Zoltán: Here is an analogy to consider for lower and higher brain regions: Let's say that you have an old radio station (hypothalamus), which is talking to people (cortex). You can have several hundred people listening to it (mouse), or you can have 2 million people (humans), but it's still the same radio station providing the crucial information.

Joy: I like that analogy. That's a very good one. Then you can have a huge audience (cortex), but it is still listening to those hypothalamic cells.

Discussion 6—22nd February 2021

1 Reviewing grant applications based on a holistic view

Tamas: I was going to bring up my experience in this grant review panel. My overall way of reviewing has been the same for a long period of time. You might say that this is because I'm lazy. But actually, I think it's not the case. So, when I'm assigned to grants, I usually look at the title, look at the person, and then eventually read the abstract. Then I read the biosketch of the individual, where this person is from, and then I look at the specific aims and the approach. Then I put it away, and I don't look at it until the last moment. I always submit my grant reviews in the last moment. I do not write too much, just some strengths and weaknesses of the 6 or so categories. I don't spend too much time on any of them. I write maybe four sentences in the summary of my review, and then, I assign a score. You might think that I'm superficially making an assessment, but then that night, when it's due, after midnight, the scores of all reviewers become available. Usually, my score matches the scores of other reviewers. Where I am heading with this rambling is that without going through sentence-by-sentence or being analytical in its classical sense, you can get the whole picture of a grant application by taking in key aspects, letting them simmer in you, and then providing an evaluation. Actually, I do the same with the review of research papers, as well.

Joy: I do the same thing, and I think most people do. I write the paragraph about my evaluation, and I assign the score. Then I do the bits and pieces of significance, the environment, and the parts that they want us to fill out. I find that I put in the information that follows or is consistent with my score, but my score comes from my overall integration of the whole thing.

2 Fast and slow decision-making

Tamas: I did the same when I had to make decisions about whether I should move to a new job, or whether I should marry this woman. I'm just making these examples. You know, there is a school of thought that you should sit down, take a piece of paper, and do two columns—the pros and cons. If I ever do anything like that, I get lost, and I can't make a decision. More frequently than not, when you pose the question, you inherently know the answer to that. You may force yourself to think about it, but you already know whether you will go this way or that way.

Tamas: So, I'm just asking you about the brain. How do we know?

Joy: Well, Daniel Kahneman got the Nobel Prize in Economics for discovering that the brain isn't rational. He has written a lot about this. The fast brain, the blink, the slow brain, and decisions, that are made rapidly, and those that are made with more thoughtful, contemplative use of information. He's a great proponent of the blink, the fast brain processes in decision-making, and no one quite understands how they come about. But people have given that type of decision-making great credibility. One model is that it is the behavioral product of a pathway that has informative value and is often correct, at the same time. Decision-making, of course, has been studied *ad nauseum* at all levels, from the neural level, to very complex behavioral levels. One interesting idea about slow-to-mid-speed decision-making is that it requires accumulation of information. When you reach a threshold, a decision is made. I mean, this is a topic that neuroscientists, psychologists, social scientists, even economists, actually, economists particularly, are interested in. Things that require decisions and how emotional factors contribute to decision-making. For example, it is well-known in economics that if people feel they're under threat, they're frightened, or they're not secure, then they will make economic decisions to save money rather than to spend or invest it. If you give them a choice, "Will you take $10 now, or will you take $20 in two weeks?" and they are frightened, they will take $10 now 100% of the time. But if they're comfortable and well fed, and feel safe and secure, they will wait two weeks to get $20.

3 Personality and spending decisions

Tamas: I look at my wife, who is frugal. If I go and buy meat and bring home some good meat for a certain price, she is asking why I did not buy meat on sale. I don't think she's threatened or, you know, stressed. They were four sisters. Some of them are like her; some of them are like spending money freely. I don't know how you develop these kinds of things.

Joy: Well, I don't know about development either. These experiments are called "delayed-choice" or something like that, it is a category of experiments. I don't know that these experimental paradigms can capture all of the nuances of a habit or pattern of not spending any more money than you must spend. I think that these tendencies are influenced by social and economic factors, big factors, the global factors that affect us all, probably on top of all of those habits. She might be more frugal when she's feeling threatened and less frugal when she's feeling secure. It's probably relative.

4 Soap and far-away hotels

Tamas: It's probably true. On the other hand, I have a habit. I don't know, Zoltán, if you have it. I blame my background in socialist Hungary. If I go to a hotel, I take all the soaps, shampoo, whatever. In Japan or China, they also put toothbrushes. I will take those as well. I bring them home, and then my wife (who is frugal) throws them out.

Zoltán: Exactly the same in the Molnár-Pollini household! I bring these 'treasures' home from my trips and then Nadia throws them away when I do not watch.

Joy: You guys are amazing! Well, I'll tell you, whenever Jim and I go to Europe, we always take our own soap because, for some reason, there's never any soap in hotels in Europe. I don't know why.

Zoltán: They don't have soap because we take them!

Joy: Okay, so much for the communist Eastern European bashing.

Zoltán: But is eating also like take whatever you can now, because you never know what will happen later? Do you think that you might develop so you always eat as much as you can, whatever you can eat, you eat it?

Tamas: Well, I think that's basic biology, but Joy is an absolute exception from that one.

5 Multi-generational effects of food shortages

Joy: I've grown up in a land of plenty, even though my parents were products of the Depression and post-World War II, and terribly scarred by the fact that there were days when they were hungry. Dad would tell us about sleeping outside in the family garden as a child with a gun to prevent people from stealing the vegetables. People were starving. He was so scarred by that. You should have seen our garden when I was a child! We had the garden to die for, and dad and mom were so adamant that every meal, there was food there for us, and we had to clean our plate.

Tamás: But that's a very good point. My childhood experience is that if I made a comment about the food or tried not to finish it, I would have been slapped by my father. I don't think I've ever been slapped, but definitely, there was pressure that you have to eat it because, how dare you not? How dare you? How could you not eat this food that you have? You are so lucky to have it.

Joy: You never had food shortages in Hungary when you were growing up?

Zoltán: No, we didn't have them. Hungary never had food shortages, but perhaps the variety of the food was not spectacular. Tamás and I came from a small city, Nagykörös; it is in the middle of the agricultural production, so food was excellent. However, we established very unhealthy eating habits.

Tamas: Our parents' generation was exposed to the Second World War. I think there is a tendency to blame that upbringing for forcing you to eat what is on your plate, and that's true. If there is a plate of food in front of me, I feel it's not appropriate to leave food on the plate. I mean, that's just the way it is. On the other hand, when I go to a buffet-style place, even if I made a decision that today, I'm gonna be careful about how much I eat, I'm the one with the most food on my plate.

Zoltán: This thrifty trait is ingrained in us, and obviously, we have genetics to support that, and there is a developmental influence. I hate the buffet-style lunches exactly because of those reasons. I much prefer if someone is serving me with a portion that they consider normal.

6 Effects of pandemic isolation in science

Tamas: One other thing I was thinking about is this issue of what we are going through the situation of COVID. Some people (such as myself) like this isolation arrangement, but I know many people don't. We will see the long-term impact on us scientists. I do hope that we're not going to go back to the "you must travel in order for you to be part of a meeting and publish a paper" mode. I was looking through the window, and I was wondering how if another person were looking through this window, would they see something different from what I see? How is perception of the environment different between people, and how does that issue itself contribute to all the various problems that we are dealing with? I'm philosophizing here, obviously, and I have not been thinking for five months. Maybe this notion that people should live in harmony, that they should have common views and approaches to things, basically is not consistent with the way our brains are constructed.

7 Our brains during agreement and disagreement

Joy: I've heard several science writers say exactly that recently. In fact, there was a science writer who wrote about my recently published "agree/disagree" paper in the Wall Street Journal. The writer was most interested in brains during disagreement, because indeed, it is the disagreement that has the evolutionary advantage for us (she proposed). In disagreement clearly, we're more alert, we use more of our brains, and there's strategy involved. That seems to be what fascinated the lay readership. When people agree, it seems to be in a more receptive mode. It's in a mode that is sort of reading somebody else's gestures and cues and aligning with them, which is very interesting. But it doesn't have the impact; it doesn't take the brain machinery that disagreement does.

8 Disagreement versus social harmony

Tamas: My question is what impact does this have on those who constantly agree vs disagree? Driving to New York yesterday, I had a conversation with my wife. We talked about how people can have real friends, acquaintances, and friends with whom we sometimes socialize, but have zero interest in actually socializing with them. Regarding this latter group, my wife made the point that regardless of if I have zero interest in socializing with these people, I am always in agreement with them and seem to have a good time. She's blaming me of being spineless, because one of these guys, for example, maintains some really radical views on certain things, and that I should be confronting the guy with my ideas. Why should I? I mean, what's the purpose of me, in a half an hour or an hour setting, trying to enlighten this idiot or convince him about the opposite view? I have been thinking, do I gain anything because I learn about the fact that there are all these lunatics around, so I need to be aware of that? But in the end, I may actually be misleading myself based on what you just told us, Joy. It may simply be that because I have no real interest in them, and hence, agreeing is the easiest way to get through such interactions, as far as my brain is concerned.

Zoltán: I noticed that my colleagues who are slightly unpleasant and constantly disagree get fewer duties than those who go with the consensus. Disagreeing can be a good trait for a scientist, but you need to learn how to handle those situations with respect and without losing your temper. We are now electing a new president for my college. This is a highly interesting process. I have been involved in 4 of such elections in my life. We are usually stuck with the person for 5—10, or even sometimes 15 years; these are long jobs. So, it is very important to see how that person leads a larger community, chairs a meeting or reacts in a confrontational situation. Of course, it's not happening spontaneously during interviews; we have to arrange it. Usually what we do is pick a person amongst us who will intentionally make somebody angry, and you see how the candidate responds when they have to draw a boundary, and they cannot cross that line. If they do, then there is a problem. I think that's good. Thank God, I usually don't have to do that! I usually have the question "What was your single most important academic contribution or discovery, if you have to select one thing?"

Tamas: Usually, when I talk to people, I get to know them a little bit around the dinner table. I usually ask them a question: "If you were to die tomorrow, what would be your last meal?" I'm sort of interested in that. Like François Mitterrand, he arranged his last meal the day before he died. It was a very special meal with unique French cuisine. It was a bird, a small bird that you have to kill while you're eating it. It's some really old school French cuisine kind of thing, and that's why he chose some sort of delicacy that is illegal.

9 "On-line" versus face-to-face disagreements

Joy: Tamas, do you think that the fact that we're doing all of this socialization in this new format online, do you think that it is changing our socialization in any way, our inherent inhibitions, or not?

Tamas: I do believe so. For example, when I need to say things, or when I feel like I want to say things that may be contentious, I feel much more liberated to say them because there's no one next to me. As soon as I'm not focusing on the faces of the individuals, I feel like I'm talking to myself, so I don't have that sort of inhibitory feedback from somebody's body language or whatnot.

Joy: But on the other hand, there's a limit on social cues about when to talk in this format. It's something that I actually want to study, eventually. We have these very well described social cues about when I talk and you talk, I talk and you talk. There's something we don't understand very well, but there's a readiness to talk and then there's a signal. Okay, it's your turn to talk. That's not so present in this new format. There's a bit of tension because sometimes, especially if you don't know the person, you don't know whether there is a little bit of a delay. I think that's causing a problem.

Zoltán: Maybe you could play around with that delay?

Joy: Well, I think so. In my group meeting this morning, we actually had a conversation about this. This is an experiment that we've had sort of in the wings to do for a long time. I do a lot of work with dialog, and we have the conversation of with the speaker and listener switching by the experimental design. Each person gets 15 s—you talk, then you talk, then you talk, and talking and listening, talking and listening—and the experimental paradigm constrains exactly when that is, but we've always thought that we'd like to do what we call the "free speech" experiment, which is to let people have a conversation without those constraints. Then we could look at the components of the activity that occurs prior to the change in speaker and listener. There are several people who have modeled this as sort of an accumulation. You reach a threshold, and then something goes off, or some kind of a specific circuit hits a threshold, and then you jump in. It's a dynamic that has never been objectively looked at, so I thought, well, maybe now is the time to do that, when it seems that it is impaired in some way.

Tamas: For example, it is annoying when you're talking to somebody, you are devoting your time to a discussion of something with somebody, and you note that the other guy is yawning. Perhaps you see them working on their phone, or you see their eyesight on something else, and then that's kind of turning you off from the excitement of the conversation. I was talking to my mother earlier, and my mother is calling me on my iPhone next to my computer on the laptop. In the meantime, I received an email, and then I open the email believing that she can't see it, but then she says, "What are you doing?"

Zoltan: Ah, your mother got you! But it is true. It was already the case I tell you, when we have friends over for dinner. They're constantly on the phone; they're always checking their text messages. I'm not talking about etiquette or that it's rude, but it's just disrespectful. In a way, it's disrespectful to the person who you are with.

Joy: Totally. I think it's generally considered terrible manners to do that. We have to kind of learn on the spot and let the social cues be our guide here.

10 Social "change blindness" and iPhones

Tamas: I remember there was a commercial some years ago, maybe it was a European commercial, in which a family is around the dinner table. Each of them is looking at their iPhone or whatever phone, and in the background, everything is changing. They replace the furniture, the pet, and eventually the children. Nobody notices anything. But I'm sure that these kinds of technologies exacerbate mental health problems even more, so if somebody has problems getting cues about when to talk and when not to talk, I think they must be completely lost.

11 Social cues for rituals compared to novel situations

Joy: Yeah, this is one of the problems with autism, that individuals simply don't have the ability to read the social cues. Not only do they say sort of inappropriate things, they're not good at greetings or goodbyes. Those things are very scripted in our normal sort of dance that we do that's sort of automatic, in times of greeting, times of discussions, times of sharing conversations and so on. People that don't get those cues are very impaired.

Zoltán: If you have a society where these formalisms are much stronger, some Asian cultures for example, then those societies may be more beneficial for those people. Those formalisms diminish the reliance on spontaneity in daily life, easing the anxiety associated with impairments in reading social cues. If you have to read the cues in real time in a dynamic way, then you're quite impaired, which seems to be the case in autism.

Joy: You know, in depression, schizophrenia, manic episodes, or certainly anxiety, social anxiety, for sure. So that's a good example. People just simply don't have those abilities, for one reason or another, to respond in a spontaneous way to other people. I'm just thinking that here, we have all become autistic. When we are faced with these this zoom stuff, none of us quite knows what the rules are, and we're affected by different social cues in different ways.

Zoltán: I speculate that the barrier to a kind of a verbal attack would be higher, perhaps in person it would be less likely. Now our meetings are three times longer because they are online. Isn't that interesting, when we have like, 50 people on a meeting? But I'll tell you why: because everybody's working on something else.

Joy: So, they are listening, but they are doing something else. That's interesting. My meetings are longer, too, but I'm not sure that's why. I was thinking that we normally have our Monday morning lab meeting at 10 o'clock; Normally, the whole lab taken care of by, the schedule for the week on the way, and we're done by 11 o'clock. Now, we are still talking at 12 o'clock on Zoom, and I honestly don't know why.

Tamas: But again, I think it's personality trait-related because my meetings are shorter. Now, the fact of the matter is, I have more meetings, in general much more in number of meetings than before, but the meetings themselves are not longer for me. One of the things I like about our new Dean is that she does not just meet to fill time. If the discussion is over, she finishes the meeting, which I like a lot.

But writing a book is a different species than a meeting… One of these days, I was thinking about these people like the person who uncovered Watergate.

Joy: Oh, yeah, Bob Woodward.

Tamas: That guy puts together a book in 2 weeks.

Zoltán: They must have ghostwriters, because there's no other way they can do it. They probably have a team of people in an office that are just writing, writing, and writing. Then they have another office of people who edit, edit, edit, and then another office of people that get it out. You know, they've got to have a machine to do that.

Joy: But how can you think that much? I mean, good writing is good thinking, and I don't think the brain can create all day long. Thinking requires energy, and the brain gets exhausted after so much thinking. I think thinking is highly under appreciated.

Tamas: Bob Woodward is on the TV a lot. He is kind of slow. I mean, it's virtually impossible that the guy actually could write those books on his own. He's a smart guy. I'm not saying that he isn't. But I think these people have engines that help them. They're gathering information. They're organizing it, I'm sure he edits, I'm sure he directs it. Did you know (I did not know for a while) that some of our esteemed colleagues, maybe those in academic leadership positions, actually don't write any of their papers, or grants for that matter? They farm it out to these professional writers. I find that to be completely eliminating whatever I can contribute to science.

12 Writing is creative

Joy: I think they're dreaming. I mean, we don't get to where we are to do something that we can just farm out to people with half of our experience, half of our degrees, half of our inspiration, half of our energy. It doesn't work. It just doesn't compute. To write these kinds of grants, to do the kind of work that we do, you have to do it. It's about you.

Tamas: It should be 100% about us. For me, in a way, the best example for it is Watson and Crick. They did not do a single experiment but took others' data and interpreted them properly.

Joy: Exactly. I sometimes find that putting together the bricks and mortar detracts from my ability to think, but somehow, I haven't figured out very well how to separate it out, and so I just do it. I just wrote this whole big grant, and in retrospect, it was hard. It took four weeks out of my life, it was just plain hard, but at the end, I shared it with my lab, and we're all in agreement. It's an enterprise that we've got going; it comes from the heart of all of us, and you can't farm that out.

Zoltán: I agree with you. When it comes to the creative process, for me, figures are the key. When I get my figures—figure 1, figure 2, and so on—I'm home free. In our publications every figure is telling a vital part of your story. You have to present them in a logical and transparent order. Sometimes I can assign the writing of the methods to others, or I can borrow that from things I published before, but I usually write all the intro and the discussions.

13 Plagiarism or not?

Tamas: The methods are an interesting point. Twenty-some years ago, I was working with the late Tony van den Pol. I was writing methods about standard electron microscopy for a paper. Mainly because of my English, I just took whatever I published before. Tony pointed out to me that officially, this is plagiarism.

Joy: I actually had a journal editor recently come back to me with a cautionary note. There was a paper of mine in press that wasn't even published yet, in another journal, and it had a very similar Methods section to what we had included in the submitted paper, for obvious reasons. They came back to me and asked me to explain the consistency between the two sections. It had a number of words in common, phrases that were alike, like when you list areas of the brain, and you're listing them in the same order every time. They eventually realized, of course, that this was not plagiarism, but I'd been caught by their rules and their system. They have a lot of people, apparently, who submit articles to multiple journals at the same time, and they want to catch that. They want to be sure that when an article is submitted to them, it is not submitted to three other journals at the same time. But. . .really. We don't do that.

Zoltán: I understand what you said, but officially, based on the legal terms, it is plagiarism, which I find absolutely ridiculous. It's like arguing that each time you make a car, every piece needs to be remade from scratch. You cannot use an automated method because then, it's plagiarizing one car. The methods are nothing. I mean, it's describing the hardware that you use.

Joy: But the journal came back to me with a counter argument, and what they said was reasonable, I think, that if you have published it, then you don't own it anymore. Another journal owns it, and you are using their text.

Tamas: As Zoltán said, that is ridiculous!

Discussion 7—3rd March 2021

1 Core temperature and longevity

Tamas: So, we are talking about Copenhagen, and the Nordic culture, riding to work and cold exposure, and how there is long-standing data to associate lower body temperature with increased lifespan. This has been around in rodent literature, and I think there is a human correlate of that, too. The lower your body temperature, the more preserved your tissues are, which kind of makes sense because it's simple thermodynamics. I believe that the less energy you expend, the longer you last, like a car in the garage will last longer. In fact, Zoltán's father had a car from the early 60s.

Zoltán: I think it is from the late 50s. It's an East German car, Wartburg deluxe with a two-stroke engine made by the East Germans. We couldn't buy anything else in Hungary, and Wartburg was already a much superior model to Trabant. We were considered as a 'bourgeois' family because our Wartburg was "deluxe," and everything was made according to the "highest specifications" with wood panels under the windows, the steering wheel had bone-like effect, and it had a metal disk in the middle with a knight holding a flag and a shield in his hand. My father used this car for decades and then he gave it to Béla, my younger brother, who showed the most interest in cars. He still has that car, it's been in the garage for a long time, and it's still good. So, it's just a proof of the point.

Joy: It sounds like it's a wonderful antique. It's probably increasing in value. Coming back to the temperature, if you are talking about the ambient temperature, that if it's lower, then you have a longer life expectancy. . . .

Tamas: Your point is very well taken. For example, look at children with obesity who are treated by placing them in a cold room at 8°C or 10°C for a prolonged period. If you lower the ambient temperature, they will start to increase their thermogenesis. From the perspective of longevity, if your core temperature is lower than mine, then statistically, there's a likelihood that you will live longer than me. For me, it is not that surprising because you're burning less energy, which then potentially adds to the preservation of cellular and tissue integrity.

Joy: It just occurs to me, you know, that women tend to live longer than men, about five years on average. I'm just wondering, we are smaller, and we're always colder. Just ask my husband—he's like a polar bear! He is always hot, and I am always cold and we say it's the surface-to-volume ratio problem. The fact that women live longer than men, might be due to just being cold?

Tamas: How about potential sex differences regarding the impact of stress, for example, in a relationship, as a contributing factor?

Joy: Maybe it's more environmental and cultural. But anyway, I'm just throwing this out as a possibility, that it's related to the temperature hypothesis.

Tamas: But consider that when you feel cold, that means that your temperature actually is higher than normal.

Joy: Some new data that we have, because of COVID: we have a rule that whenever one comes into the laboratory, we have to take our temperature. I have these months of data, and my temperature is always low. I mean, it is so low that people ask me if I'm dead! So the data may not be very good, but relatively speaking, we have temperatures from me and all the other people in the lab, and my temperature is always lower than theirs, generally. Now, it may be an artifact because I walked to work, and it may be because I do this after I come in from the cold. I mean, the data may not be good for all kinds of reasons. We were doing this for legal reasons, not for scientific reasons, and we don't have the proper control mechanisms in place. Nonetheless, over time, it is an observation that my temperature is low.

Tamas: Do you easily get a fever when you are sick, or not?

Joy: No, no.

Tamas: There are some people who very easily get an elevation in temperature, while there are others, like myself, who don't have changes in their core body temperature under any conditions, even when sick. Temperature is one of those things that are tremendously important in tissue function.

Zoltán: By the way, Joy, are you going to analyze the data you collected on temperature from your group during Covid regulations?

Joy: No, wasn't going to analyze it. The truth is that we'd have it there just as a screening measure for COVID. We have to document for the EHS that we are following rules and regulations. My documentation requirements say that everybody who comes to my laboratory demonstrates that they have a normal temperature. I wasn't going to analyze it; I was only going to save it, but perhaps we could analyze it. But there would be all kinds of problems; there's the within-subject problem, there's the cross-subject problem of variances that are going to be different because of the environments people are coming from. I think it's probably not analyzable because there are too many covariants that are confounding.

Zoltán: it would be good to see whether there is variability in the population and whether there are sex and gender differences.

2 Bio-markers for major depression

Tamas: I was going to say something about the gender or the gender difference. Last week, I had a conversation with our friend. We all know Jonathan Flint. Remember? Zoltán knows Jonathan from Oxford before he moved to UCLA. Remember Jonathan, Joy? We were in Santorini with Jonathan; he was an organizer of that meeting. He has been interested in depression. He has gained interest in mitochondria and depression. He's interested in genetics of depression, and he was looking at cohorts in China. I asked him a question: is there a reason to believe that major depression can be the cause of single gene mutation? I'm not a psychiatrist, and I'm not well-versed in nomenclature, but major depression is a clear-cut category. Is that similar to morbid obesity, which can be referred to as somebody with a body mass index of higher than 40? While obviously not an absolute value, BMI is calculated from objectively quantifiable parameters. Regardless, is there something like this, or at least remotely similar to this, in the way major depression is diagnosed? Is there an element in depression that is black and white?

Joy: Yeah, I'm not a psychiatrist. My guess would be that if you ask those card-carrying psychiatrists who would be qualified to answer the question, they would probably say yes, and one of the reasons I say that is because the DSM has five codes for major depression. They're very well-defined, and from the point of view of a non-psychiatrist, I think that's a qualifier for a black and white, yes or no answer.

Tamas: So I asked him back then in 2013, and I asked him last week again, and if you know him, Jonathan is very categorical, and there is not much space for debate. I asked him, if there is such a thing as major depression, then would you anticipate that you might be able to find single gene mutations underlying this kind of depression? If you were to visit cultures where populations may not be touched by Western civilization, and you find people sitting in the corner for no good reason, but because that's what happens in depression, then you could find those genes. Single gene mutations of obesity were found originally by chance in remote populations in Turkey and Pakistan, and not through GWAS. It was by going through very specific subpopulations that they identified morbidly obese kids that were out of the ordinary. Just to make my point here about my angle to the brain and this book, the underlying cause of morbid obesity of these kids was overeating, which is a brain-driven behavior. Guess where the gene responsible for these phenotypes was found? The fat tissue as discovered by Jeff Friedman's lab! But he said no, there is not and there will be no such thing in major depression as a single gene mutation. But I don't know if this issue has been explored appropriately. He did make an interesting point, though, which was that depression could really be a positive adaptation in certain circumstances. For example, if you don't have a sufficient amount of food for a prolonged period of time, then to be in that mood of not moving around, because you don't have energy and you have to save energy, sitting in the corner (reducing your physical activity) could be looked at as a positive adaptation rather than running around. If you approach it from that angle, then perhaps you may be able to deal with it in a different way. Then I told him (I don't know, Zoltán, if you agree with this or not) that my recollection from growing up in communist Hungary or socialist Hungary, depression was not at all really considered a disease in medical education or in practice. Thinking back, maybe it was because the society was so screwed up that the working class didn't have the understanding of what depression might be; therefore, it did not exist.

Zoltán: Even 20–30 years ago, it was not something that was considered or talked about, even in Britain. We came a long way in acknowledging mental health issues, but back then talking about depression was a taboo. Unfortunately, in some countries mental health is still considered a topic that shouldn't be openly discussed and many still suffer in silence. What about the manuals on psychiatric conditions?

Joy: I can only tell you that they are used, and they are the standard of care. Now, how much variation is there, your guess is as good as mine. There are variations in subjective evaluation. On the other side, there's variation in symptom manifestation on the part of the patient. So my guess is there's a lot of variation in symptoms, but enough distinction from normal function and other disorders for them to be at least listed in discrete categories. Remember, the way these diagnostic categories are formed is that there's a committee, and they identify research and clinical work associated with the disorder, they argue over what should be included, and then they come up with a final list of signs and symptoms for that category.

Tamas: That makes a huge difference in my view about the biology of these disorders versus how tall you are, how fat you are, or how heavy you are. You don't require a committee to realize how tall you are, or how much you weigh, and how much you eat.

Joy: Yeah, and it came into being as part of the movement of biological psychiatry. It's relatively new, Tamas; you probably know more about this than I do. What I'm telling you, I'm really scratching the bottom of my knowledge about this, but it's relatively new. I believe it was NIH mandated.

3 How do we measure the success of neuroscience?

Tamas: Well, but the reason I'm asking these questions is because this is supposedly the Mount Everest or the Chimborazo of neuroscience, how to deal with these kinds of things. The fact of the matter is that we know very little about how to assess what is there, and since we know very little, we come back to that fundamental principle: How else can we measure the success of neuroscience?

Joy: Well, so what you're saying is that diseases that are within the purview of neuroscience need tools other than neuroscience tools to understand the mechanisms.

Tamas: Research tells us that the disease may be a neuroscience disease, but the understanding of the mechanisms requires tools and concepts from other disciplines. For example, one thing that is amazing about the aging field (in which I'm getting most of my grants right now) is that they are reluctant to fund studies about developmental elements related to these late onset issues of aging and age-related diseases.

4 The big problem of understanding long-term cause and effects

Zoltán: When it is about chronic disorders and you bring in the developmental element, some people buy into it, or they don't. Unfortunately, science is not structured in a way that you have 30 years to follow things up. Every time you may have a major viral infection, or some kind of infectious disease, you use up your reserves of a particular set of neurons and you will not get more neurons later on in life. You might have now much higher chances of getting some neurodegenerative condition later in life, but these links with developmental harms 30 years ago were never really put together. If you had an infection of some sort that used up your dopamine neuron reserves in your substantia nigra, and then it eventually led to you developing Parkinson's earlier in life, then these links are very difficult to identify and put together, but there is some evidence for these kinds of links. If we only look at the last five years of somebody's life, we'll miss some of the markers that might come up 30 years before somebody has Alzheimer's. There could be a predictor of vulnerability.

Tamas: It is also development; there is a tremendous role for development. That's why I'm not clearly understanding why, if you keep a child in isolation—which obviously you can't do—but if you can do that with a mouse, you can do that with a non-human primate, I believe, that we will have major outcomes, late onset outcomes of those things, even in pregnancy, if you play impact pregnancy.

Zoltán: Unfortunately, there was such a tragic "experiment" in the Romanian orphanages in the 1980s. If you look at the very severe sensory deprivation these kids had, there are all sorts of changes in their brain and in fact lots of studies looking at DNA methylation and all sorts of things. It's horrible; the damage you can do with this kind of isolation is phenomenal. Our brain is built by using the internal and external signals we encounter during our childhood. If you have much reduced stimulation and you develop in isolation it has an impact on the rest of your life.

Joy: I think I completely agree with that. These long-term studies would find some connections, and how and when you manifest some of these conditions later.

Tamas: Along these lines, I raise another issue that I've been struggling with, which are epigenetic modifications. I don't understand that concept, to be very honest with you. I understand the theoretical principles, and once again, I'm being provocative here. Nevertheless, I'm still confused when I read papers about the methylation of this and that, because of cigarette smoking or whatnot. It is much easier for me to understand how a particular methylation process can be highly relevant for the whole body when it occurs at the time of conception, when the egg gets fertilized by the sperm, because then you have one cell, and eventually, whatever happens at that moment will obviously impact all of the cells of the emerging organism. But when you're telling me an adult exposure to this and that chemical during a random period of an adult subject's life delivers a predictable health outcome, I'm having trouble putting that in biological perspective.

Joy: Also, it's not just one generation of people; it is about two or three or more. I have never understood it. In terms of a mechanism, I've never even been certain that it was real. I mean, in terms of the evidence that I've seen, but I've been very open about it, because it's not an area that I pay a lot of attention to. But when I have paid attention to it, I've sort of looked at the evidence and said, "Gee whiz, is this correlation, or is there a mechanism that we should attend to?" So I haven't been really convinced that how credible it is, maybe that's totally wrong.

5 Conventional wisdom versus energetic naiveté

Tamas: Let me give you an alternative explanation of developmental alterations of brain circuits. Here's an example: we found that if we play selectively with these neurons in the hypothalamus during early development, you will have a completely differently organized midbrain dopamine system, which then came with very different behavior patterns and responses, for example, to cocaine. All of these things are happening in the human during the second and third trimester of pregnancy. Any metabolic manipulation a mother is exposed to would impact the brain via these hypothalamic neurons. It doesn't require any epigenetic explanation; you can simply have an explanation of how these circuits, while they develop, impact the other circuits, brain regions, and functions.

Joy: I'm trying to be really open about this, because these are very important topics. We don't have to be experts on these things to be able to discuss them. It might be somewhat refreshing and productive to have people thinking about things from a point of view of naiveté. I like thinking of naiveté as a creative resource. That is another place where conversations really help.

Discussion 8—10th March, 2021

1 Input to the brain is gaited and selected

Zoltán: In textbooks, we usually see a classic framework of the nervous system in which you have the input, something happens in the brain, then you have the output, but you have lots of things going on inside that constitute your intrinsic brain dynamics and brain states. That brain state will determine what sensory input you let into your nervous system and what perceptual awareness you will have in that particular state. I like this simplified diagram, which shows that what the nervous system can do and how you respond to environmental signals is dependent on your brain state. In some diagrams, you can also represent these brain states. I was attending a meeting in Les Treilles close to Nice online earlier today. I'm supposed to be in Provence right now. One of the speakers showed a figure where you could see the hypothalamus, and I was glad to see that they now put some emphasis on how hypothalamic activity influences the whole perception of what brain state you are in, as cholinergic, serotoninergic, histaminergic, dopaminergic... We could go on, probably. You can see the ghrelin or leptin pathways, and some of them are actually not even from the hypothalamus; they come from the body. All of this is contributing to establishing the brain state. Some of it is not related to content specific processes, and some of it is content specific. What was interesting is that the speaker was distinguishing the two states based on pupil size.

Joy: Interesting.

Zoltán: What do you think about that?

Tamas: Well, first of all, I have strabismus; I don't know how that fits into this. I'll be very honest with you. These are identical truisms that we are walking around. So, what do we mean by all this? Can you give more specific examples?

Zoltán: In sensory physiology, to see a diagram like this is already a very big step. I've been attending meetings like this for 30 years with sensory physiologists explaining their observations, and nobody ever really articulated that you could set your own perceptual threshold with these mechanisms. What I'm saying is that today I was in an online conference, and finally, I could see all of this conceptualized and some of this even quantitatively assessed. I do not think your strabism is affecting your pupil diameter when you have arousal because of an unusual stimulus or when you are frightened.

Tamas: Let me give you another example that is very interesting. I think it speaks to the points that we were discussing last time: how relevant it is that we're going to approach it from our three different perspectives, how we perceive what the brain is, what the brain might be, and how it might be functioning. You know, I got involved a little bit with immunobiology. It has been looked at as one of the biggest accomplishments of human biology in the last 50 years, and then I get involved with immunobiology. I realized that the entire field was basically doing things like taking tissue and then FACS sorting its cells. It's like taking the brain and smashing it against the wall, then starting it up and trying to understand how the brain works from that. That's what immunobiology is. So when I was introduced to this field by a colleague of ours, a smart guy, it's like he says that the earth is a globe, and they are actually all fascinated by that idea, like they never thought of that idea.

2 Thalamus "listens" to the brain and not very much to the outside world

Zoltán: For me, it was a revelation just how much we control what we want to perceive. Already, this is a big thing for sensory physiologists to understand that the retinal input to the thalamus is only about 10% of what is going into that structure; the rest is coming from the brain. This structure is not "listening" to the outside world—it's mostly "listening" to what is going on inside the brain. The input from cortex to thalamus is 90%, so the idea that you have this flow of input, and then something comes out is oversimplification. It's our brain that is an entity that has its own world, and our brain regulates its own state. It regulates whether you are fully awake or alert or not, whether you are hungry or not, whether you feel like exploring or not. These states determine your sensory perception more than anything else does.

Tamas: These arguments go back to what I was writing in my chapter, and I'm keeping it in my part, about Cajal and Golgi.

Zoltán: But what you wrote in your chapter is that it could be misunderstood; you describe the idea of continuity and contiguity within the nervous system in different ways. You almost say that Golgi was right with the continuity or reticular theory and Cajal, who emphasized the separation of neurons and their interconnections with synapses, was wrong. That is very different from the ideas that there are very complex global interactions in the body and in the nervous system that are not mediated by neuronal circuits and synapses. I think you shouldn't mix those issues; you use the terms implying that there is an electrical synapse, and Golgi was right. You should explain this is a different way. There is no doubt that the majority of the transmission in the CNS is via chemical transmission in adults. What you want to say is that Golgi was much more aware of all of the complex modulatory influences from the entire body to the brain, and also within the brain, and he proposed that all of these elements interacted along a continuum rather than in isolated circuits. You are making an argument about widespread interactions, not necessarily on the specific method of communication (chemical or electrical).

Tamas: Definitely, it's great. I see what you mean. I think we should incorporate it into our chapters when we describe our ways of looking at the nervous system. One thing that I was starting to elaborate in my writing is my issue with the somatosensory cortex.

Zoltán: So what's your issue with the somatosensory cortex?

3 The whole brain is somatosensory cortex?

Tamas: What I'm heading into is that the whole brain is somatosensory, which is basically what this guy is showing and implying with his way of looking at it.

Joy: I'm re-reading some old electrophysiology literature that was published from Columbia University around 1997. That was when they were doing awake, behaving monkey experiments. This little note in *Nature* was published by Jackie Gottlieb, who, by the way, is still in the Neuroscience Department at Columbia. She did this interesting study where she was recording from the lateral intraparietal sulcus, which is the somatosensory cortex in monkeys. It's that dorsal lateral area. Somatosensory association cortex is exactly what it is. So she's recording from there, she has her behaving monkey, and she's measuring eye movements. The monkey would respond to a salient object, and she would get a response, but then if that same object wasn't salient anymore, it didn't mean anything to the monkey, so she got no response. What she connected to was this somatosensory cortex area that responded to stimuli that had been acquired by the visual system, which brought information into the brain and was sorted according to what was relevant. Yes. That's an example of the brain being a whole somatosensory cortex. It sorts things by what it wants, what it needs, and what it's related to.

Zoltán: There's so much of sensory cortex. Absolutely true. It is very difficult to define what the somatosensory cortex is, because it's not just your skin—it's your joints, your Golgi tendons, and all your muscles... So, proprioception is not a trivial thing. When people start talking about the barrel cortex in a mouse, they say that, "Oh, it's such a nice, simple system," I always get upset, because if you think about it, it's not even a pure sensory system. That mouse is actively whisking. It's moving its whiskers. That's not a sensory task; that's a motor task. But then you have an internal prediction: Where will that whisker touch? How will it resonate when it's finished touching the object? How will it make contact with the environment? It's not a good example of a sensory system at all because active whisking is a more complex interaction between motor cortex and the Barrell cortex within the primary somatosensory cortex. In fact, in some species, like the marsupial Monodelphis domestica, the old literature keeps talking about a "somatosensory motor amalgam," and they can't distinguish between the two areas by recording. That's also supporting your point that some of these areas provide representations all together. If you think about the definition of motor cortex, Sherrington's definition of the primary motor cortex was that it's the area where you can stick an electrode and that stimulation will elicit a motor response in the animal. I must say, Sherrington's definition is not without sufficient merit even today. If you look at the origin of the pyramidal pathway in different species, only about 60% is coming from M1, 20% comes from supplementary motor, and 20% from primary somatosensory cortex. So, the point you made, that it's not just a sensory area, is completely correct. It's not as simple as just sensory; you have to calibrate all these motor and somatosensory areas as well.

Joy: So, along those lines, where is neuroscience moving in this issue of regional specification?

4 Functional specificity versus distributed processes

Tamas: You know, there are people, for example, who really believe that there is only one area that is relevant for schizophrenia or complex cognition, and it is the prefrontal cortex, and their entire life is devoted to that particular area. I'm not joking.

Joy: Is that narrow focus on prefrontal cortex ever gonna change?

Tamas: You know, it's such a simplistic model, it's kind of a conspiracy theory. It's a simple explanation of complex phenomena. It's very comfortable, so it's hard to debunk it because the human brain likes simple explanations for complex phenomena. So, it sort of fits the bill.

Zoltán: I have just read a paper that identified some enhancers in hominid evolution that specifically enlarge certain cortical areas so they become and remain slightly bigger. This paper actually surprised me because I am very much in favor of the idea that you have a blank sheet of cerebral cortex, and depending on what you use it for, your brain will allocate cortical areas according to a general overall pattern that you can adjust according to your individual needs. The representation will be proportionate for what you use, whether you are dancing or playing tennis, or you learn languages. You will have a differently divided cerebral cortex. This paper is very interesting because it's identifying some concrete enhancers that specifically enhance certain cortical areas in human so they are slightly bigger. They were probably selected, interpreted properly.

Joy: Oh, okay. Okay, great, maybe you can read it, and we can crank it up the discussion. There is an impact of subtle population stratification and selection prior to ancestry regression. So basically, you have these enhancers that mutate, and then they make rostral anterior cingulate or other cortical areas slightly bigger. Perhaps that helped us to adopt behaviors or ways of thinking that are advantages.

Zoltán: I have to modify my original position from 30 years ago and kind of accept that the cortex does have these strong gradients. No cortical areas are the same even from the beginning. Depending on how these gradients are shifted or amplified, you will have some susceptibility for the representation of these functions.

5 Functions assigned to brain areas?

Zoltán: You have some kind of a predetermination, although it is very rough, and if you have these enhancer changes, then you get a larger area. You start using these areas for novel functions, and you eventually become even more talented and better at performing certain functions. You might be selected for these behaviors, and that genetic variant will become more common in the population. I found it very stimulating. These results indicate that noncoding genomic regions active during prenatal cortical development are involved in the evolution of human brain structure, and identify novel regulatory elements and genes impacting modern human structure. The noncoding parts surely have a function, but this is going beyond that. It's saying that it's affecting certain parts of the cerebral cortex, not all parts.

Joy: Well, you know, we have all of these Brodmann areas, over 43 areas that are cytoarchitectonically very different.

Zoltán: Originally, Brodmann made these distinctions purely by anatomy. Different areas have different patterns of layers, and the patterns of layers are dependent on the composition of the cell types, their size, somatodendritic morphology, and dendrites. The subtle variations in these patterns do reflect the different circuits, and the different circuits perform different computational functions. I think it's been a mistake that people have assigned specific functions to those specific cytoarchitectonic areas. These areas contribute to different computations, and they can interact with many different cortical and subcortical areas. In combination, their computational expertise is modulatory.

Joy: If Brodmann's Areas only represented specific functions, then our repertoire of behaviors would be limited to 43.

Zoltán: If you think about it, the cortex is relatively uniform in its appearance. Historically, it has been allocated a six-layered numbering. The structure of the cortex is relatively uniform, very boring compared to other parts of the brain where you have huge changes between areas. Maybe that's the secret for the success of the cortex; you can allocate different cortical areas which have initially relatively similar structures to others to different computational functions depending on what you are doing, depending on what you spend your time on.

6 Does the inner life of the brain need a body?

Tamas: So then, another issue that we should touch on speaks to the concerns I have about reanimating any brain from an animal that has been dead. If you talk about the existence of an inner life, then all of a sudden, you have a mind without the ability to express itself. A brain without the body can't express anything.

Zoltán: I would even push that even further, and I would propose a hypothetical experiment.

Let's say that I swap heads with you. We just transplant our heads to each other.

Tamas: Yes, yes.

Zoltán: Who is that person now? Is it the body or the brain that will determine the personality, the temperament?

Tamas: I think it's a very good question. It has implications, maybe not to this extreme.

Zoltán: But this extreme is the one that people would maybe pay attention to the most.

Tamas: But if you, say (I'm going back to my favorite organ, the liver), if you do a liver transplant, to what extent will that person change when you receive the liver from someone else? The new liver is actually going to be part of you, not only as a liver, but also as whatever that contributes to your brain functions. One of the fundamental things of Nenad's invention, if you want to call it that, he's very simply keeping that isolated pig brain alive to provide input to the brain through the vasculature. That's all he does to bring signals from the periphery. There is nothing else to provide external contact, there is nothing in his advancement that relates to actual brain structure and function other than providing fuel for the brain or providing input. It is huge. That's what I'm saying. It's alive. It's huge. Yes.

Zoltán: But that brain that is now kept alive in isolation by Nenad has developed in that pig. It was wired up using all of the sensory motor experience that the pig had during its life, correct? And all that life experience, previous perception, motor programs are kind of still embedded in the connectivity and the whole infrastructure of the brain.

Tamas: If we move even further than imagining that we swap heads, now let's say these heads end up on different bodies with different metabolism and constitution. The problem is that you, Zoltan, and I have similar body composition.

Zoltán: Yes, we are both 'slightly' overweight middle-aged males with Hungarian origin. It would be better to do this head swap experiment between people with different ages, genders, constitutions, etc., so we could examine the effect of all of that. It would be interesting, because then you would have to recalibrate the whole brain.

Tamas: Well, there are three things that actually we talked about before. One was about Siamese conjoined twins, depending on how they are connected, whether only the heads are separate, or the bodies are separate, but there is only one head. Another one was about the neurobiology studies that they've done in animals when they connect the circulation of two individuals so that they get each other's blood, and then you see how the endocrine signals influence the other one. I don't think that this has ever been done in humans. I don't believe.

Zoltán: Bill Harris did some very interesting and relevant experiments in frogs. Bill did parabiosis, because one frog was producing tetrodotoxin (TTX), a sodium channel blocker that inhibits the firing of action potentials in neurons, and he wanted to see how the other frog would develop when the tetrodotoxin blocked the generation of action potentials in the other.

Tamas: Did the TTX affect the development of the other frog?

Zoltán: It did not. The development continued surprisingly normally in the presence of TTX, even in the absence of action potentials. It's a classic paper from Bill Harris, but I think when you do this head transplant, for instance, you would have to do quite a lot of learning to calibrate that brain to be able to communicate with a different body. This calibration would be, of course, metabolic, motor, sensory. Sure, everything.

Tamas: Yeah, it could take a long time.

Zoltán: Then the person, you don't know who that person is because the structure of that brain, the connections, the cytoarchitectonics, etc., was formed according to the life of a different person, but now you will have a new body with different musculature, peripheral receptors, different liver, gut and pancreas that will have to communicate with that brain. It is developmentally determined how the brain is structured, but eventually, whoever developed that brain would be the best suited to have that particular brain. That's why I'm saying that Tamas and I are too similar to swap brains because it would not make such a huge difference. We both speak Hungarian and English and we are both male, with similar age, although I have to say that I play a much higher-level tennis. Imagine if you swap brains with someone who has a different age, gender, mother tongue, or body composition! Imagine if you swap brains with somebody who speaks another language! Would you be able to speak that language once you swap heads? Probably, yes.

Tamas: I do not think that Joy is entertained by our discussion right now.

7 Brain to brain, body to body: how separable?

Joy: Well, it's totally fascinating. I mean, in a way I'm thinking, I'm not ignoring you guys. I'm thinking about what kind of ideas this thought experiment actually generates. It is pretty interesting because this is the intimate connection between brain and body that you're talking about.

Zoltán: Correct.

Joy: One of the essential aspects of our thought here is that the brain and the body are a team. And neuroscience involves somatic activities, behaviors, and it's part of the whole big picture.

Tamas: The further study of the conjoined Siamese twins I think that would be very interesting, especially if those Siamese twins have their own opinions, own temperament?

Zoltán: How would that affect the spirit and brain state of the other person? Imagine that one is getting angry, and the other is not. Would the endocrine signals through some shared circulation make the other person tensed?

Tamas: You know, let's just think about these experiments. How would the blood of one of the twins affect the other person? I don't know, what is the prevalence of Siamese twins? I don't know if it's even possible to envision that you could involve them in any sort of experiments like this. So it's all very hypothetical. I don't think this can ever be tested.

Zoltán: The exact paradigms and questions would have to be well developed. The level how the circulation, alimentary tract, liver or the central or peripheral nervous system is shared between the conjoined twins can vary by individual cases. Selection of these cases is vital for the interpretation of the results. I like what you're saying about how important the liver is to send circulatory messages to the brain. I don't think that's well appreciated, certainly not by me. But if the Siamese twins share the same liver but they have different brains, then you would have a hypothesis that many of those autonomic regulations of the functions of those two brains would be the same or very similar.

Tamas: Yes. Exactly. So you are looking for Siamese twins with and without two or one liver(s).

Zoltan: And with shared and nonshared vascular system(s).

8 The adaptable brain

Joy: I would leave you two mad men to pursue these experiments. I've always thought the Siamese twins' issues have always been really amazing examples of the adaptability of the brain. I mean, we can hardly imagine being tethered to another body. And yet, if you were born tethered to a body that you shared with somebody else, it would probably seem perfectly normal, so a brain that can adapt to that. I mean, in the interviews that I have read (which is not many, but nonetheless, they do pop up in the literature from time to time) of people who have been in the Siamese situation, they're quite adjusted to it.

Tamas: Yeah. I think it fits a general theme about the brain's ability to adapt, because if you are grown (go back to our experience), if you born into and grew up in a dictatorship, then for you to understand that you can actually exist without necessarily much stress is an entirely foreign idea for you. If you never lived in a dictatorship or something, you just don't understand how it is for people in those situations on the news. But then they can adapt. But that's not something that we really talked about. There is no other vision there, because you're born into that, so there is no reason to adapt to it. That's what it is. But what's interesting is when you come out of that, and Jorge Galan and I had this conversation, because he's coming from Argentina, he grew up during a dictatorship. There is one thing, for example, that for both of us is a complicated aspect of our existence in the United States. You need to make choices and things, like you go to a grocery store, and then you have to pick out from 12 different types of apples. In Hungary, there was one type of apple, and then you bought the apple, and you didn't have to compute anything else. That's a much simpler and less stressful existence.

Zoltán: The competitive world. We often go to Lake Garda in Northern Italy with my wife Nadia, because her family is originally from there before her parents migrated to Switzerland after the Second World War. We went to new giant supermarket Esselunga in Desenzano to do our shopping, and the choice of things was just so much that we just could not make decisions, and it took twice as long to do the shopping than we normally dedicate to such activity. So, we decided not to go to the supermarket anymore; we are just sticking to the local small "alimentari" shop at the corner. We are happy if we can choose from three types of tomatoes. You don't need hundreds of different types of tomatoes. Tamas, do you remember the sport shops at Nagykörös in Hungary? We both played basketball throughout primary and secondary school, and our brothers Béla and Balázs did too. When you needed new basketball shoes, you could only go to one shop in town and you could buy one type of Chinese canvas basketball shoes.

Tamas: The choice was either, yes, they had it in your size, or no, they did not. If you were really very lucky, you could buy it in blue, not just in white, but then you were really standing out of the team.

Zoltán: These Siamese twins think they have two gallbladders, but only one liver.

Joy: I don't know. I'm struck by this. It's very difficult for me to process.

Zoltán: It's hard for me to process that, too. For me one of the most striking aspect is that they have two hands and these hands are controlled by the two separate brains. When you do bimanual tasks, you have to work your two hands in synchrony and the two sides of your brain can send signals to each other to coordinate things. However, in these particular conjoined twins the two hands are moved by different brains that are not connected. They both receive visual signals about what the other hand is doing, but you do not receive signals how the other brain intends to move that arm. It's an intriguing thing. When you see how well they coordinate things, when they cut up an apple, cook together, or drive a car together, that is amazing. We are talking about two separate brains interacting to coordinate motor programs of two heads that are not connected in any way. They have two separate brains, so they both have to sleep their own way. How do they synchronize when and how they go to sleep? Look, I've been married for over 27 years with Nadia, but I just can't sleep sometimes in the same bed because I sleep so lightly. Can you imagine if somebody is always with you and they move or cough? I think they must have similar circadian and sleep patterns imposed on each other.

Tamas: Thank heavens they've got the same liver!

Zoltán: I think they have one liver, but two gallbladders. They have two separate hearts and stomachs, one pair of arms and one pair of legs. They have two left kidneys, but only one on the right side. Their spines join at the pelvis and they have only one set of organs from the waist down, with one pelvis, with one pair of ovaries with one uterus and one bladder. So, if they want to perform a coordinate movement, they have to constantly watch and anticipate each other and learn to do it together. They must be extremely good in anticipating what the other would like to do, but it is not through actual neuronal interconnections.

9 The "black box" and the interaction between brains

Joy: Yeah. This topic helps us to transition from the historical view of the "black box" brain to a view that that incorporates our interpersonal interactions, and what is that brain in relation to the black box. There are a lot of playful things that one can kind of do with that idea.

Zoltán: I was talking to John Kaas and Leah Krubitzer about how to write a good review many years ago on a meeting, and they told me that if you want to write a good review, then you have to state the obvious extremely clearly.

Joy: What is obvious to you? They are absolutely right because, you know, some things are obvious to you, but they're not obvious to everybody. You really have to work on it and present it in a logical, very logical sequence and give examples and take your time. I think if we all do that, we will have the word limit.

Tamas: You know, when I came to this country 31 years ago, I started to write a paper because my supervisor wanted me to write it. I remember that I followed the Hungarian way of thinking, which is a very German influenced way of thinking, which is you put down facts, develop a line of logic, then at the very end, you make the punchline. Then he told me, and this is a very practical Hungarian guy, he said in the United States, you don't go anywhere with that kind of stuff because people don't have the attention span to follow your logic. You need to make the point at the beginning, then build your reasoning about why you are making that point. I believe, now that I've been living here for 31 years, this is actually a better way to get to a large amount of people. So, when I started the opening sentence of my book—that nobody understands the brain and neuroscience sucks.

Zoltán: Once we finished our own chapters taking on the same issues by the three of us in three different ways, it would be good to circulate them around to each other. I propose that we each write a very brief introductory chapter to explain where we were trained, where we come from and what is the scientific background that is influencing our specific thinking. At the end of the book, we could insert each a summary paragraph explaining how the discussions changed our views. The structure would be an introduction, which we all three wrote, then we have the individual chapters, then the discussions and then we have the three individual summaries on the progress we made to change our thinking.

10 Three views of the brains and conversations to bring it together

Tamas: Yeah, yeah, that's the reason I'm smiling. I honestly, I really like this notion that we take these recordings, and they're one of the reasons I don't want to finish the book. I am really enjoying these meetings.

Zoltán: Finish this and we can write the next one. Oh, by the way, I shall need more money to retire. It was just announced that Oxford University is increasing our pension contributions. We will have strikes, we will have disruptions, etc. Moreover the policy and conditions of providing accommodation to official college fellows is also changing. The tax will not be affordable and I am sure many of us will have to move out from college accommodation to make ends meet. This could have an affect on the entire community in 5—10—15 years, unless something is done.

Tamas: Which I never understand, Oxford University has been ranked always number one in the world. Why is there uncertainty about this? I'm not diminishing this; I just don't understand what's behind that thing.

Zoltán: I think this ranking is done based on some selected metrics that may or may not reflect the esteem of the university. Is it good to work there? What is the H index of the faculty? What is the reputation of the employer? And if you know the metrics, you could very easily pump up any university by making the right moves. I love Oxford because of the superb colleagues and excellent students. I love the interactions in my department and in my college. The danger is that after the changes in housing policy faculty will not be going to college for functions if they have to commute from neighboring villages where they have to move out to afford a house. However, Oxford is an excellent place to work.

Tamas: And Cambridge is always second.

Zoltan: But Tamas, you have seen my lab. You have seen my department. Is it so spectacular? Some of it is run down. We have issues with the roof at Sherrington Building, rain is dripping through. The greatest asset of a university is the quality of the faculty. Oxford Anatomy and Physiology was ranked No 1 in the QS World University Rankings by subject 2017, 2018, 2020, 2021. We shall have to do some work on that roof soon to retain this position.

Discussion 9—17th March, 2021

1 Pandemic effects: is communication altered by online (Zoom) compared to face-to-face communication?

Tamas: I'm not sure who was telling the story, that if you are in a room, and there's an audience, and they're watching an object, whatever that is, if they put a plexiglass between the object and the observers, it's already a different way of perceiving what you're watching. Just the knowledge of having the plexiglass there, and not the plexiglass itself, makes a difference in the perception of the visual cue.

Joy: If you think about the zoo, when you are in the zoo. If something is behind a barrier, you don't consider that there is a potential threat is there. It can be money, food, whatever, if there's a plexiglass, the perception of that is different. There is some similarity between these examples and an online Zoom meeting, which the pandemic has forced on us. I think that Zoom is with us forever and the benefits of it as well. But it doesn't replace the face-to-face meetings. It's just not the same. And that's what's interesting to me. Like, how is it different? Why is it different? And what does this tell us about the neural circuitry that processes real people in these different environments? Like if we were all in my office now, would this be the same thing? It's not quite the same.

Zoltán: No, it's not. You're correct. It's like the difference between reading a book versus watching the same story on a film.

Joy: That's true. The book is always better. But it's a completely different process. And maybe there is not that drastic difference between having a conversation via Zoom versus having a conversation in real, physical space together. I think that the evidence we have suggests that it's really two different systems. One is, is the social system that is quite picky about being engaged only in in-person interactions. And the reason it's picky, I think, is that it's driven by saliency detectors that are actually driving the gathering of information. The eye—brain system is tuned to acquire and process information from real faces. But we don't have that kind of information in Zoom and I suspect that the encoding is different.

Tamas: How about making also the suggestion that when you are in person, then we also consider many of those cues that we respond to as human beings: threat, territorial behavior, any of those biological principles. They become part of the conscious and subconscious part of the communication.

Joy: It's interesting. There are these circumstances and probabilities that are very different. Like, you will not be going to walk in my door. My space is different from your space and yet we can have this conversation. It's a very different context on Zoom. I would like to know how the brain and why the brain interprets it so differently? Is that a question that can be answered? And how important is it? I think it is important.

2 Student admissions procedures during the pandemic

Zoltán: At Oxford University, the medical school had the tradition that nobody was admitted without a face-to-face interview. Until COVID, everybody, even from Singapore, had to fly over for the interview in December. We give all applicants four separate interviews at two different colleges with four independent panels, each panel has different composition, but they all contain some practicing clinicians. Now, after pandemic, it is suggested that we should just continue to do online interviews in future years even when the travel restrictions are eased. Admission offices are arguing that it's less stressful for the candidate. Financially it

is less demanding and it is much easier to arrange and run and it's less chaotic. Can you imagine when we do admissions, everybody's coming just for two days? They have to show up at the right place at the right time for the interview in a city where they have never been before. It's also very stressful for us, because it is done in a very compressed time-frame. I personally still prefer the face-to-face interview. I also know that many of the candidates like to visit Oxford, candidates make friendships and they keep these for the rest of their lives whether they study together or in other universities.

Joy: Well, it's, it's hard to know, what's the right procedure? We have the same struggle here at Yale. We did all of our MD and PhD candidate interviews this year online and also most of our graduate student interviews. The problem is the outcome.

How do you measure outcomes? Is one system better than the other in terms of how well the students do?

Zoltán: We can't measure whether we indeed selected the most deserving candidates from that particular pool. We can aim to do that according to the best of our abilities, but even four independent interviews combined with all sorts of aptitude and ability tests will not guarantee that the system is perfect.

3 How do we evaluate our institutions?

Tamas: Here, I have a fundamental, and unrelated problem with our institutions. Our universities ranked amongst the best educational institutions in the world. But I am not convinced that the metrics are really there to argue for that from an educational perspective. Taking the most advanced students from around the word and providing them with a place to learn and then show how great they will become, is not, in my view, a measure of success of the institution regarding its educational quality. It is like somebody bring an almost finished Mona Lisa to the home of Leonardo, Leonardo provides a state-of-the-art facility, others finish the painting, then puts his signature on the canvas. I would like to compare institutions by how well they take mediocre students and make them outstanding versus taking outstanding students, and just basically letting them navigate through the system, they are getting the name of the institution, that's all they need. That doesn't prove that the institution is actually outstanding.

Joy: This is such a complicated problem. But how you measure success of outcome is really key here. Is it number of publications? Is it number of patients? Is it the difficulty of your specialty, is that the quality of the institution that hired you? As you start putting factors together you can develop some kind of a system and then weight the factors. And then you hand it over to the mathematicians, and they do factor analysis on the variables. And then, from my point of view, it's curtains because the concepts tend to be lost and outcome measures are longitudinal so quantifying measurement techniques and strategies is generally not possible.

Zoltán: I hope that some of the students we train will change the course of medicine in the future. They will have the critical and original thinking not to accept dogmas and will find the contradictions in the inherited practices that are not based on scientific evidence. Not all medics will be able to do that, but I hope that some of ours will come up with new directions. It is incredibly difficult to give numbers to these parameters that are truly the measure of success. If you just look at the impact factor of a paper, or the H index of a researcher, the altmetrics or whatever effect of a publication, you keep on coming up with these new factors and factorings. We like these numbers because we think they make comparisons easier, however these metrics are not necessarily meaningful, since you cannot compare scientific contributions just with these numbers and you cannot compare universities with simple metrics.

Discussion 10—7th April, 2021

1 A book review

Zoltán: Gyuri Buzsaki's book came out and I really like that he has some autobiographical components.

Joy: Well, he has a lot of autobiographical parts. I think the whole book is mostly autobiographical. As far as I could tell, just sort of skimming through it all, I went through 12 chapters.

Tamas: I started to go through it, and I was reading it on my phone. One of the things that I don't like about nonfiction books, which I recognize is their very strength, is the amount of declarative statements you find in them. The author declares that "this is the way it is," and it's not the way somebody else "declared" it for this and that reason.

Joy: Aren't most of the books like that about an idea, and they lay down a chain of reasoning that if you actually follow rigidly, then you may come to the same conclusion and buy into that argument? I mean, maybe that's the purpose of a book. I feel like if I wanted to learn about Buzsaki and his Hungarian roots, this is the book I would pick up, but if I wanted to learn about neuroscience, this is not the book I would pick up. You know, he kind of flies around the various topics. He had a nice couple of pages on Weber, the Weber-Fechner law, and logarithmic relationships between stimulus intensity and perception. He kind of weaves that into his notions of neural activity and building networks, and it goes all the way to Kahaneman and Tversky. It kind of wanders around in this way, and in the middle of it you have interesting things.

Tamas: Is he challenging anything? I did not read the book through like you did. Is he giving you a new way of looking at the brain?

Joy: That was my concern. I have enjoyed his lectures so much, and I just think he's such a smart man, a thought leader in the field of neuroscience, but my reading of this book was a disappointment. It didn't come across to me as the work of an original master. It came across to me as a pendant meandering through his memoirs. It would have been better if he had disclosed the dialogs that he had with the Rodolfo Llinas, because he refers to having these lunches where they would get into fights, and those conversations must have been really interesting.

Zoltan: I have to confess that I have not read his book yet, but I listened to Gyuri Buzsaki when I was a medical student in Hungary during the 1980s. He gave a talk, and conceptually and in overall vision he was already head and shoulders above everybody. His thinking was really outstanding. His ability to integrate ideas and to see the bigger picture was equaled by no one.

Tamas: He's an outlier. Definitely, he's a once in a generation genius.

Joy: I totally agree, but this book, well, maybe I read it too quickly. Maybe I didn't do it justice. But when I read the book, I kept looking for something new, for something that would change the way I think about things. He puts a lot of things together, but I don't feel that the time I spent has changed my thinking.

Tamas: Now the other thing that I found interesting and positive, which we definitely don't have, is his warm recognition of the editor's role in the book, how the Oxford University Press editor was really very much involved with his writing of the book.

Joy: Once they get too much involved, however, the liberty of the author may be gone, and then you have to stick to certain things. I like the idea that we spontaneously come up with ideas. Of course, if you want to sell books, then you have to select a couple of ideas that you simplify. We have lots of ideas, but we have to realize how to go for the main message.

2 The neuroscience of yesterday versus the neuroscience of tomorrow

Tamas: I have a different point of view. At the time of the creative process, I really don't care whether the book sells or not. I would like it if when we are done with the book, the three of us find it to be a great accomplishment. I've been looking at all of this and thinking if that's actually the problem of neuroscience, any brain science, or maybe many other sciences: that it's really not possible to put declarative sentences one after another to come to a defensible story that explains the brain.

Joy: I had this idea as I was reading this book, and thinking about ours. I was thinking, his book is about the neuroscience of yesterday, and our book is thinking about the neuroscience of tomorrow.

Zoltán: I would like that to be true. Maybe "neuroscience of tomorrow" should be the subtitle of the book!

3 Laboratory architecture and creativity

Joy: Just give me a minute for a sidebar. When I first came to Yale, I designed my lab by taking out all of the walls, building open spaces, and having desks that automatically elevated so we could work standing up. When I designed this lab, the architects came back to me and said, "Sorry, Dr. Hirsch, but we can't do this. It doesn't have the look and feel of Yale. It's too open and does not have the proper academic feel." That's true. They actually said those sentences. I was able to take one of our architects aside who basically loved my design, and we put together a plan to appeal. He went to the architectural board at Yale and argued on my behalf. His argument was that yes, I've violated all the principles and rules of the Yale of yesterday, but I represent the Yale architecture of the future.

Zoltán: There you go! That's very true that the structure of a building and the logic of the layout do determine the kind of research and interactions you can have. In Oxford, most of our buildings in the science area were very unfit for this purpose. Look at the labs: they have their own tissue culture hoods, their own confocal microscope, their own benches, and very little interaction outside the laboratory. Some departments decided to cut the common rooms, so we don't even have a place to accidentally meet up, which frequently is the source of new ideas. Then we go back to our individual colleges, and we have lunch and dinner with our medieval historian, linguist, engineer or philosopher friends and we can have the most stimulating and invigorating discussions . . . I love the interdisciplinary nature of the collegiate system. We would need more of this in and between our departments.

Joy: We eventually got permission for the design and I opened everything up and even added a treadmill workstation. The irony is that the architects sometimes bring guests to my lab to "show-off" the modern design. That's what this reminded me of, in our book, that we're talking about the neuroscience of tomorrow. The brain isn't just this funny thing that we study in our laboratories; this brain is with us in our life.

4 Creative synergy versus isolated intelligence

Tamas: But there is a reason we ended up with these different viewpoints from Buzsaki. At the same time, I think you're saying that he had a very similar background, but came to a very different conclusion. It might be that the background doesn't matter that much. I think this way, and you, Zoltán, fall into that group. People who actually know a lot, you know, definitely Zoltán is one of those people who knows a lot. I know much less. I don't know so much. I'm not saying that in a self-deprecating way. For me, the most interesting aspects of life, to be honest with you, have always been to synergize with other people; not to pontificate that I know everything, but to synergize, so that you may have a completely different point of view or idea emerging with the combination of interactive minds. If we are interested in understanding each other, then a new thing may come out of that as opposed to thinking, "Well, look how smart I am. I will solve the problems on my own so people can see how smart I am." I reviewed a grant twenty-some years ago. The applicant was a seemingly smart and intelligent person who had been working in isolation. He studied a protein. He stated that this protein has similarities to an automobile. The reason he thought it was a brilliant analogy was because like the protein of his interest, an automobile also has a front and a rear. I am not joking. That was his point. Aside from the obvious shortcomings of analogies, this guy had been living in scientific isolation for decades. Some people, like this applicant, could be intellectually isolated because of geography; others choose such isolation because they really think that they are so special that they do not need interaction with others at their level of thought. In reflecting on my interest in synergy and this example of mental isolation, I must say that what Joy does is a crucial element in trying to understand brain functions. How does the brain work when in communication with another? This is an absolutely indispensable part of brain research.

5 The fundamental social unit and the collective behavior of ants

Joy: I am really interested in promoting a kind of neuroscience that looks at the interactions between brains. I keep writing notes. I haven't put it together, but you might remember a few weeks ago, we talked about the question, "What is a brain?" That has stuck with me, and it's sort of resonating with me, and the little answer that I just jotted down is that a brain is half of a fundamental social unit. I thought, okay, that's kind of my starting point. A brain is the other side of the dyad. I want to talk about it—brains together in action. There's a science around that. The scientist, the person who thinks of that, is irrelevant. The idea is what we want to promote, although that's one of many ideas. You guys have ideas of equal or greater stature than that.

Tamas: We talked about it, Joy, before Zoltán joined us. This is something that I have been fascinated by ever since I witnessed ants' problem solving around a pool in St. Kitts. This is an absolutely nonscientific "study." I was next to the pool, and realized that there were dozens of black ants carrying a larger dead bug. My assumption was that they were bringing it "home." They got to the concrete dividing wall at the edge of the stones surrounding the pool. This vertical wall must have been no more than 50 cm high. When they got to the wall, they started to carry the bug up. After a few inches, the dead bug fell off back to the stone pavement around the pool. The ants rapidly went back, organized, picked up the dead bug, and started to climb again. Now they carried the bug a little farther up, but then it fell down again. They made about 40 attempts, and eventually succeeded in carrying the bug up to the top of the wall, then dropped it on the other side and moved on. What was most remarkable for me was that of those 40 attempts to bring the bug to the top of the wall, every single attempt was a little better than the one before. That means that there must have been communication among these ants that made them reorganize and readjust their approach. Whatever that was, it resulted in an amazing synergy among these animals with no cerebral cortices. For me, this experience speaks both to the fundamental principles of communications between individuals as a bottom line of successful behavior of a species and to the notion that understanding complex behaviors may not arise from the understanding of molecular genetics of the cerebral cortex.

6 Collective intelligence of multiple brains

Joy: It's such a fascinating phenomenon, actually. How ants are organized and work together may provide the simplest example of the important process of interactive brains. It clearly could not have been the result of a single brain. It had to be the communication network of several brains, you know. I think there's a lot of fascinating stuff about ants, and insects in general. But what is one individual insect thinking when they do these tasks? Well, I don't know that it thinks the way that we think. I'm sure they don't. Edward Wilson, who was the proponent of studying ant colony socialization, never answered that question.

Zoltán: Have you seen videos of the murmuration of starlings? It's amazing how they form a cloud. This cloud is highly dynamic and it is constantly changing shape, position and sometimes the entire cloud moves to different directions. I could watch this for hours. What I would want to know is what are the rules guiding the behavior of an individual starling? What is that individual starling member of the large cloud of hundreds or thousands of birds thinking? When they form this cloud, I'm not sure what they think is, "Oh, I shouldn't bump into the other guy!" Perhaps there are some individuals within that group that do most of the decisions, the others follow. There were some interesting studies done on either pigeons or Canadian geese in which they put GPS devices on them and looked at the formations they flew. They could track who's the leader, who's the follower, how they changed their line up, and how they kept an eye on each other in formation.

Tamas: Another thing that I am intrigued by is the relatively low incidence of car accidents considering the number of cars on the streets. Driving relies on highly complex sophisticated communication between sensory inputs, higher brain functions, and very delicate and precise muscle function. All of these are being integrated simultaneously in hundreds of cars arranged in very close proximity moving at speeds that are magnitudes higher than humans can achieve on their own. There are accidents, but very few considering the complexities involved. This reminds me of the ants...

7 Collective intelligence without any brain

Tamas: Well, the other thing that I was thinking about today is the enormous impact of this pandemic we are experiencing. One can look at it from the perspective of how these primitive organisms are extremely intelligent. I mean, if the virus can override everything that humanity has accomplished (and let's put vaccines aside) in the last couple of 1000s of years, these RNA molecules, then you can also ask who is more "intelligent": them or us?

Zoltán: How do you define life, and how do you define "intelligence"? Is Covid 19 virus, a small single-stranded RNA associated with a nucleoprotein within a capsid comprised of matrix protein, a living organism? Are a few KB of nucleic acid with a bit of protein that can't do anything on its own intelligent?

Joy: That is well above my pay grade. The fact that there is so much written about all of that confirms that little is known. Our whole idea is that we're searching for new concepts, something different. I want to generate some new thoughts that are not already in the textbooks. We're all scientists, and we're grounded in the history of science, but that's boring. What's not boring is the science that has not happened yet.

Zoltán: We could have some maximum number of references. I want to include around 50 references into my developmental chapter.

Tamas: That is a good number. I don't like footnotes in books, either. Your reading is constantly disrupted by some specific point that you need to look up, either at the bottom of the page, or worse, in an appendix at the end of the book.

8 The "data-dump" and micro-views

Zoltán: Every time I have to review something for leading journals, occasionally for *Nature* or *Science*, they always send me these monumental and extremely boring "data dump" papers. Sometimes huge amount of work and resources go into these studies, very modern and new methods are utilized, but not much new biology is produced. The data can be very useful for further studies, but the studies on their own as presented does not offer new discoveries. Gero Miesenböck, one of the father's of optogenetics said about some of the applications of modern optogenetic tools is the many researchers, who "voyaged into the known, and discovered the expected." Still, these papers get published in high impact journals. Now what they've started doing is combined single cell sequencing with backlabeling, various viral tracing, clonal analysis to account for various modes of migrations. Very few laboratories can do all that, but if they do not discover something unexpected than they just confirm findings that were already known with more fancy methods. Sometimes I have to say to the editor: look, it's a huge amount of data. These papers are very useful and they will get cited, but is it good science? It's just a data dump. Sometimes these papers resolve the origin of certain structures, and perhaps their evolutionary and developmental origins, but they are not illuminating the function of that particular brain region. I think there are very few people now who do mostly development research and think about the function of the brain. Years ago, you had Pasko Rakic, Colin Blakemore, Dennis O'Leary, Carla Schatz, and others; they were also thinking about the function of a particular structure. There was an eye-opening event for me when I thought, "Okay, I really need to go to different meetings from those I usually attend." I went to one of my usual meetings that had the theme, "How unique is the human neocortex?" It was mostly on evolution and development. We finished the whole three-day meeting, we finished all the excellent talks that were on fossil records, on the DNA of

Neanderthals, on comparisons between reptilian, avian and mammalian brain development, and then we had one more summary session where everybody had to say where they would go with their research to resolve the uniqueness of the human brain evolution. I was expecting a really cathartic finale for the otherwise superb meeting, and I was hoping that somebody would come up with big ideas now. There were three or four very distinguished scientists, and they said, "Oh, if we understand the nature of the outer subventricular zone of cortical development and what the outer radial glial cells are doing, then we will understand the human brain evolution." It occurred to me that we are all caught up in exploring very small pieces of the puzzle and we are not willing to step back and look at the bigger pictures. I agree that the outer subventricular zone and the outer radial glia progenitors important evolutionary landmarks, I studied them myself, but surely we need more than compartmentalization of the cortical germinal zone and generation of more diverse progenitor populations to produce a human brain. So I thought after the meeting, "Okay, there's nothing to discuss here. We have to be even more interdisciplinary and provocative to get answers, or I have to go to different meetings."

9 What is your 5-year plan?

Tamas: I always hated the question when you are being interviewed for a job or you have to interview people being hired: Where do you see yourself in five years? I hate that question as a scientist.
Zoltán: You have to ask something when you are part of an appointment committee. You have to check their ambitions and scientific visions.
Joy: I just don't see how that is a measure of anybody's qualities if they can answer that question or not. In the meantime, people want to hire people who have vision and a direction, something that was driving their curiosity, which is a reasonable thing to ask of somebody who you want to bring on your faculty.

10 To plan or not to plan

Tamas: I agree with the sentiment, but I am not sure if the best measure is to ask where you want to be in 5 years. I don't know about this afternoon, let alone tomorrow or 5 years! Wherever the results take me, I will be there. Now, if you asked me what my trajectory was, that I can tell you, and it may be constructive because you can see whether I was driven by curiosity and if I could get somewhere. You will also find that we may have very different trajectories. That would be very helpful to ask graduate students, also. They would see that there is no predictive, one-size-fits-all recipe for success in research. You need to find your own path. We have a member of our faculty who pioneered and works on a very specific intracellular glycosylation pathway. It may seem to be a dry and boring enterprise in principle, but he eventually used it, asked a multitude of questions, and pursued them. He studied the white adipose tissue, the liver, brain cells, and cancer cells, what-have-you. When we were hiring him, senior people were concerned that he is "not focused." I, on the other hand, found what he was doing very attractive. He's dealing with a pleiotropic process that is in every single cell. I feel very close to that approach. You ask a question, and you get an answer, but that answer may not answer your original question. You might drop it because you are "focused." Alternatively, you can go after your results and see where they lead you in the unknown. So, I don't find this a negative approach. That is true discovery rather than career building, if that is your personality. In the meantime, I acknowledge and tremendously appreciate people who are very focused and are good at that. They have enormous impact. I just don't see a need to force that approach on scientists.

11 Function or structure?

Joy: That's an important point that might be worth writing a little bit about, the process of science. What are the qualities that will take us ahead? Focus may be an Achilles heel here; focus may not be what we want. Because what we are talking about, all three of us in our own way, is our integrative broadening of the scope of variables that we are including in our mission. I'm thinking (as I am listening to you talk) that focus, although it has been considered a predictor of success in the past, may or may not be the predictor of success in the science of tomorrow. Specially in brain sciences, I always start by asking, "What does this thing do? How do we figure out how it does it?" Structure is a detail.

Tamas: I had little respect for psychology for a long period of time. Don't forget, I was trained as a neuroanatomist. I mean, I was raised by this notion that the brain will be explained by how one neuron connects to another. I spent 15 years of doing that, but now my attitude has flipped in the sense that I do agree with you absolutely on function. For example, if you can solve the problem of schizophrenia through psychological means, I really don't care what's going on inside of their head. I mean, if you solve the problem, you solve the problem. What's going on inside of the brain during the same process may be completely different in different people. Similarly, different circuits can underlie the same behavior.

Zoltán: Yes, nobody can see any abnormalities or unusual features in your behavior, but maybe you do have completely messed up circuits in your brain. With diet, for example, you could alter a circuit's activity and function, and mask the issues even if the circuitry has some abnormalities. So yes, I agree. That's why also development is important. Basically, evolution is not centered on changes in the adult structure. Evolution is dependent on development that produces an altered adult structure, which will manifest in different behaviors, and then those behaviors will be selected if those are advantageous for that species. What we are saying is that the connectome of our brains might have huge individual variability; nevertheless, we may respond similarly to a cue coming from the environment or within our body, and produce an undistinguishable behavior with that different circuit.

Joy: This is one of the miracles of humans, that we are all so different.

Zoltán: Individual differences are one of the really big questions for the future, and it really is a question for developmental neuroscientists. What determines that you have cognitive issues or you become normal or a mega savant with some exceptional abilities?

12 What is "normal"?

Tamas: When I started to interact with people at Yale Child Psychiatry decades ago, we talked about various diseases, such as OCD and Tourette's syndrome. Anytime we discussed them, I realized—but I did not vocalize—that I had a little bit of each of these issues. I was wondering to what extent this is specific to me, and to what extent does everybody have some behaviors that could be considered symptoms of many disorders?

Joy: Yeah, we are all on many spectrums. Nevertheless, despite all of the variations, despite all of the different wiring, most of us end up being basically "normal" members of a society; however, individual differences and how we get there are leading questions. The answer will involve many disciplines and many thought processes, and that's sort of what our book is about: bringing together ideas, disciplines, and techniques that aren't considered part of traditional neuroscience.

Zoltán: If somebody has some kind of brain damage that manifests in a movement disorder very early in life, then miracles can happen. If you are motivated, you have adequate medical care, you have proper interactions with that person, you can actually correct many of these disorders. There is an institute in Hungary, Peto Institute in Budapest that was started by Dr András Petö in 1945. The idea is to teach children who have cerebral palsy and cannot control their bodies properly to lead more independent lives by means of conductive education. This institute used to be very, very popular. They performed physical therapy that produced almost miraculous recovery, but I think people didn't realize just how much care and attention went into correcting some of those movement disorders. It doesn't matter whether the disorders are happening on the spinal cord level, brainstem, thalamus, cortex, or striatum level. That child or its family will not care. All we care about is that the therapy worked and the child can now have a better quality of life. This is how I look at our brains: we have all our own limitations, but somehow, we can compensate for that in many different levels. We do not really care on what level that happens.

Joy: And then you can call it luck or whatever else that with the same pedigree or history, one person can become a success and the other becomes homeless with potentially very similar baseline circumstances.

Zoltán: We were just talking about focus. In science, sometimes focus is a good idea when you are stubborn and you believe that you do have a very original and worthy idea to pursue.

Joy: Stubborn and focus are two different things in my view.

Zoltán: Maybe being stubborn sometimes is good. I'm sure that Cajal was very stubborn, and he kept pushing and working and focusing on the brain. Sometimes maybe if you are on the edges of the spectrum, maybe you come up with more original holistic ideas. Yes, maybe 90% of the time you are wrong, but maybe if you are 10% correct, it is sufficient to make an impact.

Tamas: I can also tell you, from my perspective, I wouldn't have accomplished anything had I started out in a neuroscience department. I would have been a complete failure. Why? Because my brain doesn't work that way. You know, I am not that competitive. My ability to regurgitate book knowledge is limited. You need to show off and say all of these things that you read in journals to be favored by the person who is running the lab. I was very lucky that I ended up in an Ob/Gyn department where I started working on the brain. That's where I grew up. Without that, I would not have become a researcher or scientist, and that's the point we have been making: different arrangements of the brain may respond completely differently under the same circumstances.

13 How do we teach creativity?

Joy: Your reference to the Hungarian Peto Institute is actually speaking to that point, that you can find a way to deal with things if you approach it the right way. All of us have been running laboratories where we're creating conditions that foster creativity, and in a way, that's what you're talking about. You found the conditions that allowed you to develop. Yes, you avoided the pitfalls of being told what to do. But there's a bigger issue here: how do we teach people? How we encourage people to be creative in the world of the new neuroscience is a serious question.

Tamas: It's very personal for me. My youngest son, who's 20 years old now, he always had a very different way of looking at the world. Just different. This classical school system we have was not really fitting him at the outset. Then we started to take him to child psychiatrists to evaluate him. Does he have dyslexia? Does he have this? Does he have these classical disabilities? He didn't have any of those classical things, but he does have something that makes him different in the way he approaches things. I think, actually, because he looks at things so differently, he should become a scientist. Eventually, we went through a year of evaluation, and they prepared a long narrative on him without any advice about how to proceed. They referred him to the best child psychiatrist in town. I'm excited, and I say to myself, okay, we go on to the next level. We go to the next level, we go into the room with the guy. He meets with our son, comes out, and we go back in with him. You know what he told us? Nothing. He just prescribed amphetamines. That was his response. I was stunned. This was not 100 years ago—it was a few years ago. I really felt underwhelmed and gullible. I thought they had made some strides in child psychiatry other than prescribing drugs. Well, they did not.

Joy: I think about it on a daily basis. I have a laboratory of people out there right now who I'm looking at this very moment, and it's my job to turn them out as the most creative people that they can be. Now, I think faculty mostly hurt people, and my first job is not to do any harm. But my second job is to, in some way, build them up so that they can find their own resources in a way that they have a joyful approach to things that make them creative.

14 The academic ladder and war

Zoltán: Well, Joy, this is very rare when we're talking about science and the lives of scientists. Unfortunately, if you think about the scientific career, I don't want to exaggerate, but it is like emerging from the trenches of the First World War. When you are a graduate student finishing your thesis then you have to run, and whoever is not shot down will get to the next step and will get a postdoc position. Then you have the next step after postdoc, then promotions and so on. There is not much mentoring or career advice or discussing alternative careers. Until recently very few take that part seriously. However, I think at Oxford, we do feel responsible for every one of our students. I try to keep in touch with them whether they have a career in science or in other fields. If I look back to my own career it was not a linear straight progression from the very beginning. My life was full of uncontrollable events, some things are the result of luck, the mentors you meet, and the rest can fall into place. My scientific career was more like the feather that was blown into different directions at the end of the Forest Gump movie.

Joy: Hmmm ... emerging from the trenches of the First World War and tenure. Hold that thought. Let's take that up next week.

Discussion 11—14th April, 2021

1 The Eureka moment

Joy: I had this really interesting experience this morning with my Yale undergraduate who's doing her thesis in my lab. Every year I have a few undergraduates doing theses in my lab, and she's been working with me for quite a while. She's been doing a lot of development, kind of derivative, figuring things out, nothing very exciting. We come to the very end, and her real experiment starts. This morning, we get the first results, and they're really cool. I've got one person looking at a movie while another person is looking at the face of that person looking at the movie, and we're comparing the brain activity. Can you share information about a face? So one stimulus is a movie, one stimulus is the face. It's very interesting. And so she didn't know what to do with it. She was like, completely flummoxed, like, "Oh, my God, my thesis is over, I can't figure this out!" And I said, "Stop. This is the first time you've done any creative work since you've been at Yale. Now, let's start thinking about what's really exciting about this. This is the best work you've ever done. Let's think about it." And she got it. But at first, because it was a completely unpredictable kind of thing, she got scared, because it was completely new. Even though these ideas we've been thinking of and talking about for a long time, as we've designed the experiment, we've developed all of the tools to get to it. Then she had to face these new data. And she got it. She's smart. But she just hadn't had that experience quite this way before. I love that moment.

Zoltán: But it's also a little bit disturbing, because well, she obviously gravitated towards you or towards research in general, because she's intrigued by the idea of research.

Joy: That's right. She's young, obviously, with very little experience. She has this naiveté from college and high school and whatnot that science is not incremental, but it builds on previous knowledge, and therefore, your new discovery is going to be coming from previous knowledge. Then, if you read a lot, you're supposed to have all the knowledge. You should not be surprised or should not be feeling lost. But this isn't true—we created something new. Now we have to make up a story about it. It was a nice moment with her because she did really get it.

2 How do we structure education for science?

Tamas: I wonder, in this regard, about the impact of schooling in general. For example, I hated daycare, I hated elementary school, I hated middle school, and I hated high school. I think it didn't fit my way of being, my way of thinking. Schools historically emerged to indoctrinate people to be part of a given society. I'm not saying this in a negative way. It is easier to navigate with them as a whole, rather than getting into chaos but it is inherently against creativity and the pursuit of the unknown.

Zoltán: During preclinical years at Oxford there is a very strong incentive to be creative, original and critical. Every medical student is doing an original piece of research in the area they select. However, when they move to clinical years, things change. You have to follow strict rules and established protocols. You cannot experiment with your patient; you have to follow the guidelines and stick to them. That requires a certain mentality and discipline to protect patients, but that can destroy creativity, originality and initiatives.

Tamas: It's similar to airplane pilot training. Pilots are screened based on whether or not they are paying attention to only one thing for 12 h. That's sensible, because they should not be distracted while they are flying a plane. But it is a trait that is not conducive for spontaneity and curiosity. When we do what we do, which is to pursue the unknown, by definition, you need to be comfortable with ambiguity and uncertainty.

Joy: You know, going back to what Zoltán was saying, this is evident in how we teach medical students, at least here at Yale, in the first two years. Those are their science years. Those are the years when students really have time to do their thesis project. They're required to do a thesis project here. But if a student doesn't finish his or her thesis project in those first two years and waits until the clinical part, it's hopeless, because the thinking is so different from your career perspective. In science, you break all the rules, you do the original work, you figure it out yourself, you pursue your own hypotheses. In clinical part of medical school, it is exactly the opposite.

Tamas: I had a student who came from finishing Columbia College. She took a year off, which many, many people do. I think most of them do now. She came to work in my lab. She was extremely good, very broad-minded. We were looking at the effects of C-section on brain development. It was a great study. She published the paper, and then she entered medical school. And you're right, you know, she became more and more rigid as she advanced in her medical studies.

Joy: But it's a very different experience. You're working in a zone with zero tolerance for errors when you're treating patients. You follow standard of care because that is what is known. But in science, you work in a maximum tolerance for error zone, because that's where the creativity is, well, you're from the perspective of the overall impact of your work on science at large.

Zoltán: It's not necessarily true as you move forward in your promotion process. Initially you do things because you have to tick the various boxes to move forward. However, later in your clinical career you can really go for the big questions again. Sometimes insight gained from clinicians can move a field forward and can influence where basic science should focus. It is important not to kill out the scientific initiatives from practicing clinicians.

Tamas: I have many fellows who come from Brazil. They are very talented and eager people. One of their traits, at the time when they join my lab, is that they read a humongous amount of literature. Then they want to pursue something that they just read about in Nature, Cell, or Science. And there's nothing wrong with that, but potentially, there are more intriguing aspects of what we do when you start asking questions that others are not asking. That is my modus operandi; that approach fits my personality the best. So, I start to tell them to read less, if at all. I advise them that if they read, regardless of the journal they are reading, take everything with a grain of salt.

Joy: Exactly. I've heard myself say a few times, "stop reading." Also, unfortunately, with the kind of reproducibility crisis, even if you read it somewhere, you would redo the experiment.

Zoltán: You're 100% correct. I cannot really trust anything. Even with the old literature, you have to redo the experiments nowadays, sometimes with better and more sensitive methods. I do a lot of anatomical research and I always want to see the key results myself.

Tamas: In fact, frequently when a new person comes to the lab taking over a direction that was started by somebody who just left, the new person, by design or by default, sometimes runs the same experiment as a control that was done before. Not infrequently, they cannot replicate them to the dot. While they are frequently frustrated because of these "failures," I'm actually quite comfortable with them. I would go further to argue that the anticipation that you should be able to replicate experiments to the dot is unrealistic. For example, we do measure how much an animal or a cohort of animals eat under different circumstances or genetic background. There is individual variability, there is within-subject variability, and there are multiple factors that impact feeding. If you were to show me an identical amount of food eaten by an animal from one day to the other, I would say that you are making up the data. What I believe is that our biological systems allow us to anticipate and replicate trends. That should be sufficient. As long as that's what you mean by reproducibility, I am comfortable with it. If you are asking that the results are replicated to the standard error, then I would say that you are delusional.

3 Objective measurements and facts versus objective measures and promise

Zoltán: Certain things are a bit easier than others. For instance, take physical development or a specialized anatomical or histological feature: a structure is either there or it's not there. Where a muscle begins and ends, that measurement is reliably similar, but if you started to look at psychology and behavior, that is different.

Joy: Well, that reminds me of a paper about the false promise of quick fix psychology. This is from the New York Times Sunday section. I read this thinking that it was talking about social psychology and the trend for people to take small amounts of data, then blow it up into some big fad. That becomes popularized. Then people think that it is science, but there's very little evidence of any kind behind it. So, this is not a new idea. We're all aware of this, and this happens in all fields. But I was thinking that one of the reasons this has been a little bit hard for me to do, the writing that I'm doing, is because I like to write research papers. I like my claims and my evidence to be hardcore, upfront, and statistically quantified. Here, where I'm writing a little bit more beyond that—I realized this when I started talking about the question, "What is a brain?" Because that's sort of a theme that I've been working on. I think, well, a brain is half of a social unit. Okay, so this is an idea. And it allows me to talk and talk and talk. But then I realized I'm moving away from my data. That makes me feel a little uncomfortable. I'm finding this, this discomfort in going beyond my papers and my data to what I say about it emerges because I do not want to put us in a situation where our discussions about these social things go beyond the data, you know what I mean?

4 The "bridge" between data, findings, and interpretation

Tamas: Yes, I hear you; I completely understand what you're saying. I think because of my upbringing, because I hated school, every single moment of it, I have much less of a concern about being a provocateur. I'll give you an example. In 2005, I was invited to a Keystone meeting. There was the emergence of interest in circuitry in the hypothalamus that I was involved with. I came to that issue with a plasticity approach. We showed how synapses were moving on and off neurons depending on the metabolic environment. We showed that it is a very rapid process. We had just published that paper, and many people were skeptical about it. Some of my peers simply wondered whether I had just fabricated that data. So, I opened my talk at Keystone with a quotation from Cervantes: "The fact is the enemy of truth." You can imagine how these skeptics responded to that talk. Perhaps not entirely surprisingly, many of those lead investigators now study this plasticity and keep on reinventing the wheel with new technologies.

Zoltán: I completely understand where you're coming from. At the same time, I think because of what you're doing, you are completely at liberty to make assertions from your information, even if it's remote from the data. You just have to make sure that you present the data clearly and rigorously, but then you can let your imagination fly and speculate. But you have to acknowledge that that is what you are doing, speculating.

5 From data to interpretation: a slippery slope

Joy: Yeah, they agree or disagree, that's a different thing. I'm writing in a different way. It's sort of like Thomas Lewis writing his essays in the middle of the night, you know, sort of talking to his subjects and elaborating on it. But I realized that I'm a little far away from my data. Like this morning, I woke up, I had this idea, and I was sort of adding to this idea of a social unit—what is it? Eye contact is a social unit; a handshake is a social unit. When do we create this dyad, what happens to our brains when we create this dyad? By interacting with another person, there's a moment of connection between the brains, the dyad engages, and then there's a following moment of disconnection. And during that dyad-on moment, we are different than we are in the dyad-off moment.

Tamas: Let me ask you this. I was thinking about you and all these things about interactive brains. Bernie Madoff died today. He ran, for decades, the biggest Ponzi scheme in Wall Street.

6 The power of a personal connection

Joy: He embezzled $20 billion. He was well-known in the philanthropy circles in New York. Everybody trusted him. He was a charlatan who everybody loved. He promised money to the major institutions, guaranteed 15%—20% return per year. He had a trajectory in the previous 20 years that he could show you: And it turned out to be entirely fabricated. He paid off people with the money that he got from the new people. But the scale was so big that actually he contributed to the collapse of the markets. It was so bad that it really contributed to the downfall of the New York economy. He had the ability to look you in the eye and persuade you to buy into what he was saying, even if he was going against common sense. What power!

Tamas: The dyad exploited! They live the system that you're studying, Joy. They instinctively exploit the system that you're studying because somehow, they figure it out. And they're successfully conning people.

Joy: It's certainly interesting how individuals are able to induce a sense of trust or, just by force of nature, convince people of things. We're very much influenced by these interactions, and it always happens within the context of the dyad. That's why I think a dyad is this very special social unit.

Tamás: You're correct. Because when they speak to you one-on-one—unless you are Hitler, who brought it to the level of the whole population, so he just made the dyad bigger, correct?

Joy: I think you are onto something.

Zoltán: You have some extreme cases for instance, when somebody was brought up without any possibility to have these kinds of dyadic interactions with anybody, like Mowgli in the jungle. I think that is an extreme case of being on the spectrum.

Joy: But it doesn't have the same machinery. The dyad moment with another human being has a *risk* of reciprocal exchange of sensory information. I see you. You see me? I hear you. You hear me? I feel you. You feel me?

7 How do we measure the ability for interpersonal interactions?

Zoltán: Do you have a neurophysiological fingerprint for that dyad moment? Let's say you put the optodes on me. Let's say that Tamas and I start to speak, do you think it's a possibility that you would be able to say, "Okay, these people are or are not connected?"

Joy: I use a dependent variable of coherence. I show the variations in coherence between the brain signals of two people during the dyadic moment, although I've never called it that. It's what happens during the interaction period versus the noninteraction period. I think there is a better marker but I am still looking.

8 The development of the ability for interpersonal interactions

Zoltán: What developmental stage or age is it when you can see the first signs?

Joy: Oh, I wish I knew. I wish I knew. My guess is it's from the first millisecond following birth, you take your first breath.

Zoltán: What does your mother do there? Your doctor is holding you upside down. Can you drive this system to say, okay, there is an empathy emerging there? If you were to interrogate those first milliseconds, probably you could come up with something that is reflected in that system.

Joy: Yeah, a theory of mind. But that's above my paygrade at the moment. Those are important questions. I feel like we're just beginning to tap in. The foundations of the fundamentals of the machinery of the dyad, and theory of mind, empathy, and so on, are qualities that are part of that interaction.

9 How are dyadic interactions affected by physiological variables?

Zoltán: It would be fascinating if you were to connect this with the measurements of certain other things like skinconductance, temperature in the palm, cortisol levels, sympathetic outflow, heart rate, and so forth. Heart rate can easily be recorded.

Joy: I am finishing up one of my autism papers now, and I'm relating the absence of the neural circuitry associated with the social condition, with a variation in eye movements. So the autistic people are looking, but as opposed to typically developing people, they do not zoom right in and hold it. What that means is that the eye is not acquiring the same information. In fact, I would propose that this jitter of the eye is obstructing the acquisition of the socially nuanced information, which may be why the brain shows that it's absent. The autistic individuals don't show activity in response to a face that's associated with typical social circuitry; they process the face, but the social system is different. I'm now trying to think about the relationship between the eye, the acquisition of information, and the neural activity. That's my job today.

10 The fear of public speaking and physiological variables

Tamas: But other variables, other physiological variables, do they correlate? I think that they would. I mean, it's my personal experience that when I was young, and would go to a meeting, and I really wanted to ask a question, but there was a large audience, it was a big deal for me to walk to the microphone. Really big deal. Then when it came to me and I was about to ask the question, my heart rate would go through the roof, and I could hardly speak.

Joy: Definitely. I always go back and forth. Even to this day, I write one word down, to remind me, just in case I forget what I wanted to say at the microphone. Just one word is all I need. Usually I don't need it, because once I start to talk, you know that it just happens.

11 How are dyadic interactions developed?

Zoltán: I know that feeling so well. If you are saying that these dyads connections have to be ready and that they are established very early after birth, then it has to be some innate mechanism that is generating it to get the substrate ready on which we can establish this dyad. Evolutionarily, it must have been a huge advantage to have this ready so you can establish it immediately after birth, or maybe even before birth. If you were dropped in the jungle, would you still maintain this ability?

Tamas: There are few somewhat anecdotal and documented cases of lost kids raised in the wild. The common theme in these cases was that they could never learn to speak and had difficulty learning to walk straight up on two legs. But the fact that they could not learn any language was an amazing thing, and it pointed to this thing about the closing of that critical period where this learning is a possibility or not. I always wondered about this language thing. Would you develop a language if you were in the woods? Alone? Would you develop a language? Apparently, you would not.

Joy: But that's been observed credibly on occasions where children have not been exposed to a language. They develop their own language. I think Stephen Pinker referred to this as the language instinct. That's a really interesting point. I think it sort of gets to what you were asking, Zoltan. is the ability to become a part of a dyad innate or not? Is there an instinct to connect?

Tamas: Maybe we do develop dyads with dogs. I told you I had dreams that our dog could speak Hungarian. . ..

Joy: These are fascinating ideas. I was just sort of playing with that idea. And then I was sort of struck by this article I read in the Sunday Times about all of this, these social psychologists just go out on a limb pretending to be credible, and they are simply not. I think there is a TED talk on this, that if you want to make an argument, you should make your argument with the strongest point and then shut up! If you keep going, then you dilute your argument, and people use this to their advantage.

12 Distractions and competing forms of input information

Tamas: Every time I watch a commercial for some kind of medication with my wife, we're surprised by the tools these adverts use. It's maybe specific to the United States that you see beautiful scenery, no matter what medication it is, and there is a couple, there is a man, and a psychotropic drug is in the background. You have this continuous talk about the various side effects, which most of the time include death. And it's actually quite loud. So you hear all these things, but in the meantime, you have this beautiful picture in the background. If you are in the right mind, and you really literally listening to what they say, why would you even bother taking that medication? The picture story "drowns" the spoken words.

Joy: We are absolutely captivated by a beautiful story. So they put this narrative—children playing games, somebody's driving a car, they're having a cup of tea, beautiful stories—and you get involved in the narrative. Your brain is engaged in that you literally do not hear the words. And the manufacturers are well aware. So they meet the letter of the law, because they must tell you about all the side effects that they are liable for: you're going to get diabetes, you're going to get cancer, you're going to get brain tumors, etc., but meanwhile, you're watching this bucolic story: a child catching a fish, apples in the apple orchard.

Tamas: And because I was dropped on my head when I was a child, I actually focus much more on this background information. Whenever I take a medication, I should really not read the side effects because I immediately have them.

Zoltán: If I know that the medication has side effects, I have them even before I take the medication. That's the medical student's dilemma. Every medical student has every disease ever learned.

Joy: But that's actually a topic of potentially huge interest. How does the placebo effect come about? This is the effect of suggestion. The placebo effect is when you think you're taking a medicine. Even if you do not, you still have the benefit, and it is a real benefit. It's not a psychological, you know, it's psychologically triggered, but it's a real benefit. To what extent does that rely on empathy?

13 Variations in dyadic connections

Zoltán: Is there data on the proportions of people in the general population who are able to and who are not able to establish dyadic connections?

Joy: Regarding personal interaction? Truth is, I don't know. I gave you the example of a developmental disorder like autism, in which probably the dyad at least is compromised. This notion of the dyad is really an idea that is under development.

Zoltán: What do you mean?

Joy: I mean, in my work, I study two people; hence, the fundamental unit is a dyad. It's not new, but has not had a large following in science. We do know that people vary a great deal in their ability to be sensitive to others. My guess is the dyad is not a black or white thing, but the truth is that there's a spectrum. There's a gradient, and also some suggestion of a gender difference.

Tamas: You need interactions between people, and as we discussed before, to have the fullest impact of the dyad, you need to be in communication face-to-face, physically, in real time. What about isolation, solitary isolation, or this time we experienced now during COVID?

14 The effects of the absence of live interactions

Zoltán: Most people would not do well in solitary confinement. That is also the worst punishment for children when you ask them to stay in their room alone, with no TV, no phone.

Joy: Yeah, you know, aloneness is what jail is. Aloneness is the worst thing that you can do to a human being. Zoltan, now you really made me think about something I hadn't thought about before. We are like magnets to each other. But when you establish this, then maybe it's easier to project it even if you don't have anybody there at the moment, an imaginary person. Once you have established this, you may find ways to protect yourself against isolation.

Zoltán: You remember the movie Cast Away, where Tom Hanks plays a FEDEX analyst who is stranded on a Pacific Island on his own after his plane crashed down. He draws a face onto a Wilson volleyball with his own smeared blood, and names the ball Wilson. Wilson is his best mate: he continuously talking to the ball during the rest of his time on the island. This this is a good example to illustrate that once we have established our ability to form dyads, we have a strong preference and desire to continue these interactions. Dyads are necessary for our cognitive functions.

Tamas: That is very interesting. I was watching an interview with Jimmy Hoffa from 1971. He was the head of the Teamsters Union, had some run-ins with the Kennedys, and was jailed for 2.5 years. He said the way to survive jail for those who are there for significant amount of time is to entirely block out their past life and their potential future life. Basically, lock out the outside world entirely or as much as you can. Then you can make it without going completely crazy.

Joy: I'm sure that somebody must have studied it in prisons. Well, I'm sure they have because it's so common. I mean, people get put into solitary for substantial amounts of time quite frequently. It's horrible. And now you have this COVID, in which the whole society was put into solitary confinement.

15 Pandemic-related changes in how we connect

Zoltán: COVID is a mixed bag though. Two years ago, before COVID, the 3 of us would not have been having these fantastic weekly meetings, even if it is via Zoom.

Tamas: To be honest, from this perspective, I actually like this COVID-19 thing. I very much enjoy the fact that I don't have to travel to see and interact with other people. I got addicted to travel because of work, collecting miles with airlines and whatnot, but now that I have not been traveling, especially overseas, I just feel much more relaxed. I hope the positive aspects of this period will become part of our routine life once we completely open up.

Zoltán: It is different from solitary confinement, because you can interact with other people: you still could walk outside and so forth. Through the internet academic life can continue. I lecture, I give my tutorials, write and review papers and grants, have my laboratory meetings. However, I miss the undirected discussions you have with your colleagues in your department. When I go to the lab, I really enjoy just stopping and talking to people.

Joy: Yes! My lab has been open for quite a while, but we're open partway, you know, still social distancing. I only have some students on one day and some students on another day. That's why the moment I had this morning with my undergraduate was more important to me because it's more precious to have interpersonal moments. It will be interesting to see how society will rebuild itself.

I wish we had ways of subjecting those questions to scientific enterprise. I do think it's important to understand because it really reflects on who we are and what we need as a species to be the best we can be. The other thing that comes with this, unfortunately, is going to be in our fields.

Zoltán: In virology, immunology, this COVID-19 has absolutely been dominating all the journals. While this is completely understandable, it will take probably 5—10—15 years' time to sort out what is correct and what is incorrect. Due to the timeliness and urgency of this topic journals published a lot of studies very fast and it will be important to see what % of these publications will stand the test of time.

Joy: That is happening in psychology, too. There are all sorts of claims from psychiatrists related to COVID. An isolation of this magnitude will induce changes in our social structure, and it will affect our behavior.

Discussion 12—28th April, 2021

1 The making of brains

Zoltán: I have just read in a paper that during neurogenesis in the human brain, leptin receptors are expressed in the cerebral cortical progenitors. That's incredible. Due to the changes in the nutrition of the mother, the leptin level can have a profound effect on what kind of brain you end up with as an infant, and could even have implications in adulthood.

Tamas: Definitely.

Zoltán: For me, it is incredible that these receptors are expressed in the neuronal progenitors. As you will remember, I expressed so many times that I'm fascinated that these circuits in the brain we always associate with the highest cognitive functions, such as cerebral cortex and higher order thalamic nuclei, express these receptors like leptin or orexin in a very specific manner. These receptors are in a position to change the state of the brain, change the context of the sensory perceptions; how sensory stimuli might be perceived.

Tamas: Respectfully, this is the point that I, on the other hand, don't find surprising at all. A primary reason for you to look around is to find food and maintain your ability to survive, so for me, it's completely logical.

Zoltán: Tamas, are you not surprised to hear that leptin receptors are in the progenitors!? When the brain is generated, these receptors are already orchestrating what kind of brain that person will have!

Tamas: Well, on that I have two points. One is that I'm not surprised about any of these things, in general. But the second thing, which we had a conversation yesterday with somebody, after a presentation related to Alzheimer's disease. Somebody raised the point that these days, people are doing this and that and then paper after paper, which reminded me of my experience in the Alzheimer's meetings a few years ago in Paris, where in every single talk, they somehow solved Alzheimer's disease through a very specific mechanism. They published papers in South Asia, blah, blah, blah, but nothing really has happened in the last 20 years to address Alzheimer's disease. So, my biggest problem with papers like this is okay, so what if you have the receptor there? Well, conceptually, the metabolic state of the mother affects the brain development of the embryo. It's been there forever, and this was published on top of everything.

Zoltan: I have just read a review in *Progress in Neurobiology* from Vasistha and Khodosevich (2021) https://doi.org/10.1016/j.pneurobio.2021.102054, who discussed some of these studies. To be honest, I didn't really follow this literature. Of course, I know very well that you have cytokines, which act on the progenitors of the developing brain, then you also have maternal stress that is acting on the developing brain, then you have malnutrition, under- or over-nutrition that also (such as toxic substances, drug abuse, alcohol, smoking) act on the brain. I also published in these areas previously. I was however really shocked to hear that the leptin receptors are on the progenitors themselves.

Tamas: Do you have estrogen receptors in the progenitors? What are the other luminaries?

Zoltán: Nobody knows. Nobody looked at this specifically, but it would be very easy to look these things up from the existing single cell transcriptomic datasets. Tamas, let's collaborate on something like this!

Tamas: Well, we've been talking about that also for a couple of decades. It's like the book, like Waiting for Godot.

Zoltán: That will be completely uncharted territory. You could make amazing progress by looking at the single transcriptomic analysis, which is now available on these progenitors, and all you have to do is to look at what kind of receptors are expressed. This is the approach Arnold Kriegstein and his group used to study AXL expression in progenitors (Nowakowski et al., 2016 doi: https://doi.org/10.1016/j.stem.2016.03.012). AXL is a candidate Zika virus entry receptor in neural stem cells. Knowing where Zika receptors are expressed is important to understand why microcephaly develops. You could do exactly the same with estrogen, leptin, orexin...

Tamas: Interesting. Yeah.

Zoltán: The brain is so willing to compensate for all the harmful effects it is exposed to during development that it doesn't always manifest in the phenotype.

2 Nature, compensation, and nurture: separate or a mix?

Tamas: Point well taken, and that's my personal problem with learning anything new about what impacts brain development in a very artificial system of a mouse. Probably many of us have been exposed to those things during our early development and childhood. Chemicals, traumatic, I mean, I grew up with my parents smoking in the car in the winter, and driving, you know, two, three hours, for example. Nevertheless, we didn't end up in the kind of situation that many of these studies would predict or projected we should have.

Zoltán: The brain has so many compensatory mechanisms, and many, many factors can contribute in almost chaotic ways.

Tamas: Correct. And, you know, simple model. Good. Yeah, that's an incredible problem for a reductionist. I had an observation that I made this last week that is related to something that we talked about. It's just something that you said about fetal alcohol syndrome and the particular facial characteristics—the flat face, the widened eyes that are very typical of fetal alcohol syndrome—and that about 60% of the males in prison, this statistic is quite amazing, seem to fall into that category.

Joy: Here in the United States, and I'm sure you followed it there. We've just gone through this incredible trial of the officer who killed George Floyd. Do we have any way of understanding the brain of someone who can actually do such cruel and unthinkable things to another person?

Tamas: So something just occurred to me, and again, I'm going to make a point here that may be completely inappropriate or wrong. Isn't that typical for me? There is some similarity between the way he was just isolating everything and focusing on nothing really by just keeping his knee on this man's neck without considering the issues associated with that, but executed the job. He was executing what he was trained to do. Like all the soldiers in the entire German nation during Hitler. They just did it. He didn't even bother stepping outside of the boundaries that they set up for themselves, and they would just go along with that step-by-step, minute by minute, second by second, they just did it. I mean, I'm telling you I'm making a bit of a jump. But that guy seemed to be in a trance in that time.

Zoltán: Officers in that particular state of the United States are trained to restrain people with their knee on the neck. They should monitor the breathing of the person, but this particular practice has to change.

Tamas: No question about that. But then, and I'm a naive person. I don't think the police just beat you with this stick, but I've never seen these kinds of things before until I saw this. To be honest with you, I was not aware of these things.

Zoltán: Maybe, Joy, you are right, that maybe there are certain jobs in society that attract people who have certain innate characteristics. That is true, may be, for military members. Maybe some people like the physical demands; they like the structure; they like to get and follow orders.

Tamas: Absolutely.

Zoltan: If they don't have that, then maybe they don't find their place in society. Maybe if you would look at certain jobs, you would also find surprises.

Tamas: Yeah, I think that that'd be an interesting hypothesis to get some information on. I sort of followed the trial, which was remarkable for me, for multiple reasons. I was mainly listening to it when I was driving to New York from New Haven. When you listen to the prosecution, his reasoning was very obviously straightforward and to the point. Then you listen to the defense, who took an entirely different part of the conflict, and would not refute the prosecution. If we started reasoning from a completely different part of the whole thing, then it would also be logical for us to put forward alternative ideas. That's not only obvious for this particular case, but in general, I find it very intriguing to listen to the prosecution and then to a defense. When you have good lawyers on both sides, they are actually making very intellectually intriguing arguments regardless of the actual facts. I am fascinated by how easy it is to confuse the mind.

3 Many competing points of views about brain and behavior

Zoltan: There are some deep issues here regarding mental health. You have a kind of a hardware that you inherited; you have the cards that are mixed up from your mother and father with all your genes; and then you have all these environmental influences during your upbringing. Then you have a certain limit. If you start your life with fetal alcohol syndrome or drug abuse, you have no chance in integrating to society without some confrontations. No chance, because to succeed in our society you need a certain level of cognitive function to consider and respect others. Then, when you can't fit in or you have a brush with the law, you still get prosecuted because your disability has not been diagnosed. That particular condition can be subtle, and only a specialist examination could reveal its existence.

Joy: One thing that was made clear in the Floyd trial Is that the police officer who killed him did have a history of reprimands in the police force. Apparently, he had been a troublemaker of this kind periodically throughout his time in the police force. There had been multiple complaints against him for excessive violence. And one statistic that I heard was that on only three of them was he ever reprimanded. That seems like a lot. Why was he still in the police force? So there seems clearly, there was a great deal of tolerance for this.

Zoltan: That's just maybe some further evidence for what I was thinking—that he came to the police force as a damaged individual who had a proclivity for these kinds of poor judgments, violence and bad judgment. Of course, all this is wild speculation, we do not have actual evidence. We do not know very much about fetal alcohol syndrome, but it just seemed like that's one of the behaviors consistent with it. I am sure an expert could check that, perhaps using AI. AI is very powerful in linking certain traits to some complex physical alterations, such as altered facial structure and proportions. It would be relatively easy to scan certain parameters of faces and relate these to complex cognitive conditions. These experiments might have been already done.

4 Predictions for criminality and a mechanism for fetal alcohol syndrome

Tamas: Cesare Lombroso was an Italian physician and criminologist who came up with this facial feature-based prediction of future criminality. This was eventually used in different ways, and not necessarily in positive ways.

Joy: Yeah. Okay, but fetal alcohol syndrome implies a mechanism and damage to the frontal lobe. It includes a syndrome with a phenotype, so it's developmentally associated.

Tamas: When you talk about fetal alcohol syndrome, what are we actually talking about mechanistically? Is that known?

Zoltán: Fetal alcohol syndrome is only characterized by fetal morphological and neurological abnormalities. We actually do not know the molecular mechanisms that lead to these abnormalities. What is sure is that alcohol interferes with neurogenesis from neuronal progenitors or neuronal migration and with facial development.

Tamas: No, no—it's just a phenomenology that we know of. It's caused by a high level of alcohol, but we don't know really how it happens.

Joy: That's a question for Zoltan.

Zoltan: Fetal alcohol syndrome (or also called FAS) went largely unrecognized until 1973, when it was described by Jones and Smith at University of Washington Medical School in Seattle. They noticed that heavy alcohol consumption by pregnant mothers during the first trimester caused pre- and postnatal growth deficiencies, minor facial abnormalities, and smaller and less convoluted brains in the infants. Where the pregnant mother drank alcohol heavily, it led to severe developmental damage to the developing brain, and that resulted in behavioral, learning, and cognitive abnormalities. Jones and Smith recognized that babies with FAS had characteristic facial features. FAS is on the most severe end of the FASD spectrum. It describes people with the greatest alcohol-related effects. The diagnosis is based on special measurements and findings, such as smooth philtrum, thin upper lip, small palpebral fissures, growth deficits, and central nervous system abnormalities. There are neuronal migration and neurogenesis program problems.

Joy: Do these have different severity levels so that you might use to them to estimate the seriousness of these conditions?

Zoltan: Yes, clinicians use these facial anomalies to estimate severity of the conditions.

Subsequently researchers did monkey experiments with alcohol. When alcohol was given to pregnant monkeys, damage to the frontal lobe of the infants was noticed. They tested frontal lobe function by showing an object, such as a plastic ball, that they hid under the table from normal and FAS monkey infants. Then they waited three seconds, and asked the monkey to find the ball. A normal monkey would not have problems looking for the ball, but the monkey with FAS did not show much interest or curiosity for the hidden object. It seemingly had no idea that something had just been hidden from it.

5 Placental effects on development

Tamas: That is assumed to be directly the effect of alcohol on the developing brain, or the placental effects?

Zoltán: We always assume that alcohol exerts its effect directly on the developing brain, but it's much more complex than that. It can affect the brain directly, including progenitors and migrating neurons, but it can also affect barrier functions, vasculature in the brain or other parts of the body. For instance, when I studied congenital Zika syndrome in a mouse model with Patricia Garcez and Helen Stolp, the brain, the retina, and the placenta vasculature were just as damaged as the brain. Structures such as the placenta or blood brain barrier are always ignored by most developmental neurobiologists.

Tamas: Are they ignored by you? I mean, not by you. I'm involved with an intriguing study with this guy to cross different species to generate hybrids. We've been doing it for 10 years now, and one of the things that we are looking at is how the placenta has a tremendous impact on whatever happens. So why is that?

Zoltan: We brain development people never look at the gene expression in the placenta. We never look at the barrier functions, although they could be responsible for certain abnormalities. There are only handful of brain developmental laboratories that extend their research on vasculature and barrier functions. Even if you find some laboratories that study the placenta, you can be sure that they do not study the brain.

Tamas: Let me come yet again from out of left field here. I was originally trained in the OB/GYN field, which I eventually learned is greatly underappreciated by academic medicine. Do you think that has something to do with the fact that the placenta is not considered in any of these issues, that OB/GYN is not something that is considered relevant?

Zoltan: I teach a class of 160 medical students, and I've done that for maybe 25 years. If you ask an exam question about whether the placenta is formed according to the genome of the mother or the baby, very few students will know the answer. The placenta is formed according to the baby's genome. Basically, if you have all sorts of genetic association studies looking for genes that could explain the pathomechanism, you must include the placenta in your study, and for a long time, we all ignored it.

Joy: Excuse my naïveté here, but that means that the placenta would be a readout of a huge amount of information about the development of the baby. Even if it were smart, even a history of the emergence of the development, I don't know if that is true. That is amazing. As a potential source of understanding developmental processes, I would think, someone who doesn't know really anything about it, but it just seems like that's fertile ground.

Tamas: I assume that there is actually a strong literature base on that from the people who work on the placenta, which is isolated from many of the other research areas.

6 The placenta to brain connection

Zoltán: We don't invite people to brain development conferences to talk about the placenta, but we should. These fields are completely separate. If you look at brain development, very few publications deal with vascular development, blood brain barrier, changes in the composition of the cerebrospinal fluid, metabolism, etc.

Joy: Isn't this the overarching problem that we are trying to address in this book? Important big chunks of information are not integrated.

Zoltán: Between the mother and the baby, you have certain metabolic pathways where some enzymes are in the mother, and some in the baby. There is some hormonal synthesis, which is why you need both organisms. It's not enough to have just the baby or the mother; they have to work together.

Tamas: I don't remember, but we can look into it. These are fascinating subjects.

Zoltán: When I was a third-year medical student, and we studied these metabolic interactions between the mother and the baby, I was really surprised that some hormonal synthesis shared between the bodies of the mother and the baby. Certain steps are done in the mother, and then transported to the baby to finish the synthesis. So basically, you have two organisms, and they are doing all these metabolic jobs together in symbiosis during pregnancy.

7 Social interactions begin in utero

Joy: Interesting. The beginning of social interaction, isn't it?

Zoltan: We need a much more holistic approach for brain development, and brain neuroscience, in effect.

Joy: Yeah. But that is really the idea of the book, right? I mean, one of the things that this book could do is focus on questions that could be addressed much more fruitfully by an interaction. I mean, this is a perfect example, across disciplines. It's obvious to say it that way, but again, one of the ideas here is that our disciplines are "siloed," and that one can find ways, such as conversations, that are a way to break the barriers of the silo. When you do that, there are new insights that come about, and one of the things that we can do in this book that we have through our conversations is identify those specific questions. This is one of them. It's a good example. Maybe conversations would be good way to break down some of the traditional barriers. I like this idea.

Tamas: Or the whole body, for example, from the perspective of glucose metabolism. You can assume that these higher cortical functions will affect your beta cells in the pancreas. That's true, but maybe if you listen to good music, it could decrease your diabetes.

Zoltán: Do you think that there is such correlation?

Tamas: Absolutely.

Tamas: That has been my argument that when I have a liver disease, if Joy is correct (which she is) her brain is impacting my brain, then her liver is impacting her brain, your brain is impacting your liver, and definitely that's the route.

Zoltan: Are immunology, metabolism, social and social interactive psychology all one discipline?

Tamas: Yeah. I mean, basically, probably, if one goes into this, you will realize that many approaches to life are based on experiences, based on these kinds of communications between various parts of the body and the different interactions with the brain.

8 Sleep, brain, and lots of other

Zoltán: This is also true with the brain and sleep. Sleep also depends on metabolism and the other way around. If you don't sleep, you will have all sorts of neuroendocrine issues.

Tamas: Yesterday, I was on the radio in New York, and they wanted me to participate in some XM Sirius radio program. They asked me about cannabis, but then we ended up talking about sleep. And absolutely, I think sleep is so fundamentally important for brain functions, but also for any other tissue function, and good sleep is dependent on environment. I've been a horrible sleeper ever since I was a child. But it's very important that people sleep well, and it seems like that is the one part of our human physiology or just mammalian physiology that is simply not well understood.

Zoltán: Many people have different takes on that, but there are emerging ideas about sleep that are based on; for example, cellular overload of certain byproducts and how you would like to shift them and have a dynamic fluctuation.

Tamas: I think there are many Hungarian researchers in this field—Borbely from Zurich or Obál from the Physiology Department from Szeged.

Zoltán: Obál and Jancsó, Benedek, Sáry, they all have made great contributions to this field, along with Borbély and Tobler; they were very influential. I think Balázs, your brother, worked with them while he was a medical student at Szeged.

Tamas: Yes, I remember him talking about them.

Zoltán: Sleep research is also changing how we look at sleep. Initially, everybody was looking at sleep as a phenomenon that is regulated by these deep structures in the hypothalamus that kind of tell the brain when and how to sleep. These sleep switch centers have been established from the time of Konstantin von Economo more than 100 years ago. Von Economo practiced in Viena and diagnosed many patients with lethargic encephalitis, an epidemic brain inflammation, which left some individuals insomniac and others somnolent. Economo performed thorough clinical examinations, and others performed autopsy and concluded that the wake and sleep centers were in the anterior hypothalamus and the junction of forebrain and brainstem. A switch mechanism between these centers is telling you when to go to bed. Now it's becoming very clear that the higher cognitive areas, including cortex, are also communicating with the hypothalamus and influencing these centers so they know when to switch from sleep to waking. A much more global picture is emerging. For instance, how you use your cerebral cortex during the day can have an impact on how you sleep. We cortical enthusiasts have to respect the hypothalamus, but the hypothalamus fans also have to consider the cortical inflences on sleep regulations.

Tamas: Absolutely. I agree with you 100% on that one.

Zoltán: If you ask people now or look at textbooks to find out what is regulating the main switches that control sleep, the cortex does not feature. Recently, I worked with Vladyslav Vyazovskiy's group on cortical sleep regulation. Vlad is an amazing sleep researcher who worked with Borbely and Tobler, and then with Tonnoni.

Tamas: I know them.

Zoltán: They also had contributions to the study of consciousness. With Vlad, they proposed that there is local sleep in certain cortical areas even while you are not asleep.

Tamas: How did you show that the cortex is regulating sleep?

Zoltán: In fact, we use the mechanisms that Michael (Wilson) and Jim (Rothman) and others discovered about synaptic vesicular release. I worked with Michael Wilson who discovered Snap25, a molecule that is a member of the SNARE complex and regulates synaptic vesicular release in synchrony with Ca^{2+}. In our experiments, we genetically removed Snap25 from a selected group of layer 5 pyramidal neurons across the entire mantle of the cerebral cortex. These cortical cells cannot release synaptic vesicles in an organized fashion. They are there, they developed connections, they fire action potentials, but they can't talk to their neighbors in a meaningful fashion. They have some unregulated spontaneous synaptic vesicle release. So, these mice that contain "silenced" layer 5 cortical neurons sleep 1/3 less than they would normally sleep (https://doi.org/10.1038/s41593-021-00894-6). Vlad told me that this is the greatest influence on sleep that he has ever seen during his career. So, now we want to find out what is causing this change. Is it because of short or long-range connectivity? Is it because of the lack of influence of these layer 5 projections to the hypothalamus? I have to emphasize that we don't touch the hypothalamus in our experiments, just the cortex.

Tamas: You don't touch anything in hypothalamus, and they sleep much less. You can argue that maybe the cells project down to the hypothalamus. From a very practical perspective, this is not necessarily shocking and surprising. I mean, when we are preoccupied with things in our head, we have tremendous difficulty falling asleep and stay asleep, especially if we're anxious, and it's more complex. Sure, but that does not necessarily hinge on the hypothalamus. I mean, I can evoke the single thought that I won't go to sleep, and keep myself awake. I can trigger a few thoughts.

Zoltán: We did not trigger thoughts; all we did was silence a portion of layer five projection neurons. Layer 5 is a major output layer in the cortex, and we silenced it across the entire cortical mantle. These animals sleep much less now. Even if they're sleep deprived, they don't need to sleep. I have to emphasize that the reduction is huge. Our mice now stayed awake for at least three hours more every single day! To put this in perspective, an average mouse lives approximately 2 years, which means that it gained three full months of wake time over their lifespan. In human terms it would equate about 10 years!

9 Sleep and longevity

Tamas: In the long run, you need to maintain those animals for a long period of time. Because in the long run, if you don't sleep enough, it's related to problems like chronic diseases, diabetes, cancer propagation. Are you doing those long-term studies?

Tamas: I'm talking about lifespan studies. Are you doing any?

Zoltán: Eight months. Is it long enough?

Tamas: That's a good start, but you should keep going longer. Mice can live for three years.

Zoltán: We can't keep the animals alive longer than one year in Oxford even if we do not do any manipulations to them.

Tamas: And why? Why is that?

Zoltán: That's the Home Office requirement. It has been recently extended to 15 months.

Tamas: Are you joking? You need to extend all these studies longer, especially if you are interested in the aging brains.

Zoltán: You brought up a very interesting issue about the age of the experimental animals we use. If you look at neuroscience publications, almost every study claims that adult is postnatal day 56.

Tamas: That's not adult. That's a very pubertal age.

Zoltán: The Allen Institute for Brain Research decided that's the age that they will use for adult gene expression, and now everybody is using P56; however, patch clampers can use as early at P24 for their work because beyond that age, patching might be more challenging for certain neuronal populations.

Tamas: Yeah, I know. This is ridiculous.

Zoltán: The whole electrophysiology literature is based on male P24, or if you're lucky, P56 animals, which are almost weaned and prepubertal. One person who is very critical about this situation is Pasko. He has been talking about this very passionately. He's making fun of all, almost all these papers.

Tamas: And he's quite right about it. I think things are changing though the neurosciences a bit more globally now, but I just wonder why it is that the Home Office doesn't allow you to have animals longer than a year. What's the problem?

Zoltán: I am sure that the intention is good and it is based on animal welfare, but because of this limitation, we can't bring some studies to a conclusion. What's the lifespan of a mouse?

Tamas: Three years, and we do have lifespan studies. I mean, it's a lot of work and a lot of money and it frequently goes nowhere because they don't die earlier or later. But to argue that you have to limit the lifespan to one year is not sufficient. My prediction is that if you perform this manipulation that has such a profound influence on sleep, then it will also have impact on the lifespan of the animals.

Zoltán: We do relatively little to follow early developmental insults all the way to adulthood. It would be interesting to study how, for example, maternal infections in these particular silenced animals would accelerate decline. When you have some kind of susceptibility, and you combine this with an insult, the outcome could be very different. For example, you might have to go through life with lower brain cell reserves and as you age some conditions might manifest earlier. This is why it would be so important to be able to follow up the ageing of these "silenced layer 5" mice.

Tamas: Did you look at that, by the way, the size of the brain and everything else?

Zoltán: We looked at the size, myelination, synapse numbers, soma numbers, degeneration, cell death and distribution and activity of microglia cells age of one year at the latest, not beyond, but I completely agree with you Tamas, we should in the future.

Discussion 13—10th May 2021

1 Bad things matter in our lives

Tamas: So you said something that reminded me of something that I heard recently from the disgraced director Roman Polanski. But back in 1972, he was saying how he grew up in Poland, and during the Nazi regime, he was in the ghetto, Krakow. Eventually, he was taken by a family to the countryside. He said that he was out in the fields doing some work, then all of a sudden, he realized that there were two Nazi soldiers shooting at him. The bullets were just going by him. In the long run, it didn't really have any impact on his life. On the other hand, when he was in the ghetto, he saw the Nazis taking an old lady out of the ghetto and beating her on the street, and that had a tremendous impact, a long-term effect on his way of thinking about life. It's the sense that something has happened that threatened your life, like you were in a plane crash, but you were not clearly aware of it, as opposed to you missing the flight, but watching the crash as it happens in front of you. Perhaps it has a different overall impact on you and how you deal with it. You would have seen the pilot crash the plane, but you would not be on the plane yourself. Right?

Joy: Right.

Tamas: When you conceptualize things like COVID-19, what happens when you have something like this pandemic, or when you have a mass shooting, when some people have died and others have observed? The impact of that is unpredictable. Then you start to think about how you could have prevented it, but preventing an event doesn't make you aware of the impact it might have had. Say you uncover the plot of a terrorist attack. If it's uncovered and prevented, it does not have the same impact on you and society as it would if it had actually happened. It doesn't change how you view life. You don't really learn a lesson if the bad thing doesn't happen to you. Do you know what I'm trying to say?

Joy: Yeah.

Tamas: So if you don't experience that event—and this goes back to earlier points we talked about some time ago—then intellectualization of a difficult topic doesn't really have a major impact on society or our own way of thinking.

Joy: Yeah, yeah. How we learn and what truly provides the Pavlovian type of reinforcers for us are questions that psychologists have sort of grappled with for a long time.

2 A well-hidden dark side: is it predictable?

Tamas: Yesterday, I was watching this show on a TV channel that talks about criminal history via reenactment. They talked about some of these serial killers who, in their normal lives, appeared to be absolutely innocent. They seemed very unassuming, and you would have never have thought they were capable of what they did. This guy was a baker in Alaska, and he murdered 20-some women over, I don't know, 20 years. When people do these kind of things, other people try to put together a picture of their lives from the past and look for clues in their childhood to see how they turned out that way, but they never prospectively identify serial killers that way.

Joy: No, it's all done afterward.

Tamas: So what is that? What is the ability to learn from those things? We cannot learn from the psychological reenactment of over mind. Why is it impossible to predict who ends up with certain traits or commits certain acts in their lives, such as a mass murderer, or who will be just a normal person?

Zoltán: When we do interviews for medical student admissions for the university, we have to test the suitability of the candidate for the medical profession. We have to indicate that we did not see any signs that would question their suitability. Now, how do you do that in 30 min? Unless there is some really substantial psychological issue, it is difficult to assess it.

Joy: Good question. What is the evidence that the interviews help?

Zoltán: We have a discussion and there is a practicing physician in the panel who is asking about some ethical issues, and then you probe the answers to those ethical questions and explore their reaction. But, you can learn how to respond to those ethical questions and you can influence the panel.

Joy: Yeah.

Zoltán: Let's come back to your original point on posttraumatic stress disorder. When Eric Kandel came and gave a talk in Oxford, he was asked about how a single event can generate everlasting memories. In his book, In Search of Memory (2006) he is talking about his own memories from his childhood and the biological mechanisms that store memories. He said that he still remembers when, as a child, they had to leave Vienna, and he is still traumatized now, over 80 years later.

3 Memory enhancement during traumatic events

Joy: I think that Tamas's question is maybe not well-posed, but really important. I think it's a very important question about those nurture events that modify our behavior, and how, for example, when I get to Eric's point, he has always known that emotional intensity enhances the laying down of memory. He'll talk about that, he uses that example, but never knowing what the mechanism for amplification memory is based on emotional events. I mean, you could probably look back on your own life, when you were scared to death, or there was something awful happening, you remember to this day every single thing that was going on around you at that time.

4 Trauma in medical school and in academics

Tamas: In fact, I was talking about this when I had to present somebody to a large Yale committee from our department for promotion a couple of weeks ago. The way it has always been is that I have to wait outside of a room, and this time, it was in a Zoom meeting. But if I'm present physically, I have to wait outside the room while the committee is discussing the case. Then I have to enter the room, I'm sitting in front of 12 people, they're all looking at me, and they raise questions. I had some very bad previous experience with such events during my studies at university. When I have to wait for such interviews, it is awkward because it brings back bad memories. I'm now the Chair of the Department, and it is not even about me; I am there to speak for the person, but I'm not provided with the questions ahead of time. All of this waiting the other day brought back some bad memories, and I told the committee afterwards. I've provided feedback to them. In the medical school system in Hungary, you had oral exams in this format every month. You wait outside of the room of the Professor, you go in, you get a question randomly, you have to work that out, you have to provide an answer, and then you get your grade immediately. The predominant approach of a Professor was to make you feel horrible about yourself. It was a very provincial enterprise. Some of the professors really loved this situation. Zoltán, in your case, you knew everything, especially anatomy, so you didn't have any problems.

Zoltán: I was also traumatized by the oral exams during my university years. I remember that we had to draw a card from hundreds that contained the exam questions you had to answer. You had to struggle a bit and make notes, then you had to discuss those particular questions with your Professor. I loved anatomy! Did you know that I studied from your Mother's old anatomy books during medical school? Some of the older editions of anatomy textbooks had better descriptions and more details, and they contained better illustrations than the new ones, so this is why I used them. Since they were out of print, I asked your mother, and she very kindly lent them to me. She underlined important sentences and also made some notes in the books and anatomical atlases. I loved your mother's notes in her books, and I also made some further notes that your brother Balázs, who also studied medicine at Szeged, inherited from me. He told me that he found my notes extremely helpful!

Tamas: Great, but you know these exams were very stressful for me. Were you not anxious when you went for these oral exams?

Zoltán: By the end of the 6th year, we got used to this stress.

Tamas: You know, I think that it's due to my upbringing.

Zoltán: We now have that same kind of anxiety waiting for a Wellcome Trust interview, or when you talk to an ethical review committee. The situation is exactly the same when you apply for a home office animal licence. You write up your 150 pages, and it's prereviewed, then you submit and you have to see an ethical panel. You have all sorts of people in that panel, including scientists, then you are interviewed by 15 people, and they can ask anything.

Tamas: Why do you want to do these ad hoc?

Zoltán: In that situation, they can ask you anything, such as, "Why is it necessary to do these experiments? Why do you want to use animals? How did you calculate the number of experiments needed? What is the benefit to patients from all of this?" You have to answer these questions.

Tamas: At least in our case, we got written questions if we do an animal or human protocol.

5 Pros and cons of peer review

Tamas: With this, I would like to then segway, ad hoc, to something I think we previously covered (or maybe not). I'm going to say something that you will not like, or you may not agree with. I don't entirely agree with peer review the way it is right now. The reason I'm not entirely on the side of peer review is because it's a subjective matter. There is no paper, as far as I'm concerned, that could not be scrutinized to the extent that it should be rejected. You can always line up arguments against the paper. The way it is right now, today, I am not a huge fan of.

Joy: That's an old issue that's been around for a long, long time, ever since people began publishing. Every time we publish a paper, we have a battle with the reviewers, there's just no question about that. It does take up a lot of unnecessary time. That's my problem with it now, how much time I have to spend dealing with reviewers when I could be allocating that time to other projects, when the matters that I deal with to satisfy the reviewers usually are sufficiently trivial that it doesn't make any difference.

Tamas: I am making the point, just as a conversation amongst us, that I don't know to what extent this process properly propagates new knowledge—and I'm talking very specifically about when they assign a paper of yours or ours—to potential random reviewers out of a field of maybe 1000, or 500, or 50,000. They each have their personal take. Again, our journal club as an enterprise is proof that there is no perfect paper, which is, it's huge. I think we all need to understand that none of our work is perfect.

Zoltán: On the other hand, if there wasn't some kind of peer review, some gateway, there wouldn't be any way of sorting through the garbage. It's bad enough even with peer and editorial review.

Tamas: That is true. I agree with you on that, as well. Not only do I agree with you, I think that now I'm gonna say something that maybe I have said before, or maybe I didn't. If you look at HHMI, people don't have to write a grant. They have a very specific goal, which is to publish four to six high-impact papers, Nature or

Science papers, in a five- to seven-year period. That's how they get their renewal. Now, of course, in the field, let's say I am there with multiple people. When they receive a paper to review from another HHMI investigator, by default, is their bias to support it? Yeah, they know that they are gonna get the support back. When you do these kinds of things, I don't want to say it's always quid pro quo, because it's not, but it is in a form; from a different perspective, you know, 30,000 feet up, it's a quid pro quo system. And the editors do the same thing.

Joy: Yeah, I think the big question that you're discussing is how do scientists promote creativity, future directions and productivity? What are the things we do that favor progress and advances versus things we do conventionally that impair it?

Tamas: I am raising the question that maybe peer review is on the negative side of that ledger relative to the more creative things of just being a little bit freer in what we publish when you discover or you come up with something new. The question is, to what extent does it propagate innovation and advancement in a discipline, and to what extent does it actually propagate the status quo?

Joy: Well, it's hard to say. When people have had legitimate arguments, I've found that my fighting back has helped me think it through a little more, or helped me provide evidence that has better face validity than what I felt that I needed. You know, a lot of what we do in science requires public approval. I mean, you're not going to get anywhere in life if other scientists around you don't think you're doing really good work. So validating our claims with evidence that others feel is both necessary and sufficient is an important part of the process. Then greed is a little social, as opposed to objective, but nonetheless, we are as we say, social.

Tamas: Yeah. Yes, I hear you. I agree, of course, I agree. I'm having this discussion myself.

Zoltán: Did you just have a paper rejected?

Tamas: No, no. In fact, I'm a coauthor on this paper. The author got it out in a decent journal. I have to tell you it's one of those things that if you read the paper, there's so much information there, a sophisticated approach, sophisticated words, but if you actually scratch the surface, there is nothing else. This is one of the things about science, that if you throw all sorts of technology, recent technology, a lot of information in there... First of all, a single reviewer is incapable of actually doing a thorough job, which is fine, there's nothing wrong with that—

Joy: But frequently, I think the more sophisticated the approach, the more sophisticated the combination of approaches, the more likely that reviewers do not understand.

Tamas: For me, that's just a further indication that much of the contemporary science, including neuroscience, is closer to art than to the classical idea of science.

Joy: Yeah, this is the way of science, you know, bad stuff gets left behind because it is not right. Darwin eventually has his way in this process. That is a good thing. The system has a way of self-correction.

6 Imagine if editors competed for papers to publish

Zoltán: 20 years ago, I had a vision about how to reform publishing. Things are going in that direction, but it's not quite as I imagined. My vision was that everybody can put their publications online as they produce them and when they are ready. This part is now reality in preprint servers such as bioRxiv. Then my vision was the journal editors have to select from these online deposited papers. They have to search these repositories themselves and have to decide which ones they want to publish in their journal. This would turn our position around. The journal editors would be competing for the papers. They would not just do the correspondence to invite reviewers and then decide based on the reviews. If they find a great study, they will have to fight for the paper, and the authors will decide which bid to accept. This would speed up publishing; the faster the editors look at a promising study, and the faster they submit their offer for publishing, the more likely that the authors of that excellent study will go with them. I think that would change the whole culture of publishing.

Joy: It is a very good idea. I never thought of that.

Tamas: I don't know. I did post some of our papers on bioRxiv, which I thought for a long time, I couldn't care less about. Then at one point, I started to realize it actually has value because when you put the thing out, people can comment.

Zoltán: It is really great what is happening with bioRxiv.org. If you post your study, it is out for your colleagues to criticize, and your competitors have a chance to criticize it, but they can't block or delay the publication.

Tamas: But your vision 20 years ago was going further!

Zoltán: What I would like to see is that the editors have to make a move and examine these deposited manuscripts and compete for their publication with other editors for other journals. That way, the authors are no longer in a submissive position where we have to approach journals one by one and wait for their decision which can take years. There are now certain journals that are in that system, such as *eLife*. Their attitude is closest to my original vision. You can transfer directly from bioRxiv.

Tamas: I know. One of our papers I really thought was good, but I could not publish it. We used the plant model to mimic the behavior of animals. We wanted to come up with schizophrenia modeling plants. That has not been being accepted. But I think what you're proposing is a really good idea.

Zoltán: This would solve all of our problems. I would still maintain the veto for authors. The author would say to the editor, "Sorry, I'm not giving my paper to you. I want Acta Physiologica Hungarica or Nature Neuroscience or Cerebral Cortex because it would get more visibility there."

Joy: That's a great idea.

Zoltán: The author would be in the leading position, and the editors would come and try to win our approval rather than the other way around.

7 Peer review: pros/cons and practice

Tamas: To be honest with you, it's an injustice or abuse of power, how we are doing all this reviewing for all these journals for free. They're making money. They're living off of us.

Joy: You know, it's really true. And we do all this reviewing for nothing. We all say, oh, it's for the greater good of the field. Yeah, exactly standard in the field.

Zoltán: You can spend your entire life doing nothing but reviewing for the community. Now I don't accept as many as I used to because I just do not have time to do them. And you keep getting these harsh automated e-mails that you are 2 days late.

Joy: I have the same problem.

Zoltán: You can get credit for your reviews on the website Publons. This is just a psychological credit; you can link it to your CV and have a warm feeling for having contributed to the progress of science—https://publons.com/researcher/1175051/zoltan-molnar/peer-review/.

Tamas: I think it's called Publons. Our goodwill is exploited.

Zoltán: When I received a note that I was the highest reviewing person from University of Oxford for a particular month in a particular year, then I just said that, okay, I should reduce my activity, because it is taking up a lot of time. Moreover, I am on the board of 15—20 journals, and you usually have to review for them when you are asked.

Tamas: Maybe your colleagues don't enter their activities into this system.

Zoltán: Most journals now have a link with Publons, and you can click whether you would like to get credit for the review.

Tamas: No, I never click that.

Zoltán: I try to regulate how many reviews I do a week. I only review what I'm really interested in. I don't want to review things I would not read if it is published.

Tamas: You see that the bias comes in immediately. Then you want to review people whom you know, you like, or who you don't like, for that matter?

Zoltán: Then some of these journals offer you to be identified at the end of the process.

Tamas: Yeah, but I am not interested in being identified.

Zoltán: I don't mind being identified; in fact, I prefer that. Sometimes my reviews are longer than the paper itself, and I work very hard to make constructive suggestions for making that paper better. I do not like that most journals just reject a paper without making genuine attempts to improve the study. Sometimes journals keep sending the paper back with very little editorial insight. Once, I wrote to a journal editor that if you send me this paper again for re-review, then I want to be a coauthor!

Tamas: That's the other thing that I have to tell you as a reviewer, I'm conflicted. Sometimes when you give the authors a fundamentally important idea, you're doing it out of the goodness of your heart, but you get no credit. So that's an interesting point. They should figure out whether the data set that you have is appropriate for the argument that you're making, but it's all up to the individual reviewers, how they implement that mandate. There's a lot of variation in that. So I stick to my principles. When I review, I write a paragraph, I don't get into this 24 points of things, because that's not my thing. I don't feel that it's my responsibility.

Zoltán: How do you get the specific response to your specific points?

Tamas: In a paragraph. If there are some major flaws, I identify them; otherwise, I either support the paper, or if there's one thing missing, from my perspective, I put it there. But I almost always make the editors give the author the choice to do whatever he/she wants to improve the paper. Yeah, unless it's really that I was the only one who thought it should not be published there. I said no. And nevertheless, they published it, which is fine. I actually am fine with it.

8 A paper review: the most popular brain areas

Joy: Well, those are all kind of difficult decisions that come with the territory. Zoltán, you sent a very interesting paper about uncovering previously overlooked brain regions. You said that you wanted to discuss it with us.

Zoltán: This paper is by Simpson et al., published in *Frontiers in Systems Neuroscience* in April 2021 (Volume 15, Article 595507; doi: 10.3389/fnsys.2021.595507).

Tamas: I wanted to read it in a bit more detail, but I just wondered, what is the main message of the paper?

Joy: Okay. The paper is a mini review. And maybe Zoltán can explain better than I can why he wanted it. It's an interesting paper that includes a *meta*-analysis of what brain regions are actually talked about, and the bottom line is that about 75% of the brain regions discussed in neuroscience have to do with the lower brain structures as opposed to the cortex. This paper claims that most of the brain regions that are discussed have to do with hippocampus, hypothalamus, striatum, cerebellum, brainstem, medulla, thalamus, substantia nigra, and the pons.

Tamas: And that's not bad. I don't think that's not necessarily a bad thing.

Joy: Why should all brain areas be represented equally? I thought we were really more interested in the mechanisms, not really the specific brain areas. Anyway, Zoltán, you sent this around. What are your thoughts?

Zoltán: Sometimes a brain area is fashionable because you have a charismatic person working on that area and publishing great papers. Due to these publications, this area is suddenly becoming fashionable, and then everybody runs to study, for example prefrontal cortex or claustrum or habenula.

Joy: Yes, and all of a sudden everyone is working on subthalamic nucleus.

Tamas: Yeah, correct. You are absolutely right. That is true.

Zoltán: When I read this paper, my immediate impression was that it is very recent that neuroimaging and computational neuroscience have made it possible to make unbiased assessments of whole-brain function and identify particular regions of interest that may have been previously overlooked. Scientists are like chickens—we run wherever there are seeds dropped into the field. There are very few scientists who will not be phased by that; they will keep digging in the same corner and stay with the same set of scientific questions that they consider very important.

9 Dejavu all over again

Tamas: I just had a conversation with Marcello a couple of hours ago, and he referred to a paper that is basically an overview of what I personally experienced in my own work over the last couple of decades: that we don't realize that people reinvent things that were known 10 years ago, 25 years ago, and repackage it and sell it as a new entity, or a new discovery. That has been going on forever, and perhaps will go on forever.

Zoltán: We avoid all of this by focusing on behavior.

Tamas: And, what does the brain do? Then that sort of leads us to the bigger brain idea that it's not just about brain, it's the brain and body, brain and environment, in context. The truth of the matter is, the brain is just a gigantic toolbox of little algorithms. It's quite skilled at bringing these little algorithms together to do interesting and novel things, and even old and habitual things. The brain pulls items from its toolbox together in an adaptable, flexible way to do things, things that we need to do—to eat, socialize, procreate, whatever.

Zoltán: For instance, in this paper, if you look at figure two, you have some kind of a behavior you observe, then you look at what cells have immediate early gene expression, you look at those groups or, if you're lucky, you can image it with other functional methods. You hone into those circuits. By creating different behavioral paradigms like stress, pain, and reward, you can look at what areas overlap and what areas don't overlap, and you can identify the anatomical substrate of that particular behavior. That is our current way of thinking.

Tamas: You know, it's very interesting. You made that point all the time for me, because when I was growing up here working on the hypothalamus, my biggest issue was that from day one, I knew the circuit that I was looking at was subserving multiple functions, like thyroid, reproduction, and stress, so actually, my entrainment is that all of them are doing multiple things. So I'm coming from a very different background, very nonneuronal circuit oriented. That's why I guess I'm coming to looking at things in a different fashion because you have the same circuit subserving multiple and parallel functions.

Zoltán: If you tweak some kind of a receptor stimulation, the very same circuit can have a very different state. You endocrinologists, you don't really get hooked on a particular neuronal assembly because you can change the state of these circuits within a few milliseconds, for instance, by changing some receptor activation of a particular element of that circuit.

Tamas: Yeah. But, you know, in a way, this all flies in the face of the main development in neuroscience, which has been based on the principle of functional specificity, that specific parts do specific things, very specific brain areas do specific things. And yet, what we're talking about here are ways in which specific brain areas don't do specific things, they do very general things. The thinking is, of course, that it has to do with what it's connected to. At the moment, this principle is so deeply embedded in the foundations of neuroscience that you see it in undergraduate textbooks.

10 Isolated parts versus the integrated whole

Joy: In fact, this can be referred to as the real estate principle, which is that parts of the brain do specific things. Examples are Wernicke's area, Broca's area, the primary visual cortex, you know, the color area, the memory area, and yet, it is such a primitive way of looking at the brain. This idea is very much related to the paper you passed around today saying that people don't look at enough areas of the brain.

11 The symphony analogy

Tamas: Correct, and coming to an analogy here, it's like listening to a symphony. You can start to say, what makes that symphony great? Is it that the person sitting there playing that instrument, or is it that they're playing with the musicians?

Joy: No, it's obviously the summation of all of those, and you may have a symphony sounding the same regardless of how the orchestra is assembled. I think we talked about the same thing recently. Zoltán is the one who brings the dogmatism with the divine in a positive way, including the idea that there is some sort of regulatory or whatever principle that allows these properties to emerge.

Zoltán: Yes, which is extremely important.

Joy: So, what would you call it, the principle of adaptability? A phrase that has been applied to this relates to "constructionist models": the brain can construct various sorts of orchestral outputs by adding a little bit of language, a little bit of math, a little bit of visual association, a little action, and you put it all together and get a constructed (or orchestral) output. But the word "construction" is a process in the brain. I mean, when we put the brain together, your brain and my brain and anyone's brain, we had a predictable algorithm or process of development.

Zoltán: You have a particular shape of the particular lobes of your brain; you have the thalamus, and the cortex. Initially, it's connected in a very predictable manner; there are some developmental constraints, and there are some kind of capacity and limits to what you can do or cannot do with your brain depending on the substrate; the genetic program will produce us spontaneously. Our brain evolved to be able to progress further by changing these developmental algorithms. Your entire life experience will be reflected in the brain that you produce. If you think about it, there is very little concrete specification of these behavioral algorithms if you go on to higher levels.

12 We are sculpters of our own brains

Zoltán: You have to carve out the particular organization yourself from your own brain during your own life. It was Santiago Ramon y Cajal who said that, "Every man, if he so desires, becomes sculptor of his own brain." This is true, to some extent. Everybody can shape their brain considerably during early development, and you can sculpt your nervous system according to your own individual circumstances, ambitions, and interests.

Joy: So, you are *actually* doing it, not that you *can* do that. It is part of our life.

13 Asperger's and brain plasticity

Tamas: So, look, can I ask something very basic? Over the weekend, Saturday Night Live was hosted by Elon Musk. I didn't watch it. I'm sleeping before that time. But, you know, not surprisingly, to probably most people who know of him a little bit, he announced that he has Asperger's. That's my question. I know some people with Asperger's. Do you know if that can be transitioned into non-Asperger's, so they're not masking it? I just wonder how, because one of their biggest issues is Joy's home turf, social interactions. Can they interface with the rest of us? Can they understand or not understand cues from the rest of us? My question is if there are any successful approaches to diminish or alter that state of these people and see how they can reflect on how he was back then compared to now.

Joy: There are people here at Yale Child Study Center who would provide a more credible answer than I can. What I think they might say is that early on, you can do behavioral therapy. That helps. But there is no way to undo the basic problem you are born with. You can assist developmentally, behaviorally, you can teach people how to be aware, but you don't change it. That dovetails a little bit with the kind of work we're doing. We've got a paper just going out now on autism where we've looked at the neural face-processing system during live interactions with another human and eye movements. The hypothesis is that the way information is gathered by the active sensing system of the eye drives upstream processes that may be defective in ASD. The suggestion in this paper is that this fundamental social deficit in autism, and by extension, Asperger's, is a low-level sensory disorder in which the information is not gathered, or if it is gathered, it is corrupted and that corruption is caried throughout the processing pathway

Tamas: So, it's low level?

14 High versus low pathways leading to social malfunctions

Joy: By "low level," I'm referring to visual acquisition. But when I say high level, I generally mean neural processes up here, low level being at the peripheral, the distal end of the information gathering chain. The lowest level for visual information is what hits the retina. But the retina isn't just a passive gather of information. There are the control mechanisms for smooth pursuit and saccades that are just darting around the visual environment all the time, they're under control. We know that a quantum of information is processed as it lands on the retina, and it's sent upstream. So the suggestion is that the information gathering system, it may be part of a bigger problem in autism, but seems to be defective. At least, that's what we're finding and what is consistent with our working hypotheses.

Tamas: What you're saying is that no matter what you do to these individuals, let it be exposure to drugs or alcohol, it will not move them out of that very particular state, no matter what they do. Is that so?

Joy: Well I do not know about ultimate plasticity. What I'm saying is the symptomatology can be related to a neurological process, one quality of which is gathering information. It may be more than that, but at least gathering of information is a component.

Tamas: But I don't think that it is, it is just a lower-level peripheral. If you look at the entire nervous system, like almost in all levels, you have regulation of information gathering and information processing, including up to the cortex.

Joy: So yes, I like your idea, and I would extend my hypothesis to higher levels as well. I use the word low level to describe the visual input level, but it is controlled by higher processes that have been associated with stimulus saliency.

15 Empathy: an exaggerated social function

Tamas: But let me say something again. I have been, over the last couple of months, revealing all of my psychological shortcomings. One of them, which I think I mentioned earlier, I do believe that by now you know, whether it's a trait, either genetics or acquired, I have an overt ability to feel empathy. And it can be beneficial, but it can be actually detrimental depending on how far you go. I'll give you an example, which is neither detrimental nor beneficial. I'm watching a movie where somebody falls on their butt, but as I'm watching the movie, I physically have that feeling as if I fell on my butt. It sort of speaks to both of your points, because it has to be a visual perception. It's a visual cue. It may or may not relate to my experience, perhaps I fell on my bike as a child, and it's a very particular view to me. Or if you go on a roller coaster, that's a different thing. But to have that feeling emerging in you by simply watching something.

Zoltán: Everybody has that. I have to disappoint you; you are not special.

Tamas: No, no, I'm not trying to say I'm special. But some people have more. This is something that maybe Elon Musk doesn't have, you know.

Zoltán: Exactly.

Tamas: If these people don't have it, why don't they have it? That's a very interesting question. Can you explain everything by the blueprint of the neuronal network?

17 Narcisism

Joy: One of the favorite disorders of the day has been narcissism. There's been a great deal of focus in psychiatry on narcissism, and what is the missing module in the brain? People that are totally without a sense of empathy or caring for others. Of course, the traditional thing to do is to put people with this diagnosis in the scanner, and you have them do things like watching people fall or whatever. Then you discover that yes, there are some missing pieces in the neural circuitry, there are some weaknesses in the neural circuitry. So regardless of the details of that, the bottom line is that these individual traits, some of which can be gathered together in psychiatric disorders, are the product of neural systems that are either misaligned or not aligned or dysfunctional, hypofunctional, hyperfunctional—something is different. I don't know if that's a sufficient explanation, but the effort is to seek the biological underpinnings that are associated with the mental condition. This is the linking hypothesis that I think we need to examine. But these neurological mechanisms have qualities in terms of the amplitude of response or whatever that relate to the behavior.

Tamas: So, going back to these two people I have in my head right now; Donald Trump and Elon Musk have overlapping impairments. Why is this one a successful contributor, the other is interesting, I don't know. But I'm gonna look into Elon Musk's palette and food interest. Trump has very primitive, interesting food habits. Every day, he eats an overdone steak with ketchup and french fries. The other thing he likes is cheeseburger. It's not overtly sophisticated. Again, these are all connected, and I'm going to look into Elon Musk's eating habits. It would kind of make sense that they sort of have predictors, one behavior might be a predictor for other behaviors.

18 Anatomical substrates of psychiatric conditions

Zoltán: What is frustrating is that we know so little about the anatomical substrates of these cognitive conditions and almost any structure you touch in the brain. You will find publications describing subtle, but significant alterations. Alterations in interneurons in certain structures, but not in others. You have alterations in the white matter; distribution and number of interstitial white matter cells; hyperconnectivity; alterations in myelination; microglia activity; etc. At the end, you have nothing that you can point to explain the complex cognitive alterations.

Joy: That's a wonderful finishing sentence to a book.

Zoltán: The word "nothing" is a great ending.

Tamas: Yeah, but you have no anatomical underpinning of a single circuit that is altered and responsible for the condition. You have lots of altered circuits that each contributes.

Joy: You mean when you're talking about these kinds of behaviors or the cognitive conditions?

Tamas: Yeah, they are less simple to actually describe. If you go further down, you talk about breathing in and exhaling, for example, those are somewhat more primitive, but they actually have very specific circuits.

Joy: So yeah, I think it depends on the complexity of the thing. The brain is just such an enigma. Because it's so changeable, it's so flexible, it's so adaptable, it's so individualistic. There are big things about it that are common, we have all the lobes, you know, the brain is this thing that's sort of crumpled up in the skull, it has the same common anatomy. You have a temporal lobe, the frontal lobe, parietal lobe, and it basically looks the same. The Brodmann areas are relatively consistent. You can poke around and see that the cytoarchitecture varies, but it's reasonably constant, the layers are relatively constant. But how it works and how it interfaces with behavior is hard to grapple with. That's the part of the inspiration for our book, is that people grapple with it in different ways.

Tamas: I want to make a point when Nenead puts these dead animals through those pumps, they're regaining some function of the brain in the absence of behavior. What information will that give you about anything? The real output of the brain is the behavior! What do you learn from anything else in the absence of behavior?

19 Is the isolated brain conscious?

Joy: Is this an Important question?

Tamas: Zoltán, Nenad is your friend. So why don't you say something?

Zoltán: We always have these discussions about whether an isolated brain is conscious or not, or if you take a bunch of induced pluripotent stem cells that were cultured in isolation and assumed organ-like appearance with some degree of layering and connections, does it have consciousness or not?

Tamas: Do they know?

Zoltán: Organoids can develop spontaneous neuronal activities, but that bunch of neurons in the dish that constitutes the organoid has never received any meaningful sensory input that they could process or analyze. I'm absolutely sure that right now, we are very far away from that activity having any relevance to consciousness or anything meaningful. However, when you cut the head of a dead pig and you perfuse it with solutions to keep it alive, you get some kind of an activity in that brain. I think it is a rather different matter, because that circuitry you have there, that you resurrected with the perfusion of the solutions, could replay some of the experience that was established, captured, and still embedded in that particular brain during the life of that animal.

Tamas: So you think that there might be some meaningful neuronal activity in those brains?

Zoltán: Let me tell you why I think it is different. During the French Revolution, the guillotine was introduced for executions, and it was in use until 1977! This apparatus was specifically designed for efficiently carrying out executions by beheading. There are anecdotes about some facial movements occurring for a few seconds after the heads were severed. One of these stories claimed that when an executioner smacked the cheek of Charlotte Corday (who assassinated politician Jean-Paul Marat), Corday's face showed marked disgust. Whether these were just reflex actions or actions of expressions of cognitive functions is difficult to tell. Some publications describe EEG (electroencephalography) brain activity several seconds after decapitation in mice. The tales about severed heads that blink, change expressions, or even attempting to speak might be just reflex actions after the severe shock of isolation of the head. However, Nenad showed that he can keep some of these dead pig brains alive with perfusion, these brains have activity patterns.

Tamas: But my point here is that in the absence of having the ability to replicate a behavior or sensory perception, motor output, whatever, of that animal or whatever, what do you know about what's going on in their head, other than recording some things?

Zoltán: That brain was wired up and processed sensory input, and that brain was involved in the planning and execution of movements just a few seconds ago. Those connections that were established by using spontaneous and sensory driven activities also elicited movements and thoughts. These could, in theory, still exist in isolated brains. That is very different from the organoids, which were just grown in an incubator and were never really part of any functional neuronal networks.

Tamas: Correct.

Zoltán: When you isolate the brain, and you only have the supplying arteries and veins, no neuronal connections for sensory input and no motor output, it's still a brain. It could have some activity that is kind of related to the previous experience that brain had. For instance, if you could work out the details of how to swap heads with someone, you could probably transfer the knowledge of a language or motor function. Maybe you couldn't speak because you'd have to calibrate your facial muscles, but with practice and calibration, you might be able to do it. You have dozens and dozens of muscles that you have to coordinate when you open your mouth and change the shape of your oral cavity, position of your tongue, etc. But maybe just swapping your head could still allow you to at least understand that language that this particular brain was brought up with.

Tamas: What do you think about these ideas?

Zoltán: I still believe that isolated brain remains a brain.

Joy: Then you have everything still connected, and you could see whether you can read, activate outputs and inputs to the brain, perform a behavior, and so forth.

Zoltán: That's a very reductionist point of view, to reduce the brain to nothing, and then bring it back is its fundamental function.

Tamas: It has happened in many instances to individuals who died and fell into ice. There are some people who did that, and then they were reanimated. So it happened in the past, they were in comas for several months.

Joy: Yeah, there are examples of people that have been vegetative, and then they reemerge.

Zoltán: They can still speak, think and move after years of coma, so all these circuits were retained during the coma.

Tamas: So it's been done.

Joy: Basically, well to some extent. The loss of consciousness and then recovery is a very important medical issue, and another major topic for discussion.

20 The neuroscience department of the future

Joy: Interesting. Following up part of our conversation from last week, I was thinking that maybe we ought to think more systematically about what a neuroscience department of the future would be like, if we had our way to build a neuroscience department. How would we do that? What would we focus on?

Zoltán: Would you have a separate neuroscience department, or would you mingle it into other departments?

Tamas: Okay, those are good questions. Would you just dissolve the "Neuroscience Department"? I mean, to be very honest, it's a very simple thing to mix it with physiology. In a way, we have talked about integrative physiology.

Joy: I think the operative word is "integrative" and would include expertise in behavior and cognition, physiology, and social interactions. Whether you keep the name "neuroscience," and whether "neuroscience" is even the right term for referring to brain research, I don't know.

Tamas: Yeah. Yeah.

Zoltán: "Brain and behavior" or Life and mind institute." Or, one part of our department (Department of Physiology, Anatomy and Genetics at Oxford) is called the "Centre for Integrative Neuroscience."

Joy: Correct. What is the basic question of neuroscience? We have the integration of brain, normal and abnormal behavior, the body, and that's where the physiology comes in, and social interactions across individuals which is included in behavior. So, how do you bring the components of these disciplines together in some kind of cross-fertilization? How do you bring these disciplines together in a way that catapults, and ignites creativity?

Tamas: Yeah. I was thinking the BBB. Um, yeah, I mean, that's sort of like the brain behavior body. Three points of view, but the "body, brain, behavior" is what people are interested in anywhere in life.

Joy: Right, I was sort of inspired by thinking about bringing that home in terms of how one designs a department to train students. Do we continue to train them in little individual silos, or do we send them over to physiology, for example, to get trained in something that I don't know that much about, but they need to have.

21 What about this emerging book?

Zoltán: So I think all of us should write a one-page summary on what you think the brain is.

Joy: Yeah. Oh My God. Start with the brain. What a novel idea.

Tamas: Okay. That's good.

Zoltán: But we should do is that for the next time, we should have a summary of your chapter in one page or something like 500 words? How do you see the brain and what do you what to explain in your chapter? So it's kind of a primer?

22 The parabol of many blind people and the elephant

Zoltan: We could start the whole book with the tale of the blind and the elephant. We should each have another 500 words of kind of a confession on how did we end up where we are in our views. It's obvious to me that we begin to learn from each other, but we are still very much influenced by what we did what we learned, and this is a very big influence for the rest of our life. How we were brought up, how we were taught; how we look at the brain.

Joy: I agree with that and I think, it can be helpful for people to understand where we are, or where we are coming from, to the whole thing.

Zoltán: I agree. Tamas, you explained that you looked at the hypothalamus, and you have this kind of endocrine view of the neuronal circuits and the brain.

Tamas: Definitely, I think it's a good idea.

Joy: Okay, so, I think the brain is only one half of a communication system. Last week I started my part with some basic biographical details, and my trajectory that has led me to the study of the neural systems that manage cross-brain interpersonal interactions from a biological point of view. I think of this as the telephone communication system between brains. I only wrote about three paragraphs, a page and a half, and, maybe that is enough.

Zoltán: We have this opening chapter where we talk about how we ended up where we are, and what our views on the brain are.

23 The person in the science

Joy: Yeah, I hope the reader can sort of bond with us as individuals. Mm hmm. That's what I think is different about this book. It is that the conversation not just with us, but we're inviting our readers in a way to know us as authors and as thinkers, and in some way to join us. There's more than just science, there's a personal side. I realized as I'm writing biographical details, that this is the part that just scares me to death. Using the word "I" is terrifying. And that's why my part is so short. But, I think it's poignant and to the point. If I were a reader, I would be more interested in the people really, then necessarily what was said. But that is my "social" side.

Discussion 14—Monday, 17th May, 2021

1 On talking about interdisciplinary departments

Zoltán: Sometimes I visit institutes in Europe or in the United States, and I have the impression that everyone is working on mice. They put channel rhodopsin in different parts of the brain, and they zap these cells and look at what happens. These are considered top institutes in the world, but intellectually and conceptually, these institutes are not at the top. They might produce high-impact, flashy papers, but they might not open up new fields or change how we look at the brain. Perhaps more integrative and haphazard arrangements are better for creativity.

Joy: My appointment at UCL sort of fits our model. I am in the Department of Medical Physics and Biomedical Engineering. These engineers decided that they wanted a neuroscientist to help them integrate what they're doing into the world of science, and it is a model that is working just enormously well. You know, we share students, we publish papers, we share grants, I mean, it works because neither of us really understands what the other does, but we really appreciate it and incorporate it. It's really good.

Tamas: I have a very specific question that drives directly to Joy's work on social brains. I'm not trying to get some free counseling here, but when you say that brains exist and interact together and how fundamentally crucial that interaction is in order for two brains to function as a single unit, then we could ask, "How do you then put that into the concept of "marriage"?"

Joy: Oh my god. The scares me. "Marriage" doesn't have a scientific meaning. Maybe we should use the word "companionship." It's companionship because you have a joined up brain then.

2 Two brains and one dyad

Tamas: Yeah, you live with your companion, you have a brain with your companion, and you have another brain when you are in a different setting.

Zoltán: I would go even further. I think my wife has three brains or more in different settings. Nadia's parents were Italian migrants to the French part of Switzerland, and then she lived in the United Kingdom for the last 30 years. In different environments, she can perfectly switch between different languages, and I noticed that her gestures and attitude can also change in these different contexts. It is almost like she has three different brains to interact with. It is like living in polygamy.

Joy: My argument is that it's not about a single brain, and your single brain can be half of many different dyads. Your brain is different when you're with your wife than it is when you're with your colleague, Zoltán, for example. That is a notion that didn't originate with me. That's the notion of the old psychiatrists. There's this philosopher, Vygotsky, who was quoted all the time, who says that you know yourself by who you're with, meaning that you are different depending on who you're with. So it means that it's not about single brains anymore; it's about the combination that shares information.

Tamas: The reason I brought up the partnership, or marriage, is that one could also assume that the more time you spend with one individual, the more obviously that shapes your own brain. I was watching something over the weekend, just because it was on TV, about the life of Jim Jones. Have you ever heard of Jim Jones?

Joy: I remember to news reports of the mass suicides.

3 Leader and follower brains

Tamas: That's the biggest mass murderer on the planet. As many as 906 people committed mass murder/suicide within an hour. Jim Jones was from a very rural area. He became a very big advocate for integration of races and was fascinated by communism. If you actually look at his trajectory up until 1971 or 1972, he seemed to be somebody who you would think is a great guy or potentially is doing great things and addressing needs of the people. He had tremendous success in basically brainwashing an enormous amount of people to the extent that they committed mass suicide. So the brain-to-brain impact on those other brains was enormous. It's very intriguing, and I think it relates to your work, Joy.

Joy: It does, and you know, you've taken my ideas farther than I have. It provides a structure and a model for how one brain can dominate the other, or subsume the other, or influence the other. It may actually be a model for things like parent-child relationships and nonpathological things.

Tamas: Correct, but also pathological cases.

Joy: Exactly, but the impact of this fusion that I talked about, one brain being half of a social unit, has many implications far beyond where I've taken it. One of the reasons that I haven't been particularly thoughtful about it is because either my limitation or my strength is, as a scientist, reductionism. I stay really close to my data, and when I tell you something, it's because I think I have data that support it. That limits what I can say, but it also makes what I say more credible.

Zoltán: Joy, you could take your research on interacting brains much further. Imagine that you could do your recordings in members of a mass in a church or amongst supporters on a football pitch. You might be surprised.

Joy: Oh, yes. I think you're right. There would be surprises.

Tamas: The other thing that I find fascinating in your work is it applies to our own interactions here. Yeah, I mean, the way we are-
Joy: Exactly.

4 Interpersonal interactions and the autonomic nervous system

Tamas: I argue that there is a tremendous role of the autonomic nervous system in this, because when we start to talk, we start to synchronize, we start to engage, and we start to feed off each other, that has tremendous influence on the autonomic nervous system, which then eventually feeds back to the brain by various means.
Zoltán: When I read your personal statement, I wish that I had included more of that particular topic into my discussion. I wish I had talked more about the inclusion of physiology and how important the body is for the brain.
Joy: But I think in a way, our thoughts are independent, and then when we have these interactions and discussions we can come to a conclusion that actually fuses many of the things that we all brought to the table and developed further.
Zoltán: We wrote our initial statements on our background. I did not read yours before I finished writing mine. Reading yours, probably, I would have said more about the impact of general physiology on brain function.
Joy: That's cool that you didn't, because then we did it independently, and through the whole thing, we are going to be clear about how we might change our own ways of looking at things.
Zoltán: This way we are more original, and the differences in our way of thinking will be apparent.

5 Our background statements: Zoltán started as a neurosurgeon in Hungary

Joy: So, Zoltán, I had a question about your background statement. You're a medic; you started training as a neurosurgeon, right?
Zoltán: Yes, I studied medicine at Albert Szent-Györgyi Medical School at Szeged, and after finishing medical school, I joined Prof Mihály Bodosi's Neurosurgery Department. I practiced only for a year.
Joy: But you were trained in surgery.
Zoltán: I started my residency and I was very happy that I could start in Prof Bodosi's department. I was always research oriented, but when I finished my clinical training at medical school, at the last minute, I decided to pursue a clinical career. I did not know whether I should do research or practice clinical medicine. I liked both. My older brother, Elek, was already working at the Biochemistry Institute at the corner of Dome Square in Szeged. His passion for science and interest in basic research

also influenced me greatly. At the end, I was offered a job at both the Physiology and the Neurosurgery Departments in Szeged, and initially I wanted to study physiology.

Joy: Awesome.

Zoltán: Then at the last minute, I opted to become a clinician. I was pretty good. I even passed the basic component of the UCMLA at the time to open up possibilities for future training in the US. I had to travel to Belgrade to do that. I also loved the operating theater. I am ambidextrous, and also inherited some excellent manual skills from my parents. My father was a sculptor and ceramicist artist and my mother was a drawing and history of art teacher in the local secondary school. When I started my residency, I loved being in the hospital and being part of a larger team. What I didn't like is that you rely on the money you receive (unofficially) directly from the patients.

Joy: What?

Tamas: This is something that's probably strange for you, Joy. I mean, the medical enterprise in Hungary, I don't know to what extent it's limited to Hungary, I think it's East Europe-

Zoltán: Probably.

Tamas: I'm not sure, but in Hungary, I think it was one of the worst cases at the time. So anytime you go to see a doctor for any reason, there is an expectation by the doctor that you bring some extra money (a tip) and put it in the pocket of the guy or the woman. Yeah. That eventually became an industry; they even started taxing this unofficial income. There was a price, for example, if you give birth, you would know by word of mouth how much money you need to tip your doctor.

Zoltán: Radiologists, didn't get anything. Anesthesiologists, zero. If you were a surgeon, you got a lot of money, and specialists, like my younger brother Béla who started as an ENT surgeon in Semmelweis Medical School at Budapest were frequently given tips. The turnover of the patients is fast, so you can examine somebody and make them feel much better in just 15 minutes. Patients are in and out fast, and your money adds up, so those positions used to pay very well. Obstetrics and gynecology was one of the best paid jobs, and pediatricians were one of the worst, because the idea was that you do not pay parasolvencia (tip) for your children. There was not much logic in all of this; it was a corrupt and horribly humiliating system for everyone involved. I took this part very badly; morally, I found it wrong to accept money, but your salary was not enough to get to the end of the month.

Joy: Those things are really interesting, but I think your clinical medical past ought to be included in this because it's a very important part of your history. That wasn't where I was going, though. I was thinking, you are a medical physician, a neurosurgeon. You've been inside people's brains mucking around with instruments. You pull this part out, you push this together, you stitch this, you fix that—you've fixed brains, and this is important. This is something that's different from what Tamas and I have done; you have been mucking around in the human brain as a surgeon. I think that's a really important perspective. It's a physical perspective. You've understood the behavioral consequences of what you're doing. You fix the medical side, and you expect the behavior to return somehow. That's kind of a connection to me, in a way, because I started functional imaging using fMR, doing neurosurgical planning for surgeons so that they could do the surgery in a safe way. I don't know if we'd ever put that together. But I just feel like there's an opportunity we have between the three of us to bring in the medical culture here, the medical culture that works on brains, and you have that. Why did you select neurosurgery?

Zoltán: I selected that discipline because I always thought that it's the top of the medical profession. As a medical student, I thought that if I become a neurosurgeon, it's a well-respected profession. In Hungary, not all doctors are respected equally. Some doctors are more respected than others, and this was kind of like the top. Of course, I loved physiology, pathophysiology, and neurology. When I finally started doing neurosurgery, I realized that most of the patients on our ward were not getting better. Neurosurgery can be a very destructive branch of medicine. Most of the time, when you remove tumors, you have to plan how not to hurt patients, because to get to many tumors, you have to go through the healthy brain tissue. It is very common that you lose certain functions after surgery, and neurosurgery is not a reconstructive surgery. It is not like traumatology when you can use all sorts of tricks to put the bones and flesh of the patients together. Honestly, when I did a very brief period at neurosurgery and had to spend some time in the trauma department in the evening, I preferred trauma, because those patients were walking out much better than they came in. They were happy that we fixed them, and they recovered. I can't say the same after many tumor surgeries. As a medical student, I had the opportunity to visit some other neurosurgery departments in Holland (Professor Jan Jacob Moy) and Finland (Professor Matti Vapalahti), but even there, I was more attracted to the vascular aspects of neurosurgery, such as aneurisms and endarterectomy, and even became interested in why bypass does not work for the brain.

Joy: Bypass for the brain?

Zoltán: This surgery is also called an extracranial—intracranial bypass. You might remember, Joy, that about 30—40 years ago, there was a large multinational study in which they tried to put the extra-intracranial anastomosis in place to prevent ischemic stroke. The idea was that an artery from outside the skull (usually the superficial temporal artery) is connected to an artery inside the skull (usually a branch of the middle cerebral artery) through a craniotomy. The rationale of this cerebral bypass surgery was to restore blood flow to the brain. Professor Bodosi, my mentor in neurosurgery, was involved from Pécs and (after his move to Albert Szent-Györgyi Medical School) later from Szeged.

Joy: Mmhmm, exactly.

Zoltán: This multinational study showed that there is no clear-cut benefit of this procedure in ischemic stroke. There is some benefit in special cases of vascular developmental conditions, like moyamoya. Moyamoya are rare conditions in which the internal carotid arteries that supply blood to the brain become progressively narrowed. This limits the flow of blood to the brain, and puts the person at risk for stroke. A cerebral bypass is the brain's equivalent of a coronary bypass in the heart. It was around 1985 when the results came out, and that multinational study showed that there was no clear-cut beneficial effect. They had to stop the whole procedure because there was no justification for it. This was at the time when I talked to Professor Bodosi a lot, and we had discussions about what the way forward could be in this area. So, I liked neurosurgery, I liked my mentors, I liked my colleagues, and I particularly liked working in trauma because I could do a lot, and those patients recovered. But I didn't like the deconstructive side of neurosurgery when you actually damage the brain.

Joy: But you know, your story could include some of that. Because it's another facet of this business of the brain.

Zoltán: I have superb memories, and they will stay with me forever. I remember when I did the first lumbar puncture, first suturing of skin ruptures on the skull, watching the clipping of a berry aneurism. These were very big things for me. I enjoyed all of this as a medical student and as a young doctor. I have very fond memories. But neurosurgery is not like, you know, trauma surgery that somebody walks in with huge damage, and then you inject some stem cells, and then they recover. Unfortunately, this is not the case.

Joy: But I would like to elaborate a little bit on this topic.

Tamas: As you say, most of neurosurgery is destructive. It's not a reparative approach; it's a destructive approach. I'm going to offer my simplified way of looking at things: it's overhyped. There are all these individuals around who do things that are way beyond what neurosurgeons actually do. The reason I'm saying that is because those doctors, in my view, have a tremendous amount of potential responsibility or potential input, but there are internists who do more and are completely mistreated by the entire medical enterprise. If you become an internist, the whole system looks at you as you are a loser, and when you enter medical school, that's not the goal that you're imagining. It is unfortunate, I mean, specialties like dermatology make money with very little intervention, and automatic neurology is one of those. There are many myths about medicine, and I think neurological surgery fits into them. Again, I'm not trying to diminish anyone today.

Zoltán: Did you read Henry Marsh's book "Do No Harm" or "Admissions"? Henry Marsh is one of the UK's most famous neurosurgeons. He appeared in television in documentary films and he gave unforgettable insight into the highs and lows of being a leading neurosurgeon and operating on human brains. We also invited him to talk to our medical student society at St John's College where I am tutor of human anatomy. In his books and in his lectures Henry March speaks openly about the drama of surgery and the agonizing decisions to select the least harm for the patients. He entered neurosurgery because he thought that he would understand the human mind, he would understand how people think, and how the brain functions. He confessed that having worked in neurological surgery for decades, he has seen a lot of human suffering and tragedies.

Joy: But surely, one learns a lot about the human brain when you operate on them. That connects to Wilder Penfield. Penfield tried to understand the human mind through his experiments while he was doing neurosurgery.

Zoltán: Penfield mapped out the motor and somatosensory representation of the body in humans. In fact, Penfield spent some time with Sherrington at the Physiology Laboratory as a graduate student here at Oxford. He was a Rhodes Scholar at Merton College. Sherrington did a lot of similar mapping in monkeys back in Liverpool with Grünbaum then later at Oxford. Also, Cushing worked with Sherrington and was a regular visitor at Oxford throughout his career.

Joy: How interesting.

Zoltán: In fact, you have an amazing setup in Yale now, The Cushing Center at Yale's Harvey Cushing/John Hay Whitney Medical Library. Tamas took me there to visit a few years ago. There are hundreds of brains in pots exhibited with various tumors and lesions. Each of these preserved brains are telling a sad story of a patient. Of course, one can learn a lot from human cases and neurosurgeons are using the triad of brain exploration by examining the effects of lesion, stimulation and recording.

Joy: My fMRI center at Columbia was in the same place where Penfield was before moving to Montreal in Canada. It's amazing that they all came through the same kind of school, and I think it was also very important in those days.

Tamas: Yes, exactly.

Joy: But because it was important, then it might be kind of important for us to somehow link to it as we think about our antecedents and scientific heritage that brings us to where we are. After reading the introductory statements, I could see some significant common threads. For example, both Tamas and Zoltan proposed a completely—very similar—but avant-garde style for teaching neuroscience. I mean, each of us suggested essentially disbanding the departments as they are now and setting them in rich environments where people from multiple disciplines talk to each other and start to create.

6 Neurons aren't even the major part of the brain

Tamas: I even don't like the word "neuroscience." Of course, I got accustomed to it, and I know what it covers, but it's biased and misleading, in a way. You know, I'd never thought about that until I read your paragraph, Zoltán, about how neurons aren't even the major part of the brain.

Zoltán: Yes, I wanted to make that point.

Tamas: The fact of the matter is (and I think this is nothing new in science, as we discussed before) that there are popular areas, and eventually everybody jumps in, and everybody's working on building the given moment.

Zoltán: The same thing has happened with microglia. Initially, there were very few groups working on microglia. Everyone knew about them, but nobody was interested. Now, suddenly everyone is into microglia and microglia will solve all the problems of humanity.

Tamas: They may, but it's unlikely. After reading our introductory sections, it's interesting that all of us are saying the same things.

Joy: You know, we've been interacting with each other, although we did not read each other's sections.

7 Friends and science

Tamas: That's true, but I think that each of us, in our own way, is a bit out of the box. I'm sort of thinking beyond the mainstream, and I think we've earned the right to ask what this means in terms of the future of these directions. What it means is institutionalized integration, which is an oxymoron, actually. Yeah. When you think about it, it has to be structured. I like my idea of discussing friendship. I like friendship.

Joy: I almost erased that, and then I said, "No, I love that!"

Tamas: It is important. You know, people have to like each other in order to work. I only work with people I like. I've said that too much before.

Zoltán: I like all the people I collaborate with. There has to be trust and friendship when you do science.

Tamas: I mean, the truth of the matter is I like some more than others, of course.

Zoltán: That's absolutely normal.

Joy: You know, my collaborations with people that I like are much more productive. The happiness is sort of like, the seeds of creativity. If you can be happy with somebody, you're much more likely to be productive.

8 When friendship fails

Zoltán: However, if you have a big argument or scientific fight with someone, it could also be very destructive. I am not talking about intellectual fight. I am talking about manipulative and not trustworthy individuals in our fields.

Tamas: I agree with the big fight. But when it's an intellectual disagreement, it can become personal, emotional hatred. I gave a talk in Munich, and the person next to me was a guy who does cutting edge optogenetic techniques, and my talk was about questioning the doctrine. I was walking away from the podium, and he says, "My whole life is about the Cajal doctrine. How dare you!" I didn't say anything about his life, and what he does is probably very useful. But I'm not talking about the experience; I'm talking about the conceptualization of the thing. But he hates me because 14 years ago, he wanted to show me something, and I agreed with him. I showed him something different that, in my view, actually did not go against what he showed me. Then he started to say, "I should have taken my medication this morning, I knew you're gonna piss me off!"

Zoltán: Who is this person?

Tamas: Just some Professor at Harvard; a guy who was also an MD, trained like you. To his credit, he is a person like many of us, and I'm not speaking about myself, but generally speaking, many scientists end up being scientists because they have awkward personalities. Yeah, it's true. Many people on the spectrum end up being very, very narrowly focused, and they make very good scientists. He really applies cutting edge tools, as they contemporarily predict the questions and answer them according to that anticipation very nicely. These people sincerely believe that this is the only way you can do science. This is the way you should be doing science. That is where the interaction becomes problematic, because you don't speak the same language. I mean, I can understand them, but when I reply, it just upsets them because it moves them out of their comfort zone.

Joy: Yeah. Well, I don't do well when I don't like the people I'm with, and when they don't like me, it doesn't help me.

Tamas: Then as soon as you enter the room, or in the auditorium, immediately there is an autonomic nervous system response to that. Even just a spatial visual cue. Anyway, I was just writing, and the word friendship came out. I looked at that again. No, that's what I really mean. We are never ever asked, "Who do you want to be next to in the department?"

Joy: If I were a department Chair, and I was going to try to set up these little discovery pods with people in different departments from different disciplines, how would I do it? And the answer is, I wouldn't do it at all. I would ask people who they want to be with. I would give them choices about the departments. You know, you're a behavioral neuroscientist. I want you to talk to somebody in engineering or robotics. So go over and find somebody you like. A little directive, but not a whole lot.

Tamas: You know, maybe about 10 years ago, I was very idealistic and excited about these exact ideas and thoughts. I was living in a part of New Haven around people who worked at the School of Management, the Department of Economics, and so forth, and I was socially interacting with them a lot. I always thought that if there is a good social connection, then you can exploit that to do something innovative in your own field or the other person's field. I wanted to develop ideas with these guys in those other departments. There was one guy who had a little bit of an interest, but the majority of them were completely dismissive of this suggestion, and just would not even listen. At the same time, I have to give some credit to the late Tom Jessel, because he came around the time when they were about to launch that institute. Pasko, Pietro, Tom, and I had lunch together. Tom and I were sitting in front of each other, and he talked about collaborating to do innovative work in other fields, and I think there was some aspect of this that they were bringing to the Zuckerberg Institute at Columbia. I don't know to what extent they've succeeded, but the others were not enthusiastic about these collaborations. I came to the conclusion that there has not been any benefit of such communication between such disparate disciplines.

9 Cross-disciplinary collaborations: strengths and weaknesses

Joy: Yeah. You know, there is a very strong mainstream in our science. I see it on tenure committees, when I sit on the committee and people come through, one of the most damning criticisms of a faculty member is that they haven't been focused, which means they've collaborated, they've gone out to work with people in other disciplines. In some way, it is perceived as weakness.

Zoltán: Such committees note that the PI is "not last author on the paper," or "there are too many collaborative authors" on the paper. How can one step outside one's own field without collaboration?

Joy: Yeah, but that attitude exists, and I think it's a prevailing thing in the assessment of tenure, assessment of awards, assessment of anything intellectual. Often times, I've thought that this is where I've been vulnerable during tenure decisions. I've been through three or four of them, and every one of them was brutal and traumatic. After my final tenure decision here, I said, "Thank heavens, I'm never going through another tenure decision again as long as I live, and I'm just gonna jolly well do what I want to do!" I just published a paper with the guys in computer science and robotics, I picked up my engineering team at UCL, and we pulled things together to make it work.

Tamas: And this, I'll tell you, what makes this place work is not so much due to my effort. I pulled together a whole lot of disciplines, and I'm not saying anything negative about their particular approaches that they advocate for, because those are very important. I might have said it before, but I sincerely believe if you look at Columbus, Magellan, or United States history, there were people who ended up going to the west to explore what they did not know, and they did not have to go and deal with building every single house, village, sewer system, and whatnot while on their quest to discover something. There were the people who were exploring and interacting with others, but then they definitely needed the bricklayers. You need the high-tech bricklayers to actually do the job. So, I looked at all these optogenetics or single cell transcriptomic guys, and all of them who do that kind of thing are the high-tech bricklayers, and they are extremely important. But conceptually, more often than not, they add little to the larger picture. There's nothing wrong with that, but I'm saying we need each other to try to bring all of this together and make it a reality, to go from theory to physical substance.

Joy: I agree, and the reason that it is used to evaluate people is that it's been successful. It offers predictability; you can almost predict how successful someone will be by how much stick-to-it-iveness they've shown, how persistent they have been in solving their problems, and if they were eventually successful. Then the question is how you measure success, of course—how the individual succeeds and what impact he has on the discipline—but these two parameters do not actually show the impact of that person's work on the fundamental questions of biology. For example, in my field (and I'm sure in some of yours, too), there's a tremendous number of Houses of Cards built through this successful approach, and eventually, time will tell if they stand or they don't stand. By the time that happens, these guys are gone, and they had a successful life.

Zoltán: In my field, some of the fundamental questions require a large team of 20−30 people that to resolve. For instance, in developmental neurobiology, what cell groups were born together is a very big question. Are they clonally related? How do they disperse? What is the phenotype of the cells in that clone, in that lineage relation? Now, you can address this very easily; you go in with a retrovirus mediated gene transfer, you insert a DNA barcode into the progenitor, and then you follow the lineage by identifying the barcode in the progeny. To do that, however, you have to let the animal grow up, collect all of the target cells from the region of interest (or the entire brain, if you have enough money), perform single cell sequencing, and combine that data with the clonal analysis. This way, you know where the cells came from, where they ended up, their phenotype, and how they are clonally related. You do it thousands and thousands of times with thousands of clones, and you will know all the important principles. You then realize that Pasko Rakic predicted all these 60 years ago just by looking at birth dating in monkeys. He proposed that the majority of cells were migrating radially, and then you have some tangential migration, and then they follow inside first, outside last kinds of patterns. We knew some of it a long time ago, but of course, we also discovered some new streams of cells that change migration in between macaque and human or mouse and human. As an individual, it was possible to make some fundamental discoveries a few decades ago in this field, but a group like mine cannot do such large-scale experimental analysis. Very, very few laboratories can do this kind of scale and resolution.

Tamas: Are the studies important?

10 How do we know what is an important question?

Zoltán: I consider them very important. Why would you not want to understand where the cells were coming from?

Tamas: But is it important to understand where the cells were coming from? Can I use an absolutely wrong analogy? So let's say there is a scientific question that needs to be addressed, or some function that needs to be resolved by my laboratory that includes people from all over the world. Marcello (a guy I used to work with who is now independent) and I get together, we solve the problem, and it's an absolute success. He came from Brazil, and I came from Hungary. It made no difference where we came from, in the end.

Zoltán: The fate determination of neurons is very different. You used a wrong analogy. The fate of a neuron is largely determined at the site of the last S-phase in the germinal zone. Germinal zones can have very different combinatorial transcription factor influences on the progenitors depending on the time and location of the particular segment of the neuroepithelium. After birth of that neuron, there is very little that can happen to drastically change the fate of that cell.

Tamas: Oh, so you mean that origin, temperament, and emotionality may or may not be related.

Zoltán: That's another question. If you mix up the neurons of the brain in a different way by changing the clonal dispersion, you would end up with a different brain. If you have different clones with different dispersions, you might not end up with a functioning nervous system at all.

Tamas: So this is why people are obsessed with development! People play around with some cadherins and stuff, observe how sticky the cells are with the others, how well they disperse, and I'm sure that will lead to some subtle phenotype.

Zoltán: If you look at the brain structure in somebody with autism spectrum disorder, you hardly see any difference. The anatomical phenotype is very subtle.

Joy: There have been some claims for subtle disparities in fractional anisotropy, but for the most part, the brain is not distinguishably different in any way from a typical brain.

Tamas: Maybe you have a slight change in the dispersion of the cells, a difference in how they disperse among other clonally related cells, and that might result in this manifestation. So is it important?

Zoltán: People are working on these kinds of questions and examining not only neurons, but also the other cell types that you have in the brain. I was very interested in the tiling of progenitors and the possible pathologies that arise when the regular tiling of radial glia progenitors is disturbed (https://doi.org/10.1016/j.neuron.2019.05.049). There are very big questions about the origin and clonal relationships of excitatory and inhibitory cells. Do they come from the same clones? Do they come from separate clones and then meet? Are there other differences in the human or in the mouse with these proportions?

11 The compass going forward: uncharted territory

Joy: I think that what we're talking about is the last chapter of the book, because really, what we want to talk about is what are the future important questions that we have from our domains? I can see taking what we started here and adding another section on it asking what are the important questions that come from this discipline that we're talking about? What would be the ways to solve them? Somehow, I don't have any clear answers. I think it's sort of hard writing because it requires some creative thinking, and I don't have a clear answer to guide me. I admitted to Tamas that I hadn't thought deeply about what the two-person paradigm really means, but maybe that would be a service that we could provide in the book, some creative thinking about where we want to go.

Zoltán: Some of the papers you see published in *Nature or Science* now, maybe, don't reflect the directions in which we want to go.

Joy: Well, that's kind of the point.

12 Most important papers and the life span of a scientist

Tamas: There is this Hungarian guy named Laszlo Barabasi from Budapest who lives in Boston. He published, I don't know, dozens and dozens of influential papers in *Nature* and *Science* in the last 20 years or so about network analysis. One of the interesting things he produced (which may or may not be directly related to this topic) was an analysis of when scientists' most impactful papers were published in relation to the timeline of their entire careers. It turned out to be absolutely unpredictable. There's no predictability of whether somebody's most impactful paper would be published during their postdoc years, earlier or later. I think it's a very interesting paper. There were zero predictors. It was not related to professional development. Obviously, you can assess a scientist's impact on their field in other ways, but this was absolutely random. You don't really know whether your most influential work is behind you or yet to come.

13 Scientists and transitions

Joy: I hope it's coming up. I always feel like I'm just beginning. I started all over with a new question and new technology when I came to Yale. It was both thrilling and a scary transition. The only difference from past transitions was that I had the startup package and experience of a senior recruit but I still had the energy and motivation to do the job. My husband refers to me as the "ever-ready" battery. I guess I am lucky about that.

Tamas: This is the topic about which I sent both of you: a link to a speech or to an interview of Albert Szent-Györgyi. Did you read you listen to this Joy?

14 Vitamin C, the Nobel Prize, and the Hungarian crown (you can't make this up!)

Tamas: You should, because the guy basically says that he discovered actin back in the 1930s or something, after he got the Nobel Prize 1937 for vitamin C. He worked out how muscles move, discovered actin and myosin, and then he was absolutely fascinated. He continued working, but after 30 years, he realized that he did not accomplish anything. He believed that he accomplished nothing. Then he moved to another field. For me, it was very encouraging, because I had the same feeling.

Zoltán: Did you know that it was Albert Szent-Györgyi who led the delegation that brought back the Hungarian crown by the order of US President Jimmy Carter on January 6, 1978? During World War II, the crown and the regalia were buried by the guards in the Buda Castle in October 1944. After Ferenc Szálasi came into power, he unearthed it and swore to the crown. Then it was taken to Austria to save it from the Soviet troops, and eventually given to the United States Army. For decades, the crown was held in Kentucky alongside the bulk of America's gold reserves. Albert Szent-Györgyi was in the US delegation that brought it back on a big jumbo jet from the US, and I even remember that they had to extend the Ferihegy Airport at Budapest because it was not big enough for landing a jet at the time. Szent-Györgyi gave an interview to the Hungarian television, and I remember it was on a special channel (channel 2!) that we did not have in my home, so I asked Tamas' mum to let me watch it in their home. Your mum was also very interested in it, but your father slept through it because he was a veterinarian, and he worked from very early in the morning until late at night.

15 Books not brains are repositories of information

Tamas: I remember when he brought back the crown, and then he gave several lectures, and one of the things that I always liked was that he emphasized that it's not crucial to memorize things; what's crucial is to understand the conceptual framework and know exactly know which book to go to and where in the book you can find the information you need. He said, "Books are there to keep the knowledge while we use our heads for something better. Books may also be a better place for such knowledge. In my own head, any book-knowledge has a half-life of a few weeks, so I leave knowledge, for safe-keeping, to books and libraries, and go fishing, sometimes for fish, sometimes for new knowledge."

Zoltán: Unfortunately, medicine was not always taught like that. Half of what you learn in medical school will be shown to be either wrong or out-of-date within 5 years of your graduation; the trouble is that nobody can tell you which half. This is why you have to teach creative and critical thinking so that doctors can have lifelong learning.

Tamas: Well, you never know, that is true. I really hated passionately anatomy because you need to memorize in detail the origin of the muscle, where they go, the bones, etc. In the veterinary school, we had to study five different species, and each has their own anatomy slightly but significantly different. I hated it. But in retrospect, I had to recognize that in medicine, that's the only objective discipline. Everything else is constantly changing, but anatomy has a real ground truth.

Joy: I am terrible at memorizing things. I used to write songs about lists and details that I thought would be on a test. I could remember the song and that help me remember the list item. However, I have a really good spatial memory. When I started working with the neurosurgeons doing the neurosurgery planning, I realized that I could retain extraordinary detail about specific brains. The brains were so amazing.

Zoltán: The human body is just beautiful. I liked anatomy not because of what we had to memorize, but because I liked to understand how things work and how evolution and development produced that arrangement for a particular function. I did not study veterinary anatomy, just human, but I would have liked it. When I visit in museums, it is so interesting to compare some stuffed animals or some skeletons of animals, like a giraffe dinosaur or other species to human. Tamas, you were taught to know how to compare different animals.

Tamas: Yeah.

Zoltán: Did you know that the giraffe has also seven cervical vertebrae?

Tamas: Yes, of course, for me, that was that's a given. I knew that.

Joy: You know, Zoltán. What you just said about the human body or bodies in general are very beautiful. I think that's kind of an entry-level part of your section and something that drives you, and it puts you where you are now and how you got started. I can see that you like anatomy.

16 Evolutionary influences on the human brain

Zoltán: Anatomy is fascinating. If you just look at your hand, you will see an incredible sophistication; this structure evolved into an incredible organ. Neuroanatomy is even more fascinating. Current brain structure reflects its evolutionary path. You are thinking about how to design such structures. But these were not designed. Nobody designed these intricate brains. We were selected over millions and millions of years and evolution was grinding this out.

Joy: I found this statistic recently . . . 400 million years?

Zoltán: I thought it was 500.

Joy: My source is from a flyer from the American Museum of Natural History, but you may be right . . .. what's a hundred million years on this scale? It took a long time. Any ideas about the next 400 million years?

Zoltán: It would be very interesting to study the evolution of body brain interactions. I do not think that the evolution of these interactions has been compared. I would like to know what the vagus nerve signals in various species.

Tamas: I don't know much about it, to be honest with you, I would not dare to say yes or no. And there may be some, you know, fundamental works on that in journals that we never read. But that's definitely something that is crucial in a way for them to conceptualize these kinds of communications, and definitely for model organisms, what we are doing and what we are not.

17 The "zero" cranial nerve

Zoltán: Did you know that we have one more cranial nerve, the zero cranial nerve? The current nomenclature for the 12 pairs of cranial nerves was established by German anatomist Samuel Soemmering in 1778, and it is generally accepted today. Zero cranial nerve was described originally in the brains of sharks; it was first found in humans in 1913. They didn't want to modify the numbering of the cranial nerves. Now they call it zero cranial nerve. The core cranial nerve zero is also called the terminal nerve. It was discovered by German anatomist and anthropologist Gustav Fritsch. With Eduard Hitzig, he did the very first localization studies in the dog motor cortex in Berlin in 1870. Ferrier, Sherrington and Penfield continued this type of research later.

Tamas: And where does it come from?

Zoltán: The terminal nerve appears as a microscopic plexus of unmyelinated peripheral nerve fascicles covering the gyrus rectus near the cribriform plate and travels posteriorly toward the olfactory trigone, medial olfactory gyrus, and lamina terminalis. I think it is sensing pheromones or perhaps humidity changes.

Tamas: But what can you tell me about it, I tell you why I'm extremely sensitive to humidity.

Zoltán: This is the area you have to study. I am interested in it actually. Paul Manger a friend of mine who is at the University of the Witwatersrand, Johannesburg, developed the theory that this nerve, is signaling us when to fall asleep by sensing changes in humidity.

Tamas: Oh, so when it's drier then you would fall asleep.

Zoltán: His idea is that when the sun goes down the humidity is changing. This might generate an important signal to go to sleep. Paul would like to exploit this knowledge by generating a device that can change the humidity in rooms and use it to help to fall asleep. We might have a device that makes us sleep better. I think he even patented this.

Tamas: Can you send me that reference because that would have a revolutionary impact on my marriage because I always fight with my wife, I can't sleep in humid weather, and she can't stand the air conditioning in the room.

Zoltán: I am reading more on the nerve: "the terminal nerve is a common finding in the adult human brain. It is very close to and often confused with a branch of the olfactory nerve. The terminal nerve is not connected to the olfactory bulb where smells are analyzed." In fact, this suggests that the nerves are either vestigial or maybe related to the sensing of pheromones. "The hypothesis of sensing pheromones is further supported by the fact that the terminal nerve projects to the medial and lateral septal nuclei, and the pre optic areas which are involved in regular sexual behavior. A 1987 study finding reports that mating in hamsters is reduced when the terminal nerve is severed." Coming to your family life. . .

Discussion 15—20th May 2021

1 Various roads to becoming a neuroscientist

Joy: We all ended up doing neuroscience, but we arrived through different routes. In life all of us have different opportunities put in front of us. The route that we take is very individualistic, in that if we just keep going forward doing what we want to do, making choices every day, we eventually get there and accomplish our bigger goals. In retrospect it might look like a direct path, even if it was not planned that way.

2 Introduction to the hypothalamus

Tamas: And again, in that regard, I feel lucky that I started doing my research on part of a brain in a department that really had nothing to do with brain research in general, but gynecology. Working on the hypothalamus was also a plus, because that's the part of the brain that no "contemporary" neuroscientist was really interested in. But recently, all this is turning around. This perception is now corrected, which is good. But turning around for the moment, in my view, which I think I wrote in my chapter is that it's turning around by introducing the tools that people have been using in cortex and hippocampus and thalamus and whatnot. And they are using the same tools to implement those concepts that were derived or delivered in the cortex trying to apply to the hypothalamus, which may not work. The functionality of those hypothalamic circuits is much more complicated than the cortical circuits. Individual neurons and circuits carry information on multiple homeostatic functions and impact multitude of behaviors. So, it's very different. I appreciate that many people are coming into the hypothalamus with the cortical centric philosophy. And what I'm trying to do is do the opposite to take the hypothalamic centered philosophy and bring it to the cortex.

Joy: I really like that. When I was working at Columbia we did a functional imaging study of the hypothalamus. I didn't know very much about the hypothalamus. I knew that it had not been functionally imaged before, and that it was a difficult structure to image because of the surrounding artifact generating tissues. The hypothalamus is essentially a puddle of water nestled in between a pile of bones and an air pocket. I referred to it as a swamp. Yeah, it was a swamp sitting between air and bone and everything that made the functional imaging just almost impossible. And yet, I was able to show activity in that area in relation to leptin infusion. And it occurred to me that because it worked on this part of the brain, it worked on sort of a diffusion principle, like things just oozed out of the blood into this area. Chemicals came in and chemicals came out. It wasn't like a neural system; it was a flow of liquid going in and out. I don't think it's a structure that really lends itself to typical neural investigations that well.

Tamas: I agree. And so, my question really is, how does the interface between those two systems work? You've got this chemical system that's washing something with something and something changes. I mean, you're talking about the blood flow coming from the periphery through the Willis circle. In principle, it shouldn't be that different from any parts of the brain, except that you're right. It's more concentrated. There's more of it. But the problem with those questions is the tools are really not good enough to answer a question in a very rigorous way.

Zoltán: I don't think it's just this sort of neurosecretory "swamp way" how the hypothalamus is interacting with the rest of the brain. For instance, if you look at some of the actions of the lateral hypothalamus on the thalamus and cortex, there is an extremely precise and targeted action of these lateral hypothalamic peptides like orexin/hypocretin, and, these can have very specific action on the entire brain that can change the alertness, arousal, attention or general state of that brain. I am looking at least three waves of actions here. One is the kind of endocrine action that your peptide is secreted to the bloodstream and then it is acting on a distant receptor of a neural or nonneural structure. The other way of communicating is through direct neural pathways from the lateral hypothalamus to other parts of the hypothalamus, to parts of the thalamus, claustrum or cerebral cortex. These pathways are extremely precise because of the actual neuronal connectivity. For instance, orexinergic lateral hypothalamic cells have direct projections to the prefrontal cortex and higher order thalamic nuclei and to the claustrum. Some of these projections either directly contact cells with synapses or release the peptide close to these targets. We recently did some monosynaptic rabies virus tracing from selected populations of layer 6b neurons from the mouse prefrontal cortex and we could reveal that some orexin immunoreactive cells from the lateral hypothalamus directly innervated these layer 6b neurons. The lateral hypothalamic input could elicit a fast transmission on a very specific cell population. In addition, these orexinergic projections might also release these peptides to the cells in the neighborhood and then the cells expressing the receptors will respond. Here comes the surprise: these higher executive centers of our brain, cortex, thalamus, claustrum have extremely specific receptor distribution to these peptides. This is now just a coincidence. These phylogenetically new structures like cortex, higher order thalamic nuclei evolved to make use of the preexisting hypothalamic signaling and evolved mechanisms that are highly dependent on these specific interactions. There are at least three ways how the lateral hypothalamus can act on specific cerebral cortical cells in a very intricate fashion. We have the endocrine action through vasculature, and you have the fast synaptic and slow paracrine action through these direct neuronal projections. What is even more surprising is that the cerebral cortex has a lot of projections directly back to lateral hypothalamus. The two structures are very intricately linked! We have seen some direct cortical projections from layer 5 pyramidal cells to the lateral hypothalamus[1,2]. Moreover, William Wisden at Imperial College found some long-range GABA cells from the prefrontal cortex with direct projections to the preoptic area and back to the lateral hypothalamus. Chemogenetic stimulation of these particular cortical GABAergic cells reduce the sleep need. Vlad Vyazovskiy and my laboratory showed that if you silence some cortical layer five pyramidal and dentate gyrus granular cells from postnatal development in the mouse, then you can reduce the sleep need of an animal by 1/3! I was told by Vlad that this is one of the greatest reductions in sleep that the field has ever seen!

Tamas: So, what do you mean by sleep need?

3 Cellular physiology of sleep

Zoltán: I learned from my colleagues Vladyslav Vyazovskiy and Lukas Krone with whom I have been collaborating on this issue for over 5 years. They always talk about a clock and an hourglass model to explain sleep need. The clock helps us to keep a 24-hour biological rhythm. The clock is controlled by the suprachiasmatic nucleus that guides the rest of our body to sleep at night and wake up in the morning. The other regulator of sleep is the "hourglass." This keeps track of the accumulated amount of sleep we've had. This hourglass slowly empties while we're awake and refills while we're asleep. Many centers that can switch the brain between wakefulness and sleep have been identified—but no hourglass center has been found. Vlad Vyazovskiy discovered that parts of the cerebral cortex can go to sleep while the animal is awake. We were interested to examine whether manipulating the activity of cortical neurons can also activate or inactivate the main sleep centers. So far, we studied the effect of silencing a layer 5 projection neuron population. We measured the time the mice spend in slow wave sleep, and we measured the sleep after sleep deprivation. These layer 5-silenced animals sleep less, and they recover from sleep deprivation much better.

Tamas: Did you measure stress hormones?

Zoltan: Yes, we measured steroids in this paradigm, but they appeared normal.

Tamas: So, sleep and oscillations are impacted or whatever. So that's great. One of the questions I have is how does it impact that that animal in the long run? What is the impact on physiology and chronic disease development?

4 Ablation of layer 5

Zoltán: The transgenic mouse model we used to silence layer 5, layer 6, and layer 6b neurons in the entire cortical mantle in my laboratory is to delete a protein Snap25 that is part of the SNARE complex. Without this protein no synaptic vesicle can be released in an ordered manner from that particular cell population after the recombination, which occurs around birth. We made comprehensive analysis in these brains, and we examined the developing connectivity, synapse formation and maintenance and myelination. Depending on the exact cell group that we silence, these silenced neurons die earlier. Layer 5 is the most sensitive. We don't know why they die.

Tamas: So, you're saying that these animals have less sleep need?

Zoltán: Yes. We observed this when you silence a particular layer 5 neuronal population. The cell population that we silence (Rbp4-Cre) is a mixed layer 5 neuronal population that has projections to other ipsilateral and contralateral cortical areas, to higher-order thalamic nuclei, to striatum and to basal pons, spinal cord. We also examined layer 6b silenced mice, but they did not show such change. A normal mouse is nocturnal, sleeping during the day and active during the night, but they do have some naps even then. If you just look at the sleep recordings from our layer 5-silenced animals, they sleep less during the day, they are a bit more active during the night. In the homozygous knockout a fraction of the layer five cells was silenced, and this also has an effect on sleep deprivation. If you keep placing new objects into the cage of the mice, they keep awake and you can sleep deprive them. The layer 5-silenced mice don't mind this sleep deprivation, the sleep deprivation is not catching up with them, whereas the control animals have a rebound. We don't know why.

Tamas: And then, you said, they die a bit earlier.

Zoltán: They live for a year, but their layer 5 might degenerate around eighth month of life. These mice have all sorts of subtle changes in synapse maintenance, maintenance of myelination as well. Moreover, microglia are also accumulating around layer 5 neurons.

Tamas: So, do they have circadian problems?

Zoltán: No, interestingly these layer 5-silenced mice have no circadian changes at all. We examined them while keeping them in normal light cycle, or when you keep them constantly in the dark. Also, we gave them 6 h of phase advancement, kind of giving them jetlag, they respond the same way as normal controls. However, not everything is normal with these mice. If you examine the morphology of the layer 5 neurons, they begin to have serious degeneration from around 8 months, at later stages of life. Initially, these neurons have normal connectivity, normal synapses and they also have normal myelination. The only difference we could see with Shuichi Hayashi is that the specialized layer 5 projections to the higher order thalamic nuclei never developed their characteristic specialized structures. Normally the thalamic neuronal dendrites protrude into the layer 5 gyant terminal, but not if the layer 5 neurons have been silenced. They form synapses, but they do not develop these protrusions. The layer 5 synapses in thalamus are also smaller than in control, they are beginning to lose the synapses, their myelin and eventually even the cell body is affected. All this occurs because they can't talk to the other cells in the brain. This might have an impact on what signals they receive from the other cells in return. The microglia activation is also different in this model. Microglia accumulates in layer 5 and in the target areas of the particular layer 5 projections. Probably the microglia are chewing these silent synapses and projections and eventually the cell bodies. I really don't want to bore you with the details, but this conditional Snap25 model is a very complex chronic degenerative condition. Only Joy's husband, Jim, would understand what is going on in a cell without a functional SNARE complex. If you remove Snap25 protein from a population of cortical neurons, then this cells population initially will develop fine. They even form synapses and myelinate their projections, but eventually they will lose their synapses and myelin and with time, some of these neurons will begin to die.

Tamas: What I would also like to know whether the animals themselves have shorter life. They have less sleep. Do you know how exactly and of what they die? When you mentioned the silencing of the different cell populations and differential death, was it about the cells or the animals?

Zoltán: Sorry, I was talking about the death of the different silenced cell populations, not the animals. We have so far silenced some cerebral cortical layer 5, layer 6, and layer 6b subtypes. The silencing of the cells in these models starts from around birth, since the recombination comes in shortly before or after birth. You begin to see the first signs that something is wrong with the silenced neurons around the end of the third postnatal week, and then you have a reduction in the terminal, and then about eight months of age, they start to die. In the knockouts there is no Snap25 protein, the neurons cannot talk to the other neurons with regulated synaptic vesicle release triggered by an action potential, but they can still release vesicles spontaneously. Probably, there are some other issues with these neurons without Snap25 effecting their vesicle transport.

Tamas: Is there a difference in their life expectancy?

Zoltan: According to our current licence we can't keep them longer than a year, but you are right, it would be great idea to compare life expectancy in these different models and also use more refined and timed manipulations for the silencing or activation of the different projection neuron populations.

Tamas: Did you silence or stimulate these very same cells in the adult? Did you have the same phenotype?

Zoltán: That's exactly what we are doing now in my laboratory. We use chemogenetic approach. We express artificial receptors genetically in the same cerebral cortical layer 5 pyramidal neurons and we activate them with a chemical substance that is only acting on that particular receptor. We stimulate or excite these layer 5 neurons. Basically, what we noticed was that if you stimulate layer 5 neurons, the animals really slow down. If you give them a very low doses, you already see that. If you give a bit higher dose than they have seizures.

Tamas: Okay, but the original effect was by silencing this layer 5 neuronal population.

Zoltan: This is why we recently used inhibitory chemogenetic manipulations in the adult and the preliminary experiments suggest less sleep, so there is an effect in the same way as we expected. If you excite layer 5 neurons them, they slow down and show sleep –like behavior, but higher doses can elicit seizure. We recently also used inhibitory chemogenetic manipulations in the adult and the preliminary experiments suggest less sleep. Even if you excite them, or inhibit them, there is a slowing down of behavior. You need EEG to see what's happening exactly.

Tamas: All this is observed by manipulating cortical layer 5 neurons. You were saying that you are now looking at mainly in layer 6.

5 Layer 6b neurons

Zoltán: I am most interested in these layer 6b neurons, but silencing these had the least effect so far. The reason why I am particularly interested in this layer 6b cerebral cortical neuronal population because they are the only cortical neuronal population that has orexin receptors. As you very well know orexin comes from the lateral hypothalamus and without orexin the arousal can be instable. Just think about the human patients with cataplexy without orexin. Since orexin receptors are selectively on the layer 6b neurons, if I silence these neurons, the cortical effect of orexin will not be mediated. This did not happen so far. However, layer five had the biggest effect on sleep.

Tamas: But those cortical layers, what do they do anyhow, originally?

Zoltán: Apparently, it's a wrong idea to think about brain function according to cortical layers, but layer 6b is so special that I like to emphasize their uniqueness. Some of layer 6b neurons have very widespread connectivity to prefrontal cortex, distant cortical areas, and they also have projections to higher order thalamic nuclei. As you will remember I mentioned our own data that monosynaptic rabies tracing showed that the lateral hypothalamus is sending projections directly to these layer 6b neurons.

Tamas: When we worked on this lateral hypothalamic system back in 1998/99, the way I started to think about this system is very specifically, the wrong word to use is that they are responsible for nonfocused attention.

Zoltán: That's exactly that we want to study further in this model. We would like to see the response to the so called "oddball paradigm." We silence selected population of layer 6b neurons and we look at how they respond to unusual auditory signals. It is like you hear a noise like "tee, tee, tee, tee," and suddenly "too" and then again "tee, tee," etc. We want to examine how they respond to this unusual noise. Do they pay more or less attention to these unusual stimuli?

Tamas: And have you tried any smell related signals?

Zoltan: No, but it would be interesting, since smell is not mediated through the thalamus. I don't know how to do those types of smell testing experiments.

Tamas: Why don't we do it? Even just the food cue and stuff like that, you know, hiding food in the cage?

6 Inhibitory signals

Zoltán: The inhibitory chemogenetic experiment is working very well and safe as an acute manipulation in a circuit that developed normally. Let's think about these experiments. I know that you don't like the idea that we silence these neuronal populations in a chronic manner from birth. You are right, that is changing the whole development from early stages. Of course, it is also interesting on its own right to examine the differences between the acute and chronic silencing protocols. If we want to manipulate a normal adult circuit, we need to use chemogenetic approaches in the adult.

Tamas: Yes.

Zoltán: When Jeff Friedman gave his Krebs Lecture at Oxford last year he showed a couple of interesting slides where he had some kind of unexplained central effect of leptin. These might be mediated through cortex, hippocampus or amygdala. Leptin had an effect on anticipation, anxiety, exploration.

7 What does the cortex do?

Tamas: And, as you just mentioned earlier, the way how this can communicate directly with the cortex, all of these systems will eventually utilize the cortex for output. As I indicated earlier, in my view, the cortex is a tool, it's definitely not the "brain."

Zoltán: Cortex has to execute, since nobody else is projecting out of your skull, just your layer five projection neurons. If you want to move a finger, or move your body, you either have to have an active ventral horn motor neuron, or a cranial nerve motor neuron. There is no other way to move your body. That is the final executive pathway. All those arise from descending cortical pathways. Some of this is coming from layer five.

Tamas: But if you decerebrate an animal, some localized functions are retained quite well.

Zoltán: Depending on the particular animal, the brainstem and spinal cord can retain some complex functions. They can do certain things, even though there is no continuity between the cortex and the brainstem or spinal cord. You will remember that I showed you a video of Sherrington's decerebrate cat preparation a few months ago. This cat could keep and adjust its gate extremely well without control above thalamus. Also, you have these stories about chicken that can run around without a head. My point is that just because the hypothalamus is such a diffuse and complex structure, it can have all these very complex interactions with the cortex. The receptors where they are expressed and where they are acting are very specific and this suggests some very specific functional adjustments that the hypothalamus can exert on these cortical circuits. Cortical researchers might not realize just how specific and widespread these interactions are they still consider the hypothalamus as a "swamp."

Tamas: No, I definitely have no problem with that at all, from my own perspective,

Zoltán: We can go even further and suggest that the state of the cortical, thalamic, claustrum circuits can be switched by these interactions. Lateral hypothalamus can change the state of these complex circuits by oozing out a little bit of lateral hypothalamic peptide somewhere, or acting through direct and faster projections through synapses, or release close to the cell bodies. Acting on very specific elements of these higher cognitive circuits you can completely switch one brain state into another. I agree, this is extremely fascinating and not so many researchers consider the hypothalamic control of experimental subjects.

8 Physiological homeostasis

Tamas: Joy, do you ever look at the homeostatic status of your subjects when you image them or record from them?

Joy: We can and do measure systemic variables, blood pressure, temperature. You know, we can ask questions about, how hungry are you. I don't routinely do this. I have in the past for experiments where, I thought it was relevant, but I don't routinely do it.

Tamas: I believe, based on my experience that it's relevant in everything. If you were to bring in people who were fasting like you usually do for a blood test of glucose, they will perform differently, or maybe your standard deviation or error will be different.

9 Global versus local effects

Joy: Let me address that in a different way. I'm interested in specific things going on in the brain. But the whole brain has this exterior layer of cardiovascular stuff going on, that is very affected by blood pressure, by excitability, by temperament, by temperature, it's so and so computationally, I remove it computationally. And what I do is that I average across all of my signals, and I say, alright, computer, take out every signal that's common across the whole brain, and leave me only the signals that are the delta. And so, in a sense, I can say that I computationally remove the effects of the homeostasis in order to amplify the effects of the conditions that I'm running in my experiments.

Tamas: I think it's a very good answer, but I'm trying to still work on it. Because what is your readout at the end?

Joy: Okay, the readout is the amplitude of a signal, that is the result of the change in either the oxy or the deoxyhemoglobin, in response to a task. And what I take out, is the overall change, like say that one is anxious, and there's a whole lot of blood that's rushing to the brain. Well, that's a big problem for me because I have to see that little thing down there that is activate by what I'm doing.

Tamas: In your setup, can you measure subjective responses and performance?

Joy: Oh, yes. I have, depending on experiment, a response dial. And people indicate their subjective responses. Or I can do things like reaction times, I can have them do a test reaction. I can measure things like performance or correctness, might tell someone to look to the right or the left, and I can measure how often they do what I tell them to do. You know, those kinds of things. It's nothing very sophisticated.

Zoltán: What you want to see is the difference on a functional level when somebody was very hungry or just had a Mars bar. We would like to see how this would affect the brain state of that individual.

Joy: When we ran those hypothalamus experiments, we made sure that everybody's appetitive level was exactly the same. I don't control for that usually in my current experiments. Maybe I should, there have been some functional imaging studies that have tried to ramp up their signals that gave their subjects M&Ms before they went into the scanner, because they thought that both the chocolate and the glucose actually increased the hemodynamic signal. And actually, coffee also has been shown to amplify the hemodynamic signal.

Tamas: I mentioned this before, and it's in my writing, too, it is this study of these judges that are looking at the parolees, and if they made the decision before lunch, they were likely to stay at the jail. If they made the decision after the lunch, they were more likely to go home. Yeah, and I think I've been using that in my own way. As an administrator, if I have contentious issues that I have to attend in the morning, I make sure that I have a breakfast before because then all of a sudden, if you don't, you're much more tempted to be agitated or whatnot.

Zoltán: That's true that your brain state can change because of metabolic or endocrine influences. It would be so interesting to see the very same network, but in a different state. Perhaps we could even image the transition from one state into another. Or, we could explore the way it is activated. You could just pump in a bit of lactate or glucose during the MRI imaging session and compare the results. You have the tools now to look at the spontaneous network activities even if you're not giving them any stimulus, just examining what's going on in the brain. How do these networks get activated with and without a Mars bar? I'm absolutely sure, with the right methods, you could see the differences how the state of the brain is changing.

10 Resting state signals and their interpretation

Joy: Now, to your point, Zoltán, there is a technique called a resting state. Participants go in the scanner and are asked to let their minds go blank, and then you just look at the correlations of the hemodynamic signals across the brain. And what you see are specific areas usually like, for example, motor cortex, that are highly correlated across both hemispheres and in some back and frontal. These intrinsic activities are there, and you could take one of those correlated networks and ask specifically, if it would be up or down regulated during some variation and homeostasis.

Zoltán: The chap who discovered this spontaneous network is Marcus Raichle. We had a discussion about the anatomical substrate of these spontaneous networks on a meeting in Frankfurt and he and I believe that in this network we can see the lower cortical layers quite a bit. Of course, you always see the structure you want to see, but I can see the interstitial white matter cells active at the bottom of the of the cortex as a major player in this activity. The claustrum shows great similarity to these cells; they are almost brothers and sisters developmentally. They have very similar neurochemical markers, transcriptomic signatures, connectivity, and also receptors. Both subplate and claustrum has orexin receptor expression, and they are both packed with orexin immunoreactive projections coming from the lateral hypothalamus.

Joy: So interesting, that has a ring of ground truth to it, that these low-level systems are regulating managing the higher-level ones.

Tamas: I'm not sure you noticed me smiling here now because I'm entertained by this notion, which I probably don't understand properly. And I want to ask, what did this guy discover?

Joy: He discovered that if you don't receive any stimulus, that is, no external visual or auditory sensory input and you just let your brain "zone out," you will still see spontaneous network activities.

Tamas: It's different from when you have a sensory stimulus?

Joy: If you're just looking at the firing of cells all over the brain, and you tell your graduate students to go look at what's been correlated with what, and you do this grand correlation matrix, you find that there are signals that are correlated. Now, it doesn't tell you why. And I don't know of any studies that have tried to manipulate that thinking as I understand it, I don't like this type of work and don't do it. So I don't even read these type of papers. But the thinking is that if signals are correlated in their firing the on and off and on and off and off, even though it's random, that they must be connected somehow. And that's the basic idea of resting state. People have probed the observations by varying the size of the segment that they use,

Tamas: I still don't understand what's the big finding here. We all assumed that things are connected in the brain. And even if you have whatever sensory stimuli, you're talking about, that you cut off, that you will have that communication oscillations going on, regardless of sensory stimuli.

Zoltan: Not everything is connected with everything in the brain and the resting state activity is also not entirely random. Some areas coactivate more frequently. If you look at these activities without any sensory stimulation, there are resting state activity patterns, networks.

Tamas: But this is where I have a conceptual problem with your discussions, because for me, neural activity is a major sensory stimulus, and despite of eliminating "sensory input," the brain is continuously receiving dynamically fluctuating stimuli from the body.

Zoltan: They do have a differential patterns of resting state network activities under those various metabolic conditions, for example, because under these different metabolic conditions, you have different sets of neurons with different sets of receptors. Depending on which set of neurons which set of receptors are activated, you can shift how these resting state activity patterns change.

Tamas: Yes, I completely agree with that.

Zoltan: You could have 1000s of "brains" embedded into one "brain." Basically, you can change which brain you have, depending on the state that is modulated with all these hypothalamic influences.

Tamas: I agree with that. I completely agree with that.

Zoltán: I tried to get funding for such project over the last 5 years, but I keep getting rejections. My proposal was to manipulate the target cells of the orexin projections in the cortex. The lateral hypothalamus has a direct effect on the cortex. This is mediated through the layer 6b cells, which are located at the bottom of the cortex. My hypothesis is that if you could silence the cells that mediate hypothalamic cortical interactions on cortical level. We anticipated that these animals will lose the behavior which is changing these states. We expected that they will not be able to maintain sustained arousal or keep attention. When we silenced layer 6b cells along the entire cortical mantle (similarly as I explained to you about layer 5 projection neurons), we noticed that the animals were not anxious, they walked out to the open arms of the elevated classmates. They also wonder out to the bright side of a box and don't preferentially spend more time in the dark side. I am sure that one can elicit some very interesting manipulations if you block these receptors in the cortex. For instance, one could get rid of the orexin or leptin receptors just in the cortical neurons.

Tamas: Interesting.

Zoltán: Your ideas from 30 years ago, are coming of age now. People will now try to get rid of the receptors in targets in the so called higher cognitive areas and have a look at the effect. That's the future.

Joy: And that's actually some of the ideas that are bringing us together, in a way. Except, the resting state network is not particularly interesting to me.

Zoltán: I am quite interested in resting state networks and would be even interested to study them during very early development. At very early stages there is not even sensory periphery that linked into the thalamocortical circuits, so all activity is initially coming from within the developing brain.

Tamas: Well, I'm just, again, I'm arrogant in this way. But because I'm coming from a different background from endocrinology and metabolism, I don't find resting states surprising or even an entity. A lot of people have been directed to study resting state, for no good reason. We don't understand anymore about how the brain works. I know of no studies that use in resting state as a hypothesis that takes us somewhere else. But so tell me the impact that resting state had on your development?

11 The value of understanding single cell types

Tamas: Can I also ask your view of the value of the single cell transcriptomic analysis of the whole brain?

Zoltán: I do not do such studies myself in my laboratory, but I am very interested in the results. I have just participated in an online conference that was hosted by the Allen Brain Institute in Seattle. I wanted to show you some examples how useful this data can be for us. For my own research, it is imperative to know how many different cell types are in our body or in our brain. It is also very important to understand how similar these cells are to each other based on their gene expression, morphology, projection site or physiological characteristics. Without this knowledge you can't start selectively monitoring or modulating these elements of complex neurological circuits. It is like knowing the trees of the forest. It is good to know what trees constitute the forest, even if you are interested in the forest itself. If you walk into a forest, of course, within minutes, you can say, I have pine trees, I have oak trees and things like that. You do not need to do single cell tree gene expression analysis to establish that. However, there are questions that you can only answer by knowing the exact characteristics of these trees. For instance, some trees are more susceptible for a particular adverse condition or a pathogen. You need to know why. It is not enough to know the general shape of a pine or oak tree; you need more detailed knowledge. And that's exactly what happened in neuroscience. Researchers were looking at the different cell morphologies since Golgi and Cajal. Their basic classification, based on morphology, is still valid. The different cell types received their name at this stage and we still use some of this cataloging today. This clustering of neurons based on their shape and location match with their projections, physiological properties or gene expression. How can we match up firing pattern with particular types of receptors for certain substances? Knowing the gene expression on single cell level can really help the field to move forward. We are now in a stage when we have a much more objective clustering of the neurons of our brain. You might say that this type of research is industrial, it's not very intuitive and anybody with lots of money can do it, but I have to emphasize that it is very useful.

Tamas: No, I completely understand. And I think it's good idea to have these big consortia doing all these cataloging. I think they do it better high-quality control, they have enough money to do. I completely agree with that. The question I have is: what impact it had on the conceptualization of brain that shifted our current understanding of brain development, or in any shape or form that makes it somewhat more, you know, touchable?

Zoltán: This information is eventually very useful for that. You have a complex network, and you know the gene expression of their single elements. Now you can go to the single cell transcriptomic data, where all the cells in the human brain are known, and you ask the question: Which cells will change their properties after you had your Mars bar? Which cells have leptin or orexin receptors? You will have your answer from all cells of the brain, not just the selected ones that you picked for analysis because of your previous bias. You will know the position of all these neurons not just in your preferred circuit, but in the entire brain and you just tap in the receptor you are interested in and get your answer from these large high-quality databases. You begin to see neurons or other cells in specific regions of the brain where you would have never looked for them. For instance, I am looking at layer 6b or claustrum neurons because of their specific characteristics. They both have orexin receptors. With the knowledge of other characteristics of these particular neuronal populations you can specifically target your investigations to include those.

Tamas: The Allen Brain Institute Atlas probably has the receptor expression of the orexin in the entire brain in mouse and human. You already had that information.

Zoltan: You need the single cell transcriptomic analysis to go further. You knew the orexin receptor expression in layer 6b, but you did not know in exactly which cell type of the layer 6b it was expressed, you do not know what other special characteristics those cell populations might have.

Tamas: So, you are saying that potentially it will have an impact.

Zoltan: It already has a huge impact how we design our research.

Tamas: But so far, it has not. Why would it be revolutionary?

12 The development of inhibitory neurons

Zoltán: Most of the understanding of cell types and classifications in the nervous system was done surprisingly well. These early classifications match the clustering of the cells based on their transcirptome. Let me give you some potential examples where some unexpected observations were made. I told you that for us interested in brain development, it is a very big issue to know where the inhibitory neurons of our brain come from. You know, 20% of our cerebral cortex contains GABAergic interneurons.

We still debate whether the GABAergic and the Glutamatergic neurons come from the same or different progenitors.

Tamas: Okay, let me ask you an arrogant question: Why should we care?

Zoltán: Why do you need to know this?

Tamas: Yes, why is this an important question?

Zoltán: This issue has huge clinical, theoretical (evolutionary and developmental) importance. Imagine that because of some genetic or developmental reasons you have a subtle change, maybe just 5% less or more of these cortical GABA cells. A particular subtype is never generated or you generate too much of them and you can't scale them to your needs during development. You could end up spending the rest of your life in epilepsy or you could have severe learning difficulties.

Tamas: That's a good reason.

Zoltán: You can use these single cell transcriptomics in combination with lineage analysis. You can actually answer the questions related to origin and relationship of your cortical neurons easily. It's not something my laboratory will do, because we just don't have the money and resources to do it, but the Allen Brain Institute or Broad Institute or other larger and well-funded laboratories can do. I am sure someone must have already done the single cell transcriptomics on the mouse hypothalamus. Did you have a look at that?

Tamas: I did. And I was surprised that the journal even published the paper.

13 The impact of "data dump" papers

Zoltán: I agree that sometimes these "data dump" papers end up in high impact journals with no particular reasons. Of course, editors know that they will be cited well and it is good for their journal. They publish them regardless what the referees say, but conceptually they are not always up to standards.

Tamas: Yes, there is an obvious impact. But our task is not to publish in high impact journals, but to understand how the brain functions.

Zoltan: That will come. These are amazing tools just to do that.

Tamas: Yes, but now it becomes like optogenetics, you need to do it every time you ask a question, even though they may not do anything to actually answer a question. We have a lot of information that we have to figure out what it means. I'm not saying that all of it is useless, I'm not putting this kind of research down. I'm just saying that all this will have to be placed in context and used properly to answer the big questions. Not it's like this, but it just reminded me of that when you talked about the resting state. Is that really important issue?

14 Can perceptions be read from the brain?

Zoltán: Coming back to the resting state, I would like to convince you that it is. Imagine that you can image or record from someone's brain with high resolution and you can see what areas are active and in what order these activity patterns change.

You can do it in resting state or you can do that when someone is watching a particular movie or you can do it what your subject is thinking about a particular person, object or a concept. It is not far-fetched, that with time, you will be able to tell what that person is watching and what they are thinking. You record from the brain in different paradigms, and then you use AI and match up these events. Eventually you can actually predict what movie they're watching or what they freely associate from their resting state. I do not think it is science fiction.

Joy: Oh, yeah, yeah. There people that have done that. One of them is actually is Jack Gallant at Berkeley.

Zoltan: Matching the movies to brain imaging and then to be able to predict which movie the person was watching is the first step. However, even these imaging studies have problems at the moment, because they train the same individuals for weeks, usually authors on the paper and them just show half a dozen distinct movies. However, eventually you could develop methods to decode from the signals whatever movie I am watching. Some speech can be already reconstructed from the signals you record from multiple electrode arrays before epilepsy surgery. You kind of use the cortex as a transistor to break down the auditory signals, and then using AI, you reconfigure it again from the multiple electrode recordings. If they can do that, eventually we can predict more things from the recordings and imaging and eventually we should be able to read someone's thoughts.

Tamas: But I don't believe that it works that way.

Zoltán: I think it'd be possible.

Tamas: I don't know why you would do that.

Zoltán: Why not? If you develop these methods, you will not have to do polygraph or lie detector test. Currently you record several very indirect physiological indicators such as blood pressure, pulse, respiration, and skin conductivity while a person is asked and answers a series of questions. Or you can really find out what someone is really thinking of you. Or, you can detect the effects of various stimuli in different brains and from that you can judge what are the differences in the responses, where do they stand in the spectrum.

Tamas: Okay, good. So let me tell you why I thought about this a long time ago. I can see that you can reconstruct some of the thoughts eventually from the living brain. But how about if you were able to actually reconstruct from information that you can find in the dead brain? Doing a postmortem fast analysis of all the expressed RNAs or immediate early gene expression, one could reconstruct what the individual actually saw before he/she died. I see the functional elements there. But I don't think it's going to ever be possible.

15 The binding problem

Joy: Well, not the way we currently understand the brain. You know, it's interesting, my journal club today was about the binding problem. The Binding problem is a very interesting problem. The fact is that the brain puts together many, many pieces of things that are perceived as a unity or a whole. So how is it that the brain binds bits and pieces, and then perceives them as a unit? I read a paper today about a study of faces, which is something that we're very interested in, where dynamic movies were made bits and pieces of faces. The bits and pieces were sort of roving around randomly and every once in a while, they would come together, and it would be a face. And then they would separate again. The events that the investigators were interested in was the neural circuitry during those moments of face-together, as opposed to the isolated parts. The hypothesis was that it would be the gamma band or the gamma wave oscillations, that would become more synchronous during the face percept representing the "moment of binding. It was a good idea; the data weren't terribly convincing. But anyway, the binding problem is an old, venerable, and interesting. These ideas have been connected to consciousness, for example, a conscious perception is one in which perceptions are bound together to create a unified view of the world.

Zoltán: This is an, interesting idea to test the "moment of binding."

Joy: Not quite sure, exactly how relevant it is to the fundamentals that we're interested in. But it's an amazing thing. It's been around for a long time, and people, smart people have thought about it in smart ways.

Tamas: So, I have to go because my daughter has just arrived back from DC and brought a ginormous amount of clothes that she left there last year. Okay, so what are you showing on screenshare?

Zoltan: I wanted to show you this summary figure of the classification of all cells in the mouse brain from an Allen Brain Institute publication because you're mocking all this single cell data.

16 Classification of cells in the mouse brain

Tamas: Look, I have very same idea that, you know, I would never do this kind of work. I'm not talking about that, to be honest with you, I would definitely not be I couldn't even do it. I don't know how to.

Zoltán: Once we have this single cell data, we should try to use the data. I keep telling this to members of my own group. We should use this data, because it's showing so much. For instance, in the past, we classified neurons, because someone classified these neurons some way, like Cajal or Golgi. They noticed that some of them are a bit more bushy or have strange shape. It is like when you walk in the forest, you kind of classify the trees. We learn how to call these trees, but our names and classifications are still subjective. Carl von Linné, the Swedish botanist, zoologist, taxonomist formalized the modern system of naming organisms. He thought that he put things into order, but he was also subjective. Old anatomists thought the same. And now, what you do is grind up the brain of different species, different regions, and you ask the computer to cluster them, you just decide on the threshold to detect the differences and similarities. You can probably end up with slightly different clusters when you are looking at just a nuclear transcriptome or the cytosolic, transcriptome or both. When you do that, you want your program to cluster this data and see who is related to who and then you have a couple of master genes, which you have been using for decades and try to link the old data up with the new. It is like linking Peter Somogyi's inter-neuron classification to the single cell transcriptomic data of the Allen Brain Institute. You can actually see whether your previous classification is actually valid or not, or whether we should put move some cells into a different category. What is very reassuring is that astrocytes and neurons, they cluster differently, the inhibitory, etc., neurons, they cluster differently. And micro glial is completely different from neurons. That's also reassuring. It is a very useful tool. But I completely agree with you Tamas, just producing this data in more and more detail without linking it up to other aspects of the brain will not going to answer the big questions of neuroscience.

Tamas: Let me ask you something on this data. Is all this information coming from a single animal? Or, are these from super imposed individual animals? Is it just massive number of cells, regardless of who they belong to?

Zoltán: You can do this from a single animal or a single human donor. You can also do it from many different animals and even different species.

Tamas: But do they usually use one single animal or donor?

Zoltan: You can do that from a single animal or a single human donor.

Tamas: Yeah, I see. I see. So can you compare different species? That will be very interesting.

Zoltan: You can compare the clustering of individuals of the same or different species. You can compare the clustering of the cell types in various clinical conditions due to genetics or environmental insults. You can identify what cells are affected the most in those particular conditions. I believe that this will have a huge impact on how we look at molecular pathology. It already has a major impact. Comparing cell types in different species is not that simple. The developmental program, the lineage, gene expression evolved separately and these cells can retain some similarities, but diverge on others. If you want to compare cell types in turtle, mouse, monkey or human, you have to consider that these cell types evolved further apart and they only preserve some characteristics. Your clustering based on these characteristics might reflect this divergent evolution and you are shooting on a moving target to try to link the different cell types based on their transcriptomic similarities.

References

1. Hoerder-Suabedissen A, Hayashi S, Upton L, Nolan Z, Casas-Torremocha D, Grant E, et al. Subset of Cortical Layer 6b Neurons Selectively Innervates Higher Order Thalamic Nuclei in Mice. Cereb Cortex. 2018;28(5):1882—1897. Available from: https://doi.org/10.1093/cercor/bhy036.
2. Krone LB, Yamagata T, Blanco-Duque C, Guillaumin MCC, Kahn MC, van der Vinne V, et al. A role for the cortex in sleep—wake regulation. Nature Neuroscience. 2021;24:1210—1215. Available from: https://doi.org/10.1038/s41593-021-00894-6.

Discussion 16—3rd June 2021

Joy: I want to continue working on my part because it leads in to sort of all the ideas that I wanted to present about the two brains, the social system, the interactive systems and who we are as interactive beings. The idea is that other brains outside your own brain are part of your brain in a way. That was the idea that brought Tamas and me together before Zoltán was invited to join in, Tamas' idea that the other organs of the body that are outside the brain are part of the brain as well. We were both thinking that a single brain is only one component in a larger system.

Tamas: Take for example my colleague who unexpectedly exploded at a dinner party accusing a guest of being a racist. It would have been great if I had videotaped, it would have been a great example of how something social can go in the wrong direction, you know, the two-brain thing.

Joy: Yeah, you bring up a really good point. And that is, with your example of the explosion in an inappropriate social situation, somebody just had no ability to edit or control an urge to be angry. How do these kinds of antisocial behaviors come about? And how do we deal with them? I mean, is this what mental health is all about? How is antisocial behavior managed? It's sort of a lead in to the kind of stuff that I'm thinking about that involves dynamic interactions between two brains.

1 Does "normal" mean the absence of antisocial behavior? How do we separate unusual behaviors into normal or mental illness categories?

Tamas: So along those lines there was a talk in our department a week ago, maybe a few days ago, actually, a woman was talking about in utero exposures, or insults and their impact on the mental health of the adult. It was predominantly associated with metabolic impairments, or alterations and so forth. But she said something very interesting probably for you, it's well known. She can, you know, group various mental issues and disorders in let's say, six different groups arbitrarily, but also she made the point that by the time somebody is 18, almost 60% of people experience some of these conditions which raises the question, what is normal? So who is normal? Am I basically coming to believe that nobody is normal? And we all have issues that are caused by various things. The fetal alcohol syndrome that Zoltán brought up is an extreme example of that, where you sort of know what is the intervention and that is to not to drink during pregnancy. Or when you have all these other potential impacts? Food, omega three fatty acids, blah, blah, blah? How do you come up with any sort of advice about how to deal with that, and so that you predictably can avoid something? I think it's virtually impossible, when you have fetal alcohol syndrome, yes, you know the answer, but when you add these other new things being discovered constantly between impact second third trimester, third trimester, and associated with that brain development, lateral, whatever, that she was talking about various sub parts of the brain, how do you make sense of that Zoltan?

2 What role does genetics and development play in health and disease?

Zoltán: We all have a unique genetic platform, which is very different from the others. We start our life with a set of cards, our genes that we have to get through life. The kind of cards we get for our life is different, we have those particular cards, we have to get through life using those. In addition to the different cards, we also have very different environment that we live in. The interactions between our genetic background and our environment are very complex. In a way, if you have a set of decent cards and you are a good card player with your existing cards (genes) you get through life fine if you know when to play these cards. Sometimes, there is a crisis, you don't have the card and you are in a situation when you need that specific card, otherwise, you cannot produce dopamine, you cannot arrange your neurons in the appropriate manner or you can't regulate synaptic release or cell death. The lack of a particular gene can have huge consequence on your entire life. For instance, if you cannot produce androgen receptors, you might have XY chromosomes, and you have SRY gene, you have Sox9 gene and you produce testicles and they produce testosterone, but then you cannot get a male phenotype. Androgen insensitivity syndrome will then have a major impact on that individual's life. This is just one example to illustrate the impact of missing one card/gene. Or, you can have some deficiency in the nutrients and that can lead some serious issues. Just think about the neural tube closure defects such as spina bifida and the levels of folic acid. There is probably no simpler inexpensive and effective means of preventing a serious fatal or disabling congenital disorder than implementing supplementation of folic acid, yet it has not been generally introduced around the world and the consequences are very considerable to those individuals and families who are born with spina bifida with paralysis and incontinence or hydrocephalus. I only mentioned two simple conditions, one genetic and one environmental, bot dependent on a single factor. Brain developmental disorders usually depend on multiple environmental and multiple genetic conditions. I think the development and function of the brain is much more complex than any other body parts, and that's why we don't understand these cognitive disorders, because of a highly complex interplay between genetics and environmental influences. Sometimes missing a card/gene will not have so much impact because we play our existing cards differently and under normal circumstances very different nervous systems can produce reasonable behavior and mental capacities.

3 Is mental illness a common mental state? Where is the threshold between "normal" and "not-normal"?

Tamás: So but isn't it also potentially impossible to say that all of us have some sort of mental issues? I am absolutely sure that almost everyone could be triggered in some specific situation and expose these issues. So, what the question is how to cope with these different conditions, not necessarily how to prevent their existence. Most of these anomalies might be normal. Of course, you can have extreme situations, such as in extreme fetal alcohol syndrome or severe metabolic condition, otherwise, we all have some form of anomaly.

Zoltán: The challenge is that if you have a condition, which will be normally exposed, maybe you can work on therapies, which will hide these conditions. There are interesting efforts to change the particular diet in cognitive conditions, such as schizophrenia or autism and this could adjust behavior. There are many forms of training regimes that can alter conditions and this can help individuals to get through life if they recognize these limitations and they adopt a behavior to cope with that. I had lots of students over the year who had attention deficit hyperactivity disorder and they went through medical school without much problem. We had to adjust the structure of the tutorial, but then they coped with this condition. I have never been diagnosed, but I am sure I have dyslexia. When I was a child, dyslexia was not recognized in Hungary and my teachers did not understand that I could not read aloud (in fact they kept asking me to do that in front of the entire class to make fun of my reading), but when it came to the IQ tests and other metrics that did not depend on reading aloud, even just reading comprehension, I came absolutely at the top of my class. I am sure many others had similar experience from my generation, but this probably just increased our resilience. Now about 10% of Oxford medical students get extra time to write their essays for their exams if they were diagnosed with some issues, like dyslexia. I am sure they will make superb doctors, but perhaps a few decades ago they would have not had a chance to go through primary, secondary or medical school.

4 How valid is our current information about the effects of diet on development and mental illness?

Tamás: I agree with that statement on the role of genetics and environment and how environment can compensate for many things. I'm just saying that during development, when you have the pregnancy, and you say, this is what you should do, this is what you should not do. This is the food you should eat, this is the food you should not eat. Every 15 years the guidelines are changing, there is a sort of recycling of ideas, this is bad, this is good. Now, this is bad, this is good. So, I'm skeptical about the possibility to inform a mother for example, in a manner that, you know, it's really truly going to be beneficial for the child. Other than, yes, of course, don't drink, don't smoke, and those kind of things that are much more straightforward.

Zoltán: Fortification with folic acid or dietary supplementation with iodine in certain parts of the world could avoid devastating conditions.

5 Human development is so complicated, how does it ever turn out right?

Joy: You know, one of the things is quite amazing to me, is how resilient development is, how many circumstances can be overcome by biology, which is, the smart friend behind the scenes, that covers an awful lot of deficiencies. But some things one cannot overcome. For some reason, fetal alcohol syndrome seems to be one of them. I mean, malnutrition, maybe is another one. You know, various types of neurological disorders and so on. However, the fact that there are so many of us in the "normal" category, when the variability of our developmental experiences is all over the place, is remarkable. Perhaps maybe that's what we should look at, how in the world is it that we turn out normal?

Tamas: The normal? So my question to this seminar speaker at Yale was, what is normal?

6 The concept of a spectrum for behaviors is useful in describing clusters of social qualities such as is autism spectrum disorder

Joy: Well, you know, I like the way behavior in autism is structured, it's a spectrum. And more and more papers come out showing that autism traits are prevalent in typical populations. That is just how much of one thing you have, rather than whether you have it or not. So thinking about your experience with your cocktail guests who got out of hand and just sort of popped off explosively and unexpectedly: you might say this was a problem of cognitive control and that this person didn't have much. Also, you could say maybe frontal lobe mechanisms failed to truncate or diminish the more primitive emotional responses. Some models propose that frontal lobe plays a role in shutting down those inappropriate explosive behaviors. These putative control mechanisms have been identified in neuroimaging studies. So you can propose the whole thing in terms of deficiency of cognitive control mechanisms and even propose some kind of separate therapeutic intervention. Nonetheless, does this supposed failure of a cognitive control mechanism cross the threshold between "normal" and not?

7 Antisocial behaviors can be self-promoting and reinforcing for some people

Tamás: Is it possible however, that actually this individual is like Salvador Dali, who is exploiting that trait and actually uses that to succeed? Salvador Dali early on was a very special person in high school, for example, he would throw himself down on the staircase, simply just to get the attention of the other students. I mean, these are not necessarily normal behaviors. This guest who got out of control during dinner is a very successful scientist. So one discussion that came up after the incident is that this person, like Salvador Dali has narcissistic tendencies and you know, it's bringing their attention to him entirely by everybody.

8 How much of our behavior is modulated by our genes? It is an enigma that our genes are so similar and our behaviors are so diverse. Can genes give us information about susceptibilities to conditions, such as alcoholism, heart disease, and dementia?

Zoltán: We have to accept that we are all different. One of the biggest asset humanity has is that we are a huge population, very highly monitored, our differences and similarities are very well documented. Our genetics, general behavior, cognitive abilities are very similar, but we are not the same. We have variations in our genetic background, we have huge variations how we live, we impose the different behavior and our living conditions to ourselves. Just think about extreme sport, extreme climate, extreme social behaviors that we can document and link it up to genetics and other parameters. Therefore, we humans are excellent experimental animals to study everything but this has to be done in on ethical grounds that we are all equal. Your colleague who got upset during your dinner is probably different. He has a different way of looking at the word, on relationships, conflicts and resolutions. We are not the same, we can be very different. Of course, 99.9% of the genes are the same, but we have some minor variations that can make the difference. It would be good to exploit these differences to learn about our biology much more. Who is susceptible for certain conditions? Wo is susceptible for addiction? Who is susceptible for depression or heart attack or these kinds of biologies? At the end we shall have excellent monitoring and records and we can test hypotheses easily. Unfortunately, we are not there yet. For instance, until recently in the United Kingdom, if you moved from one GP practice to the other, they have no idea about your previous paperwork. In some countries the network for the health documentation is much better. Denmark is always mentioned as a positive example in this context. Do you have a good documentation system of your health status in the United States? And can they compare you with your brother or mother or grandmother?

9 The problem of sharing medical information and records

Joy: We do not have an integrated medical record system in the US. It's as in as much as dis-repair as it could be. Within a limited system, like if you're faculty working for a university, then there is the Columbia system, or the Yale system, or some other system, within that system the medical records are easily shared. But if you want to step out of it, and exchange records, say you go from the east coast to the west coast, forget it.

Zoltán: An integrated medical record system would be extremely informative to detect regional, gender or age variations of certain conditions and this information could be used to combine them with other information that we all collect on exercise, sleep or social media use, etc. Just think about all the apps we now use on our phones that constantly monitor us. I have at least five.

10 Prediction of risk factors from SNIPS can be dangerous and contribute to unnecessary concerns or worries because the probabilities are not well-applied to individuals

Tamas: An attempt is by companies like 23 and me, and ancestry.com, that are trying to combine the genetic data with the traits and potentially with some medical information, which I think is in the direction that you were talking about. On the other hand, I have to tell you, it's extremely dangerous, because I looked at my snips. And I didn't ask for the, you know, vulnerability analysis, which you can ask for, but I can look at my snips. And I have a lot of snips associated with all sorts of different things, which, you know, maybe true for certain things, but definitely not true for others. And I think it's very dangerous because of what you just said earlier that because of genetics, and then very different environmental exposures. Those genetics may or may not have anything to do with your phenotypes, those ideas that you should be able to prevent things that those genes predict are very dangerous as well, in my view.

11 Normative versus idiopathic information

Joy: yeah, I think that's a really an important "Red Flag." It is so easy to make assumptions from data that goes beyond the data and is not true for a single individual.

12 Information gathered from vaccinated people would have been very informative. Should we vaccinate children?

Zoltan: It would be good to have the data, at least, Because for instance, when you got your COVID vaccination it would have been so easy to set up something on the internet, to document the side effects. Which vaccine did you get? What's your age? All sorts of other data to add, and nobody monitored it. It would have been a free and perhaps very important way to gather information to design future immunization programs.

They only monitored the original 40,000 people that tested the vaccines.

I wonder if you're doing that with the younger ages, because the question is coming up about vaccinating children.

13 How do we understand cognitive side effects following COVID vaccination and consequences for development?

Tamas: you know, this is a very interesting point, because I think at least I definitely had side effects of the second dose of Pfizer. And most people who I knew had side effects. Mine were relatively minor, but definitely for a day I was useless. My brain was not functioning well. And there is a reason why my brain was not functioning well, which at least in part is due to effects on microglia. And I think that has major implications for development as well. I think that my microglia was activated and microglia is involved in everything.

14 When we consider the possible physiological substrates that underlie behavior, health, and disease, we have to consider nonneural mechanisms including the microglia

Zoltán: ... including sleep, plasticity, memory. In fact, if you do genetic association studies to autism, and now have this human single cell sequencing stuff, the closest association with the autism susceptibility genes are genes that are expressed in microglia and not in neurons. Microglia is considered the gardener of the nervous system. They prune the flowers and bushes. We always think about the neurons as the major issue in cognitive conditions, but it might not be the case.

15 The brain is like a garden and there are many reasons why the flowers do and do not grow

Zoltán: ... you have a garden, and something is wrong with that garden then you have to consider whether it is a problem with the gardener. The problem is that maybe the gardener is too enthusiastic, cutting everything back more than needed, or not cutting anything at all. You still have your flowers in your garden, but they do not look good. The main cause is not in the flowers or bushes. Microglia can be activated by many different ways. The gardener's attitude is key. Perhaps vaccines can change that temporarily. The gardener can cut too much or too little. If you have too much of these going on, it might lead to schizophrenia, too little then maybe autism. That's the idea now in a very simplified manner.

Tamas: And it may be different, as we discussed in different people. So I may have a microbiome feature and think that is predictive of Alzheimer's disease, and I may not have Alzheimer's Disease. That's the problem.

Joy: Yeah, that sort of the probabilistic approach to predictions of a single individual just doesn't work. I mean, in science, we work with normative properties with the bell curve as our friend. We collect samples, and we look for normative properties. And that's important for doing generalizable experiments, and so on. But when you want information about a single individual, that method tells you very little, and that's the problem here. We think we understand something from a normative point of view. And we don't appreciate our profound ignorance when it comes to the individual.

Tamas: And that's my biggest issue with medicine, with internal medicine specifically, because of the ignorance of that very fundamental point,

Joy: Right. Standard of care is largely based on probabilities of an outcome. And the probabilities and outcome influence whether you prescribe a medication for chronic heart disease or something, as opposed to a detailed understanding the micro environment of that individual patient. Of course, that's really difficult. I mean, patients come in "n"s of 1, and you can't know everything about that universe. So it's complicated, I think the more options we have, the more complicated it is to make decisions about how to employ those options.

16 Where do all the strains of mice and other animals come from? Why are there so many?

Zoltan: Even mouse strains can be very different from one another. We usually select a strain for our experiments and then we stick to that because we do not like variability when we perform a particular set of experiments. We work with inbred strains that are virtually clones of each other. Today, I was doing some cell implantation in mice. I had to wait a bit and there was a big charter, about the origin of the different mouse strains in the word from you know, 1921 onwards. C57BL/6J strain started in Miss Abbie Lathrop's "pet shop" stock and C.C. Little (1921) mating of female 57.[1] Immediately above that line I could see the word "fekete" that means black in Hungarian. I would have loved to know who gave that name to that mouse at the stem of that huge C57BL/6 tree, he or she must have spoken Hungarian. All these strains are maintained by sibling mating for hundreds of consecutive generations. By selectively breeding brother-sister pairs, choosing mice with desired characteristics like litter size, weight, color, temperament and robustness scientists created "inbred" strains of mouse. It's incredible where these mouse strains came from, and how long and how specialized they are. I can't believe it's a single species still.

Joy: it's interesting It's like dogs and cats. Different cultures have had different dogs. I just learned about the most interesting dog, the lurcher. A lurcher is the name of a breed of the dog that is a combination between a greyhound and something else. Culturally, their story is that they were the dogs that traveled with the gypsies in Europe, and that these dogs were trained to go to the hen houses of the landowners, and steal eggs and bring them back without breaking them. This is a side-bar. But anyway, to your point. How did these strains get there?

Zoltán: It is interesting that now these strains are crossed again with mice that came from distant geographical location some Japanese scientists captured some mice and they had some mice from Russia. We have all these various mouse strains everywhere, and sometimes they were brought together and they made new strains.

Joy: Interesting.

Zoltan: Coming back to your lurcher dog story. I have seen an interesting video on Facebook: a guy broke his leg and was, limping and walking with some aid. He had to take the dog to the vet, because the dog started walking with a limp. However, the dog had nothing wrong. He was limping in solidarity with his master. I can imagine that after a generation these lurcher dogs started to walk and behave like their owners.

Joy: Funny. Anyway, we digress. Okay, so I have to go to another meeting soon.

Reference

1. Staats J. Standardized Nomenclature for Inbred Strains of Mice: Seventh Listing. *Cancer Res.* 1980;40:2083−2128.

Discussion 17—17th June 2021

1 Individual differences: savants and the biology of special talents

Zoltán: It's interesting that you have this huge variability in different sensory and cognitive skills in savants. I think we talked about the face recognition and super recognizers previously. In fact, I have very bad name memory, but I am a super recognizer. I took some of the tests that are available online from University of Greenwich, Reading, University of New South Wales and most of the times I pass at the level of a super recognizer. I know that Joy, you examined some very interesting personalities with all sorts of exceptional abilities. What kind of savants did you image Joy?

Joy: One of the savants that I imaged was George Widener, who's a very famous savant who does calendar calculations. He's also a very well-known artist. He does some very beautiful things with very detailed representations, often with numbers embedded in them. George Widener is a very interesting savant, because he's autistic, as many savants are. He's also very high functioning. He came to my scanner when I was at Columbia on several occasions. The idea that I had was that savant brains are very much like everybody else's brains, except they can do these complicated things. You can get them in the scanner doing these complicated things, and their brains light up, but you don't have a control experiment; we don't have a nonsavant person who can do the task. You can never know what mechanisms the savant has that are unique. I thought what we could do was ask the savant to do things that everybody else could do and see if they did them differently.

Zoltán: Did you publish any of these observations?

Joy: In fact, I have not published this paper, because the differences we found were very minor. Again, it's an N of one, there are differences. But it's very hard to credibly (as in statistically significantly) say, "This is how this savant can calculate dates and times back, you know, 50, 40, hundreds of years, for any event." That's what George Widener does easily.

Zoltán: Was your imaging performed while he was doing those tasks?

Joy: Yes, yes. I had him imaged during his incredible calendar calculation task. I also had him imaged during the typical tasks that people did, like object naming, simple language tasks, and memory tasks. The control was an atypical person who was to trying to do the same tasks. Well, in one experiment, I did that, but you know, getting a medical student or any bright graduate student to do these calendar calculations is quite difficult. I've had students that actually practiced a lot, and they were able to do the task with difficulty, but then they would do it very differently.

Zoltán: What brain regions of the brain were active in George Widener and what regions were active in the students?

Joy: When he was doing the task, George Widener used a very small part of his brain to do those things. He did them very efficiently using the parietal lobe primarily. The student was using his whole brain from the hippocampus throughout the entire temporal lobe.

Tamas: What is this calendar calculation? I don't know much about this.

Joy: Well, these are "megasavant" individuals, and it happens that they come about in just about every culture since recorded time, that there are a few individuals who have this particular talent and variant. They are able to visualize calendars back 1000s of years (in some cases), and you can ask them to tell you what date or day would be. They do it two ways: what day of the week does a particular date fall on or what date is a particular day of the week? So, if I ask, "What day was 14 December 1943?," the calendar calculator will say Tuesday or whatever, instantaneously. Or if you ask the date of Tuesday, the third week of December in 1943, then they can tell you that it was the 14th of December, 1943. I mean, you can do it both ways. You can get them to retrieve the date or the day of the week, and they just have these calendars in their heads.

Zoltán: That ability must be very common amongst megasavants, because when I just searched savants on Google, this comes up a lot.

Joy: Dates, facts and dates, yeah. When you look at savants, you can find George Widener.

Zoltán: On Google, I found an artist named George Widener. Is there an artist, as well, or is this the same person?

Joy: Yes, he's an artist. That's correct.

Zoltán: That is very interesting that he is an artist, as well. Art can be extremely revealing about the cognitive condition of the artists. Just think about all the examples that you can see from artists who had migraine, epilepsy, stroke, intracranial tumor at various locations, multiple sclerosis, parkinsonism, dementia, schizophrenia, or hemispatial neglect. You can almost monitor the progression of the conditions by examining the artworks. What kind of art is he producing?

Joy: In fact, in my office right now, I'm looking at a painting that he painted for me. Well, they're nonfigurative. They're very detailed.

Zoltán: I found some pictures on Wikipedia. Do you want me to share my screen?

Joy: He has one that is called Blue Monday. That's very famous here. It was very interesting that George Widener was very preoccupied with the Titanic.

Zoltán: I found his picture.

Joy: There he is. That's him. That's his picture. I have his painting he gave to me as a gift. You see, lined up dates, and that's his painting behind it. That's probably something that he did. Yeah.

Tamas: Let's see. Also, he's preoccupied with numbers.

Zoltán: This is what you can find on him in Wikipedia: "...as a child growing up in Kentucky in the 60s, he exhibited exceptional arithmetical skills. He was also compulsive drawer, with a photographic memory and an interest in machines. He joined the US military at 18 to work in intelligence." Can you imagine working in the military with these abilities?

Joy: Yeah.

Zoltán: He was based in West Germany using his pattern recognition skills to analyze photos from the Stasi and the KGB? He was even flown to England to advise the British Army during the Falklands War. He left the military because of his poor social skills and enrolled in the University of Texas to study engineering, but his mind was so full of numbers and dates that he was unable to cope with the course. He ended up living in hostels and on the street.

Joy: Yes, he became homeless for a good long time.

Zoltán: This is what he writes about himself: "I began to get a bit obsessed with these things I had as a child, these numbers and stuff. They calmed me down in my times of stress. I sort of regressed. I started to retreat. I was filling my notebooks up with the dates. I had dozens of notebooks. I was keeping these notebooks in a back-pack." Joy, did he give you the impression that he was suffering because of these exceptional abilities? How was it to interact with him?

Joy: I actually felt like I had a meaningful relationship with him. I enjoyed talking with him very much. He's a very sincere, gentle person; he's limited in many respects, but just interesting, a warm person that one is naturally attracted to. Of course, he was in my lab because he wanted somebody to try to image his brain. I agreed to do that—I actually have the manuscript sitting in my desk still. I just don't know what to do with it because the data aren't very compelling. The answer to what makes him special is not in the images, really.

Tamas: Because it's in his liver!

Joy: Oh, silly me. I should have imaged his liver.

Tamas: Yeah, yeah, apparently.

Zoltán: Can you imagine that he could clear all the cash from all the casinos with these abilities?

Joy: No, because this might not work in gambling very easily. He doesn't have a crystal ball in his head.

Tamas: But what he does have is the ability to remember long strings of numbers. Well, this is what people can do in blackjack, they can count cards, and they can figure out patterns. I watched or listened to some interesting program on public TV or radio, and they were talking to different people about how they do calculations, you know, like if you must subtract, 30 from 68. How do you do that?

Joy: That's kind of interesting.

Tamas: People have proposed that there are two strategies: One is sort of the exact strategy where you exactly do the computation, and the other is the approximate strategy. Is that when they round up or down to the nearest multiple of 10, subtract the numbers, then adjust? I forgot the details of it, but it was kind of interesting to figure out how people are using their brains to calculate one way or the other.

Zoltán: Joy, did you do some structural imaging on George Widener's brain?

Joy: Yes, I have the structural images, his brain looks perfectly normal.

Zoltán: Anything abnormal?

Joy: Nothing. His brain looks perfectly normal structurally; there's nothing about his brain that is notable, given the resolution of a 3-T MRI. I believe we did a single scan on him. It will be difficult to compare because there's so much variability in the typical population. It's an N of one, but with savants, it's hard to know if they are homogeneous even though conventionally, they are distinguished by these calendar calculation and extraordinary number abilities. It is not even clear that they are even alike, although they do those things the same way.

Zoltán: Because I think "Rain Man," Kim Peek, did a similar calculation.

Joy: That's a true story of a savant. He was a good friend of George Widener. Yes, yes. And George cared very, very much for Kim Peek. When Kim Peek died, George was absolutely bereaved; it was a very major loss to him. In some of his paintings, I remember George told me that he would leave messages for Kim Peek in the form of mistakes in series of numbers. He says that he knew Kim would understand, but nobody else would.

Zoltán: He was very much aware of their special abilities, and he was very much aware the special ability of his friend. He wanted to communicate all this to his friend Kim Peek, and this does require some empathy to assume all this.

Joy: Yes.

Zoltán: Kim Peek, he was a megasavant, who had some congenital brain alterations. He was the inspiration for the character of Raymond Babbitt in the movie Rain Man in 1988. I found some stuff on him on the internet just now: "Scientific investigation in 2004 at the center of bioinformatics space Life Sciences at the NASA Ames Research Center examined Peek with a series of tests, including computer tomography and magnetic resonance imaging. The intent was to create a three-dimensional view of his brain structure to compare the images to MRI scans performed in 1988. These were the first tentative approaches in using noninvasive technology to further investigate someone's abilities. A 2008 study concluded that Peek probably had a rare X-chromosomal genetic syndrome that causes physical anomalies such as hypotonia, low muscle tone, and macrocephaly (abnormally large head). He died from a heart attack at his home on December 19, 2009. Age 58." I think he could read two different pages with his two eyes synchronously.

Tamas: I tried that, because my own eyes do not line up, but I didn't succeed.

Zoltán: We are not megasavants.

Discussion 18—17th June 2021

1 Study section blues

Tamas: I was on a study section, so I had little time to work on my chapter.

Joy: After being on a study section, I always have about three days of depression, because I feel so bad for the people who couldn't get funded. The resources are so limited, and the critics are so ingrained and powerful that it's depressing.

Zoltán: When I served on the MRC Neuroscience and Mental Health Board for 5 years, I was also depressed after the board meetings (three to four times during the year). There is such a low percentage of grants that can be funded.

Tamas: Yes, I am still recovering from my session yesterday.

2 What is a brain?

Zoltán: Let's talk about the question, "What is the brain?" When I talk about the brain, I never think about the peripheral autonomic nervous system; until I started talking to you I would associate the term brain with the tissue that is inside your skull. However, brain is much bigger than that. This is becoming more and more obvious that your brain extends to your entire body. Some even go further and they say that in addition to the autonomic nervous system we should consider the microbiota. Now they are saying that you have another type of brain that is comprised of microbes in your gut.

Tamas: I'm a little bit skeptical about many aspects of that.

3 Diversity of the microbiome

Zoltán: A friend of mine, Christine, sent me a YouTube video link about these issues, and this further increased my interest. The Stomach: Our Second Brain I ARTE Documentary: https://www.youtube.com/watch?v=UJ80OIHO0Wc. Unfortunately, I do not think you can play it in the US because of some restrictions that I do not fully understand. In this documentary, they made the point that a fat mouse has a different constitution of its microbiome than a lean mouse. If you do a transfer of feces from the fat mouse and put in the lean mouse, the lean mouse becomes fat. The reverse happens, too, and that was a leap.

Tamas: That has not been confirmed in real time. In 2007, I was in Shanghai. That was in the summer of 2007, and it was before those papers came out. I visited a lab in one of the institutes in Shanghai, and the guy was working on the microbiome before these things came to the surface here. He showed that based on whether an individual was eating a Japanese diet versus a Chinese diet versus a Western diet, they have almost nonoverlapping microbiome constitutions. The microbial species are very different. I have a very strong memory of this. What ensued was 12 years of people claiming that if you have schizophrenia, if you have depression, if you have this, if you have autism, it's all the microbiota, it will explain everything. So, they definitely went to the extreme on that. Going back to the guy's experiment, if the Chinese diet is the "champion," and you eat the American diet, which has a very different constitution of the microbiota, does that mean that if you eat Chinese food next week, you're more intelligent or something? You may be autistic on Chinese food, but not on a Japanese diet? So, there are some far-fetched claims in this whole field.

Zoltán: How about if you have a ketogenic diet?

4 Diet, behavior, energetics, and microbiota

Tamas: Definitely, I think you have a very different driver of cellular energetics, which will have a huge impact on signaling events in the cell. So, you might anticipate that would change your CNS function. Yeah, I buy into that completely.

Zoltán: When I went to a meeting in Sydney, Australia, I was talking about brain development, but the meeting was very broad, and they were talking about parabiotic supplements. Some of the speakers started talking about the transplantation of feces in autistic kids, to transplant the microbiota to them. I asked them about these probiotic yogurts and drinks, which you can buy and eat. It's a multimillion-dollar industry. They laughed at my question. They told me in front of everyone that they are all ineffective, and they don't work. They told me that if I start administering these from the other end of my alimentary tract, and then you do the transplant, yes, it'll work. But if you eat your yogurt, it's not going to help you, apparently, because nothing survives the stomach acidity. So, for instance, I love kimchi.

Tamas: Me too.

Zoltán: That's the best thing because this spicy fermented cabbage is, apparently, good for you. I don't get to eat it very often. There are not many places to get it in Oxford.

Tamas: You can get some in New Haven.

Zoltán: I watched this documentary film on YouTube that Christine recommended. It was in French, and I understand you cannot open it in the United States. They described an experiment with anxious mice and more tame mice, and when they did feces transplants, these mice changed their behavior. That's why they say your microbiota is also part of your brain.

Tamas: Again, these are very good arguments and popular things to say. There are about two kilograms of bacteria in your gut, and that can have a big effect on you.

I don't think, to be honest with you, that there are tools available that could really answer those questions today.

Zoltán: You know, there are 200 million neurons in our stomach. They say that this is approximately the same number of neurons that a dog has in its brain.

Tamas: Is that correct?

Zoltán: I don't think it's correct about the dogs. I have just read somewhere that dogs have 530 million neurons in the cortex, while cats have 250 million. So it is more like you have a cat brain in your stomach.

Tamas: The mouse has a million.

5 Neurons in the stomach: are they part of the brain?

Zoltán: Yeah. But let's say 200 million. So, there are quite a lot of neurons in your stomach. And we must teach these ganglia to medical students—the sympathetic paravertebral ganglia, intramural parasympathetic ganglia, the myenteric plexus, also known as Auerbach's plexus between the layers of the esophagus, stomach, and small and large intestines. You also have the submucosal or Meissner's plexus between the circular muscle and mucosa. I think most people are aware of the importance of the vagal system. But again, for whatever reason, we do not think about these cells as part of our brain, or we do not consider them as our second brain.

Tamas: Yes, it has an important function in regulation of the intestinal function, but surely, you cannot call it a brain. Well, it's a question of how you define brain.

Zoltán: How did the word "brain" come about? If you had 302 neurons as a nematode, then you would call it a brain, if you had 200 million neurons, as the cat has, then you would call it a brain. Human brain has around 100 billion cells, but that 200 million in the periphery outside your skull and spinal canal is still significant. Okay, so that was the first point they made in the film. The second one was that the brain you have in your skull and spinal canal and the brain that you have in your gut have to work together.

Tamas: Sure.

Zoltán: I accept that our discussions have now converted me to think that way.

Tamas: You know, that is not only true for that, it's also true for your hepatocytes. They work with your brain indirectly through the autonomic nervous system.

Zoltán: What do you make of this statement, that you have different enterotypes? You know, that some people have different autonomic nervous system components than the others.

Tamas: That is for sure. I mean, I hate to come to this very simple argument about how regular one is versus how irregular somebody else is. That's partially because of the nervous system.

Zoltán: The whole peristaltic movement is related to nerves.

Tamas: And there is a humongous amount of individual variability in that. That variability, is it linked to some other entities of your brain? This is something that I've been very much interested in—whether any traits that you have in the periphery would be correlated or functional or causally connected to behavior traits of some sort. I think because they are physically connected, you could assume that yes, you will have traits associated with gut movement or a heart function or liver function that occur along with certain traits, behaviors, or brain functions.

Zoltán: For my third-year medical school project, I was going out with a very nice girl, Nóra, also a medic, and I wanted to work with her on a scientific project. The project was very simple. We went to the abdominal surgery unit and measured the blood pressure of patients before and after putting their hands into ice cold water. This can be considered perhaps an autonomic nervous system reaction test. We measured how the blood pressure and the pulse changed. Then they were operated on, and they all had some kind of abdominal surgery, such as cystectomy or hernia repair. We followed up these patients and visited them a few days after surgery and we asked them whether their intestinal motility was fine or they experienced some difficulties. We also examined them and listened to their gut mobility to see whether their guts started to move again properly. We were interested to detect whether a small minority could develop paralytic or spastic ileus after surgery and whether this could have been predicted from the changes in blood pression after the ice-water test. The idea was that if your blood pressure changes more or less when you have this ice-water exposure, that could predict problems with intestinal motility after surgery.

Tamas: Reasonable approach.

Zoltán: We had a very low n for this after several months of hard work. We had less than 100. Also, just measuring blood pressure change and pulse might not be sensitive enough. There are other parameters that we could have included, in retrospect. Nothing significant came out of the study. But I think those are very good questions, and I assume people are pursuing many of them. Now you could probably explore these questions by measuring catecholamines, or you could perform a much more sensitive measurement of the reactions of the central and autonomic nervous systems.

6 Enterotypes and behavior

Tamas: But what do you mean by enterotypes?

Zoltán: I have no idea, but that's the term they used in the film. We have all different mutual brain relationships.

Tamas: I'm sure we do. I'm sure there is individuality and everything.

Zoltán: The other question is when watching this film was that you can diagnose neurological conditions like Parkinson's from a gut biopsy, and that's been going on for a long time. One can determine the age before the onset of the condition. How do you explain that?

Tamas: Is the person who started that from Germany? I think they were looking at people for a long time.

7 What does the intestinal autonomic system have to do with the substantia nigra?

Zoltán: How do you explain it? I usually teach the medical students that the neuroretina has a central nervous system origin, and the "retina is a window to the brain." You can diagnose all sorts of neurological disorders by examining the retina. What does your intestinal autonomic nervous system have to do with your substantia nigra? One could assume that maybe the pathology that underlies a substantial amount of dopamine cell death also occurs in other parts of the body, including in the gut.

Tamas: Actually, we have been thinking about that in many other diseases, whether you could take a muscle sample from someone and see whether that will give you some information that would be predictive of something in your brain. I think these all need to be investigated. The problem is, as I thought while was sitting on an Alzheimer study section yesterday, that when you present an out-of-the-box approach or idea, there's this tremendous reluctance to support that. You need money to start researching these kinds of things. Some of these people like Braak, they didn't care about publishing. They were just very interested in doing the work, not getting grants, and so forth.

Zoltán: That speaks to biases in contemporary biology. The film was talking about the two kilograms of bacteria in your body, and how after birth, these bacteria are colonizing your gut. Actually, I think they're probably doing it even before that, because in utero, we swallow the amniotic fluid. We also release urine into the amniotic fluid.

8 Where does the gut bacteria come from?

Tamas: I think that the amniotic fluid is assumed to not carry many of these bacteria. The bacteria are really introduced during the birthing process by traveling through the birth canal.

Zoltán: Really.

Tamas: Yes, that's the contemporary idea. Today in The New York Times, one of the discussions was about a guy who is interested in in pursuing this old idea. They call it a particular thing, whoever brought it up in the 60s. He only does things that were already available 500, 600 years ago. He eats that kind of stuff. He doesn't do things that emerged in the last 50 years. He makes the point that he doesn't use mouthwash, although it's very popular, but mouthwash removes the good bacteria and so forth. It's something that I suggest that you read. It's kind of an interesting argument, and they made the point about the mouth because it's loaded with bacteria, obviously.

Zoltán: Apparently, you have more bacteria in your mouth than at the distal end of your alimentary tract.

9 Dementia, chronic inflammation, and the mouth

Tamas: Maybe so, and if you use mouthwash, that kills all of them, or many of them. Oh, he did say very something very specific, that it has been associated with dementia. Actually, there is a set of observations that many related things have been associated with dementia.

Zoltán: I heard that oral hygiene is related to dementia, because your gingiva around your teeth is a large surface where you can have chronic inflammation, and that can have an effect on your immune status.

Tamas: A mother of my daughter's classmate had these tooth implants in her early 50s. She developed Alzheimer's and died within a few years. Now, she obviously had a genetic predisposition, and nobody connected those things, but this is a recurring theme about the mouth.

Zoltán: These issues should be studied further in Alzheimer's disease. We need solid, rather than anecdotal evidence for all this. It might take decades to get such data collected properly, but someone has to do it.

10 A science hero: Katalin Kariko

Tamas: Sometimes you must have the vision and the guts to continue research wherever you see some interesting links. That's why for me, this Hungarian biochemist, Katalin Kariko, with the mRNA vaccine is a very special case. I think she's more like a saint. Very rare, I think that she represents a very rare breed of researcher. You read about her. She invented the technology that made these vaccines available. I nominated her for an honorary degree at Yale, but it didn't go through.
Joy: Try again.
Zoltán: I was so glad to see that she got the Honorary Doctorate from Szeged University. In fact, my brother Béla got "Mecénás-dij" - Patron Award at the very same occasion and he managed to get some pictures with Professor Kariko.

11 To be a scientist …

Tamas: The point is that this is the kind of person who was never really ruined by the fact that people completely ignored her, oppressed her, and fired her. She just kept plodding along. That's a trait that very few people have. I don't have it. I'm fine, but Katalin Kariko was constantly running against the current, constantly from an early age.
Zoltán: You have to be born with those traits. You can't just become like that. There are so many similar stories about scientists who just had to go through difficult times, but they never lost confidence. Once Gyuri Buzsaki was asked what is the most important trait of a scientist. He replied: how to handle rejection.
Tamas: Yeah, I think that's right.
Joy: I think that scientists have to be prepared for very few reinforcements, because they come so few and far between. It is very important that your compass is so internal that it's not shaken by exterior opinions or conditions that alter what you want to do. Scientists that have contributed the mRNA vaccine are truly heroes.

Discussion 19—24th June 2021

1 What is dance?

Tamas: So, the point is that the dance is a set of movements mimicking a purpose, but it serves its purpose with no purpose. I was thinking about the question, what is dance?

Joy: It's a really interesting question, what movements in dance represent, and dancers think about this a lot. I listened to a tape just recently of one of the national ballroom champions, and he was talking about dance, and what distinguishes a really good dancer from one that's just kind of an amateur dancer. The description is that an amateur dancer can do exactly the same movements, but you just kind of skip through them, and there's no passion, but a good dancer... What he said was that the "good dancer will make you cry." I think that's a very telling statement: a good dancer says something in movement that you, as a watcher, incorporate. It gives you an emotional feeling somehow. That's very visceral, I think.

Zoltán: Which a good singer and a good musician will also do.

Joy: Yeah, I think with dance or music you are not limited to spoken language.

Tamas: I was wondering to what extent, and obviously, in the animal kingdom, you observe dancing, you observe movements of animals...

2 Why do we dance?

Joy: ... but it is absolutely fascinating that we dance, like what in the world does that mean? And there's a big difference between whether you're watching a ballerina, which is a solo act, and whether you're watching or actually dancing in partnership. The story in partnership is very different from the story from a ballerina or a performing artist, such as a modern dancer. For example, in partnership, the story is a conversation. You and your partner have this way of producing those movements together. That has an added attraction, I think, over just a solo dancer. It's very different.

Zoltán: If you look at the animal kingdom, you also have both types of dances, solo and in partnership. Dance is used in courtship, when various animals display a set of behaviors that males, or sometimes females, use to attract a mate. This display is key for sexual selection. These can include dances, songs, or even temporary ornaments or tidying up their display. I am sure you all watched David Attenborough's natural history documentaries. I remember a really impressive set of dancing behaviors of the birds-of-paradise from New Guinea during which they show off their impressive plumage.

Tamas: Displays are key for sexual selection in any animal, for that matter.

Zoltán: They behave in a certain manner to attract attention, and that is a very important part of how they select a partner. If you think about this, it is a fantastic test of the genome and how well the unfolding genetic program interacted with the environment to produce that particular behavior. The female will select a partner based on how well they dance, sing, and display their plumage. In a way, all of that can be the phenotypic readout of their genes, and shows how successful they are—how shiny the feathers are, how well they sing and move, and how well they can entertain. Then you also have the males and females dancing together.

Tamas: I'm pretty sure that there must be a history of each of the dances we do. What was the origin, where did each dance come from?

Joy: Absolutely, and of course, scholars have spent a lifetime studying the origin of the dances. It's really interesting that some of the Latin dances we do have their origins in Africa, and the mixing of cultures and music have contributed to the evolution of the dances we do now, such as the rumba, cha cha, and samba. Those dances have a very earthy quality, and the movements are very different. They're down to the earth, they're into the floor. It's very, very sensual, actually, explicit. The way the dances evolve goes along with the evolution of the music, and so then, if the dance gets up to Spain, it becomes more flamenco, representing the music, etc. If it goes to England, they write about it and it becomes a little bit more rigid. We refer to Jive as a Latin dance, but it originated here in New York, and melds a lot of our street cultures.

Tamas: I wanted to go and see the movie The Heights—about Washington Heights.

Joy: I want to see the movie because of the dance music and the reflection of the Latin-American Manhattan society, Samba, not so much rumba, but they do a lot of Meringue.: And a little bit of "hustle," actually.

3 Can we study the dancing brain?

Tamas: So, you could study these people dancing using your fNIRS techniques.

Joy: We tried to do that, but it was a little difficult because dancing has a little bit too much movement for our techniques, but pretty soon, I'm receiving a pair of head caps from a company that wants me to test them. The caps are wireless, and one of my ideas is to put somebody in a dance frame and see if we can image partners doing, say, a waltz.

Tamas: Let's see what you can see! You may not even need to move; you can visualize your dance in your head.

Joy: Absolutely. I do that to train as a dancer. Yeah, I close my eyes and just review the school figures for the choreography that I am doing. As you know, any athlete (I think) learns that you have to do this mental rehearsal.

Zoltán: Joy, when you do a dance, do you have a particular moment when you must decide how to move? When you make a choice and pull out a particular version from two, three, or more motor programs? I am talking about moments when you can improvise from various possibilities. The reason I'm asking is because a tennis player is also visualizing his movements, and you can see some players swinging from both sides before settling to return a serve. When they have to release a shot, they have three or four choices, but at some moment, they have to release that motor program to hit the ball a certain way. I'm interested in the brain activity that occurs when we make that choice and lock into that motor program. When you dance, you have an algorithm, a pattern that you have to follow and do in concert with the music and in concert with your partner. One movement is triggering the next one. We learn how to generate these automatisms, the choreography. That's probably why dancing is such a powerful protective mechanism for Parkinson's disease, and also against Alzheimer's disease, because you have the rhythm, the music, the movements of your partner, and you have your own movement algorithms.

4 The magic of music

Joy: Even if you have Parkinsonism, there are patients who still dance extremely well, even though they have profound motor disabilities. There are examples in the literature where people can't walk, but they can dance. Oliver Sacks describes cases. One thing that we have in dance that you don't have in tennis is music, and one of the goals in dancing well is to represent the music in its most beautiful, original form.

Richard, my dance partner, and I have the notion of "pure Latin," which is dancing what the music says. I think with that, there are fewer decisions in ballroom dancing because you're so entrained by the music. Then comes the partnership, and in Latin dancing, the best partnerships work as a single unit.

5 Social partnership and couples dancing

Tamas: Who is in charge while you dance with a partner?

Joy: In the Latin dancing, we are both in charge, but of slightly different parts. Richard is in charge of "floor craft," and I do my part to make it beautiful. Somehow, we communicate continuously during the dance. It's a kind a language. I wish I knew how it works.

Tamas: What if you were to put with another partner for your upcoming competition, so you do the same dances, and then you change partners. Would that ever work out?

Zoltán: Would that work out with a substitute partner, like Baby volunteers to stand in for Penny with Johnny in the Dirty Dancing movie?

Tamas: I tell you why I'm asking that. When Sabrina and I enrolled in tango class, I really didn't like that you constantly had to change partners, but maybe that's a regular way of doing these classes—to make everyone switch partners. You really don't develop partnerships. First of all, you go with your spouse, and you work best together, but all of a sudden, you're moved around to different partners, and you never really have a chance to develop a partnership.

Joy: I do not like that, either, but apparently, this is common when these dances are taught by the professional studios.

Zoltán: With Nadia, we went to a Salsa class at the City of Oxford Rowing Club once. At the beginning of the class, you have to form a circle. Then you move around, and you have a new partner after every 5 min, but then you can actually stay with one partner at the end. People are changing partners like this at the beginning because it gives them an opportunity to meet people. For a dancer, it must be frustrating because it is a separate thing.

Tamas: So, it's interesting, you don't even envision that you could be able to dance properly with another person.

Joy: Well, I mean, I can dance perfectly well. If you give me a rumba, I can dance a rumba with anybody, but not in a competition. Richard is a professional dancer, and I am an amateur, but we have a seasoned partnership and have been competing as a pro-Am team for many years. So a school figure is no longer just a school figure with 1—2—3—4 beats. We work off of each other; I'm into him, he's into me, and we're pulling each other. We're following and leading each other at the same time, and it involves both of us as a unit in a way that does not happen unless you've trained for that.

Tamas: Now, if you were to dance a tango, would that be different? In at least a few lessons I took, communication is hugely important, and it's driven predominantly by the man.

Joy: Yeah, that's a little different. In Argentinian tango, everything is a little bit different. There's this leg stuff, and I think it's quite easy. It's an easy dance to do because the cues are so physical. A good woman dancer can literally follow that man or make him chase her. I do a different type of Tango competitively. It is the English Tango, and it's not nearly so explicit. So I'm in a proper dance position (almost like a waltz), and although it's a little snappy, it's not nearly as embellished as the Argentinian tango.

Tamas: But I always had this problem because Sabrina just refused to be led by me. Emancipated woman, you know.

Joy: She needs to learn how to "let him think that he is leading!" There are ways to have your way while he leads.

Zoltán: I would go to the salsa classes with Nadia, but because of these two circles and constant changing of partners, it takes another half an hour until you get back to your original partner with whom you started, and would like to dance with. That's why Nadia and I stopped going. We wanted to go and dance with each other.

Joy: But all you need to do is turn on the music at home and start to dance salsa.

Zoltán: No, that's not the same—you need somebody to show you things, give instructions, correct your movement, and shout at you.

Joy: I actually I kind of like what Tamas said. I think that dancing is a part of us; my feeling is that everybody can dance, and that we kind of know how to move with music. Dancing is from the heart. It comes with the human package.

6 Everybody can dance: a Nobel dance story

Tamas: That's a very good point of view. Everybody, by default, knows how to dance or how to respond to music and rhythm, and they should just be left alone.

Joy: Can I tell you my dance story? Okay, this is at the Nobel dinner, after the Nobel Prize ceremony on the 10th of December, 2013. I was in the Grand Ballroom with my husband. It was a beautiful occasion—so many beautiful gowns, so many beautiful men and women. Everybody was dressed up, and the King and Queen of Sweden were dancing. I was standing on the side of the room with Jim, and then this very beautiful, tall, handsome man came to me and says, "Would you dance with me?" I, of course, said yes. So he takes me by the hand, and we walk out to the middle of the floor, and I get into my dance position. The music was a foxtrot, so it was a usual dance frame position, and he pauses and says to me, "Oh dear, I can't dance." I said "What!!" and thought, "Oh my God, where is the instruction book for this?" So, this was Professor John Martin. You may know John Martin. He is a dear friend, and he's British. I paused for a minute on the floor, and said, "All right, you just walk and

I'll make you look good." So he walked, and I did beautiful chasse turns, and I kept my promise. He looked good.

The next morning, Jim and I opened up the papers, and of course, Jim thinks the papers will cover him and all the other Nobel Laureates. No. The papers featured the dresses and the dancers at the Nobel Ball. First, there was the picture of the Queen and her dress. She got a top rating of five stars, and then there were the other dresses. I was next to the Queen with a four star rating. My dress was beautiful! On the next page were all of the pictures of everybody dancing, and right in front was my handsome Brit (John Martin) and me. The paparazzi got a beautiful shot of him "leading" me doing a fancy spin turn.

Tamas: There you go. You stole the show.

Joy: Well, at least the eye of the paparazzi. Jim told me that he had to read the news to discover that he had a "four star" wife. That was all such fun!

7 Learning to dance

Zoltán: Unfortunately, dancing is not part of the education nowadays. It would be good to have it in the curriculum at early stages. I remember when we were 13—14 and finishing primary school, and we could go to dancing classes at Nagykörös. Did you also go, Tamas?

Tamas: No. I know people who did, but I played basketball.

Zoltán: I did too, but I also enrolled in the dance classes. It was so liberating, because at that age, 14, you never mingle with the girls. We always had separate circles, and we did not really mix with them. It was so liberating that you could dance with girls, and meanwhile, you learned the basics of some of the classical dances. This class went on for a year.

Joy: Yeah, that's good.

Zoltán: We learned csárdás, which is a Hungarian folk dance, and many other dances. When we finished secondary school, we were 18, and we had to do a Waltz dance in front of the parents, the so-called "Szalagavató Bál." Tamas, do you remember that?

Tamas: Yeah, by that time my year was doing dances to Stevie Wonder.

Zoltán: It took us half a year to learn the Waltz steps because it was not only the dance itself we had to learn, but the couples had to dance according to various configurations. I remember I was dancing with Emöke Törös, and it was not trivial to learn all that. Dancing is not taught anymore at schools these days, and I think it's a mistake because it's a kind of cultural activity, particularly the smooth dances that are part of our culture. The Waltz, Foxtrot, Tango—they're so easy to learn early in life when you are a bit more plastic, but now it's very difficult.

Joy: Oh, yes, difficult it is. My dance competition next week is my excuse for not being able to have our meeting next Thursday.

8 Possible opinions about our book

Tamas: The book will be under control. I was kind of thinking about what our critics are going to say about it.
Joy: Um, you know, I can imagine good things and bad things about a book like this.

9 Core problems in neuroscience

Tamas: I started to think about one specific thing more and more because of our conversations and because of an incident that happened to me in which somebody told me in a paper that they're going to prove my theory wrong. My issue with these is that these technologies that are invented in neurobiology, predominantly to study circuits of the cortex and the like, they don't really generate answers in relation to these deeper circuits. The reason is because your tools are potentially limiting for the study of these parts of the brain. You can set up a question, and you can actually address that and find an answer, but it may not really be helpful or solve any issues of a function that you're trying to understand in the brain.
Joy: I think you've got such a good point. It really is the essence of problems in neuroscience; the questions are huge. Everybody studies a little part of it, and it's hard to know how that little part fits into a bigger part, let alone the whole big picture. Really, when you think about it, that's the big struggle, and what led Tamas to be inspired about this book.
Tamas: That is correct. So, let me simplify this for you. You're probing the relationship between one part of the brain and another—Schaffer collaterals, whatever, it doesn't matter. You use your tools to examine one neuronal subpopulation, using an intervention that we all understand, whether it's cellular or circuitry or of brain physiology. There's a tremendous role for the glial cells, for microglia, for astrocytes, oligodendrocytes, for endothelial cells in these events, as well. If you are limiting, and basically demanding, that a question is asked with only those tools that are available, like optogenetics, it's not going to give you the full or the right answer.
Zoltán: The problem is that we are in a situation where you have to "smack me in the face" with results, because that's what the journals publish. You know very well that we can communicate not just by smacking the face, but also by winking an eye, raising our eyebrows, whispering into someone's ears, etc.
Tamas: Exactly.
Zoltán: Then you start explaining your story that you can lip read, you can also vocalize, you can mimic facial expressions, but the editor will say, "So, which one is it? You did not resolve the questions; you have to let me know exactly which one it is."
Tamas: No, I'm trying to put that in my part. In the end, some of our conversation also covers this, because this is one of the things I really would like to throw out there

as one of the problems of neuroscience: they came up with these tools that are great for investigating certain things, but to my limited understanding, they did not actually revolutionize the way we are thinking about the brain. They keep on using the same technology, specifically the optogenetic tools, that continue to reinforce information that we all knew before.

Zoltán: Gero Miesenböck, one of the founding fathers of optogenetics, expressed it even more elegantly than what you have just articulated Tamás: When you use optogenetics for a particular scientific question avoid the fate of many researchers who "voyaged into the known, and discovered the expected."

Tamas: Yes, it is a good one.

10 Science is what we measure

Zoltán: Now we have much better methods to ask important questions. We have much better temporal and spatial tools to stimulate to record from very selective neuronal populations. You no longer stick an electrode somewhere, and although optogenetics (in my view) is the thing to use, you can now manipulate a very selective group of cells with amazing precision, but that never happens in real life, that you will naturally be only manipulating small batches of cells in your brain. The techniques of neuroscience didn't change much in the last 500 years. We use basically the same methods, which are lesioning, stimulating, and recording. These are the three tools, and you don't have anything else. Stimulation can be pharmacogenetic or by optogenetic electrode, or you give your peptides or stimulate the receptors. You can do lesion experiments by cutting out the brain, cooling or burning it, or cutting or silencing fiber pathways. Recording techniques became much more sophisticated. Look at what you can now record in great detail with huge temporal and spatial resolution! You also have the computational power to examine the brains. The categories of tools, though, the basic principles are still the same.

11 Fitting the pieces together

Joy: I completely agree that we should try to approach these questions more holistically.

However, we have to study things in a very specific, localized way; we have to get under the hood to understand mechanisms. I think it might be important to develop a style, where we actually, as investigators, fit our little pieces into the bigger pieces so that small pieces provide structure for the big pieces, and vice versa. I think one thing relating directly to that is to be able to listen to each other. I have learned so much from the two of you. . . in spite of the nonsense.

Tamas: Remember, I tell you this, which is not really conceptual; it is technical. So, I've done a lot of electron microscopy in collaboration, as well. You ask me to do something in the cortex, we will do; you ask me to do something in the cerebellum, we will work on the cerebellum; you asked me about the liver, we will do work for you on the liver with electron microscopy. About 25 20 years ago, when we didn't have electrophysiology of our own, I would ask a colleague of mine, "Can your science department record miniature (or whatever) excitatory postsynaptic potentials (EPSPs) and IPSPs?," and the guy would say, "Well, I'm actually focusing exclusively on EPSPs and in the dorsal hippocampus. We cannot help you." For me, it's like saying that with electron microscopy, "I'm only looking at the spines, and I don't care to understand anything else on the picture." It's that sort of a mentality of "I do what I do." "I don't really understand, I don't care to understand." This is probably one of the major issues historically.

Zoltán: We are almost forced to concentrate on a particular topic and claim that we specialized in that and we are the best in that particular area How many times did you receive comments from grant agencies claiming that you have no track record in one or other areas? Then in a few years, you developed those methods and now it is the strength of your laboratory. It is good to be driven by questions rather than techniques.

Tamas: I'll give you this example. In one of our conversations, we discussed this guy who came from London and talked about how he revolutionized the study of the visual system because they were previously studying anesthetized monkeys, and the animals' V1 (visual neurons or whatever) were actually not working normally under anesthesia. If you keep the animal awake and hold its head still by fixing it to a ball system, it's possible to study these neurons when they're actively in use, but then having its head fixed to this ball is not truly a normal setting, either. So, everybody kind of reinvents the wheel and says, "This is the only way you're supposed to be doing things, and I don't care how others are thinking about it." I think this is a problem.

Joy: This is an example of people changing their perspectives on where the work that we do fits into the big picture and how we communicate it.

12 The "right" of being wrong

Tamas: The other thing, the way I thought about this book is that from my perspective, it's easy to emphasize that a little humbleness is useful, but it's also important to accept and actually be proud of the fact that if you are pursuing the unknown, you can absolutely be wrong, and you should not be afraid of that. That's for sure. You should be wrong most of the time. It's the other side of being able to recognize when you're right. If you're always right, and you keep on believing you're always right, then you're delusional.

Zoltán: You have to keep an open mind to the possibility that you can be wrong. You have a set of evidence, but under different conditions, or if something is changing, you might be wrong. Did you know Ray Guillery?

Tamas: I know the name.

13 The "wrong" of being right

Zoltán: He was a very famous neuroanatomy Professor working mostly on the thalamus and thalamocortical interactions. I went to his lab meetings, and I presented my thalamocortical development work when I was a graduate student. It took five hours to give a one-hour presentation because they kept interrupting. They criticized everything and requested clarifications. One of the reasons why Ray's laboratory meetings lasted so long was because all of the bits of evidence were questioned at every step: data in the published literature, the presenter's experimental evidence, possible methodological pitfalls, and the interpretations. I did not understand why these arguments were so critical. Sometimes, we could not even get through someone's introduction because some of the illustrations were not accurate or there were too many generalizations in the opening sentences. When I presented in one of these meetings and my "feathers were ruffled," Ray Collello, who was a senior postdoc in Ray's lab at the time, gave me some advice on how to present at Ray's laboratory meetings. "Imagine that you can see a bunch of white sheep grazing on a hill. You can't just say that these sheep are white. What you should say is that the profiles of the sheep are white as shown from our current viewpoint." Anything else would be an undue generalization in Ray's laboratory meetings. If you give a talk like that, then you will be all right.

Tamas: So actually, that's a very good point. I was writing a little bit of a preface to my chapter. There, I make the point that one of the things that is irritating to me is to see repeated chains of declarative sentences in these kinds of books.

Zoltán: You always have to leave room for doubt in everything. You have to be ready to change your mind if that sheep turns and you see another color.

Joy: It's not easy once you are ingrained in something. I think that we need a fresh start.

Epilogue—Zoltám Molnár

1 What did you lean and how did you change your views after the discussions?

Before starting our discussions, the three of us had little scientific interactions or collaborations because we thought that we had limited common themes. Our early conversations represented different views. Initially we kept repeating our own individual standpoints, but as we gradually learned more about our respected fields and interests, we came up with some more unique insights. Our discussions reconfirmed some of our initial ideas to keep our interests broad and after the discussions all three of us are even more determined to extend our horizons further.

Endocrinology, immunology, individual variability, and brains state in different conditions.

None of us satisfied to look at a single brain in our research. All three of us are constantly thinking about multiple brains, but in different ways. To understand the brain, we need to understand where it came from. For me a brain is a product of a developmental process that is based on millions of years of evolution. When I think about a brain, I see the developmental process that evolved during milenea. I never think about just that single brain. Both Tamas and Joy also think about multiple brains, but in a different manner. For Tamas, there are multiple brains in an individual. There is a brain in the cranium and another in the rest of the body. Study of the interactions with autonomic nervous system, interoception, metabolic regulations, microbiota all beginning to receive increased attention. For Tamas, the liver was part of the nervous system over the last 30 years. The two must work together in harmony. If not, there are numerous pathologies emerge. These pathologies cannot be understood just by examining the brains in the cranium, without the rest of the body.

Joy believes that a single individual brain does not exist on its own. Our brains constantly interact with other brains around us. Brains evolved and developed to be able to interact with other brains. According to Joy, in all of our interactions there is a diode, comprised of your brain and the brain you are interacting with. For Joy, the two brains form a unit during this interaction. She is advocating and pioneering methods to study multiple brains together. When we know someone going through an experience, we mimic their breathing, heart rate, and overall state. We start mirroring their manners and it has impact on our brain activity. Without knowing it we register the state of others and mirror this behavior. Spending so much time discussing these ideas our three brains started to form a unit. Our discussion exposed ideas that demonstrate that there is huge added value in exchanging ideas the sum of our combined expertise is more (Fig. 1).

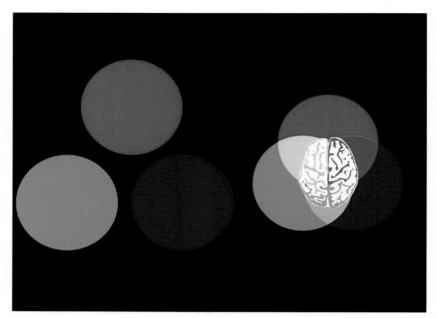

Figure 1 The advantages of multidisciplinary approaches in brain research.

The three of us all agreed that study of the nervous system requires a certain integrative level of study. If we want to understand how New York City functions, it is important to study the buildings, traffic, electricity at different times of the day and in different seasons. We cannot just spend all our attention to study the bricks that the houses are made of. Bricks are useless to understand the issues a metropolis might have during the rush hours or under extreme weather conditions. Of course, we need to study the bricks as well, but it is important to understand that we are at the level of a brick. That will not help in understanding how buildings are connected, how roads are laid out and why you might have a traffic jam at a particular time of the day. It was Winston Churchill who said about history that "The further backward you look, the further forward you can see." This could be extended to the level of the study of the nervous system as well. The more we understand about the evolution, development, metabolic regulation, and interactions between brains, the more we can understand about the true principles of brain function and disfunction.

I was surprised to see that there were certain topics that we have been going around repeatedly. We did not even notice, but we repeated some elements of our previous conversations several months later! Interestingly, we kept making similar arguments, which suggests that we are stubborn, and we still have preconceived ideas. Perhaps we should have been exposed to these ideas earlier. We advocate for the start of these interdisciplinary conversations in graduate school or even before that.

Neuroscience should be taught together with general physiology, endocrinology, anthropology, environmental studies, and even in conjunction with economy.

After our discussions I now have a new set of questions that apply to my own scientific thinking that have arisen from our conversations. These new questions demonstrate the value of nonformal conversations in expanding our thinking beyond our specific disciplines.

In my future research I shall appreciate more the endocrine interactions during brain development. Tamas published on the role of testosterone and estrogen on the developing visual system decades ago, I only recently started to consider the role of diet and metabolism during brain development. Some conditions, such as congenital Zika syndrome, cannot be understood just examining the nervous system. Metabolism, barrier functions, vasculature, placenta, and immunology all must be considered in conjunction.

The other topic that I shall pay more attention to after our discussions with Tamas and Joy is the issue of individual variability. There can be huge variations in our brains, autonomic nervous system, body, but we can still perform the basic neurological simple functions in similar manner. The individual brain connectivity or endocrine regulation might be very different in each of us and still we can perform the very same functions. We need very sophisticated tests to expose these individual differences. Also, it is very difficult to find predictor of a malfunctioning brains. Brains can cope with alterations extremely well.

We know a little of the internal workings of the brain in different brain states. Currently, understanding the workings of our brain without any specific external sensory input is considered the "holy grail" of neuroscience. We understand some principles how sensory inputs are processed and how the final elements of motor outputs regulated. However, in between, we have no comprehension. We have a black box. In one conversation, we called it the "gradient of chaos." Neurons oscillate and we believe that these oscillations are meaningful to understand computations. We wish to know how these oscillations change when we recognize a face, object, or a voice or just some noise. We study task related response and study it as localized response. We would like to predict what someone is thinking about just by observing the oscillations. Would these oscillations drastically differ if someone is hungry, frightened, or just met his/her partner? However, to predict cognitive functions from brain activity might be as difficult as predicting longer term weather predictions from the current meteorological conditions. However, there is no complete resting state. There are constant interactions between the body and the brain even if we do not hear sounds or open our eyes. Time will tell whether resting state is important or not.

After our discussions, I had to change my way of thinking of the functional localizations in the brain. I was a strong advocate of the "real estate principles of the brain." I always believed that functions are associated to particular circuits. What other cells a neuron talks to can be a strong predictor of its function. I love anatomy of neuronal circuits because I strongly believe that the possible functions can be predicted from the knowledge what other neurons a particular neurons is connected to. Talking to Tamas and Joy, I now appreciate the there is an enormous flexibly in the neuronal interactions in multiplex ways not just direct synaptic connectivity and we must consider other levels of interactions that are beyond classical neuronal networks mediated by synapses as Cajal described them over hundred years ago. Our neurons interact with distant parts of our brain and body through endocrine, paracrine, and autocrine manners. Our brain state can be changed and influenced from our entire body. The highly specific distribution of receptors for molecules that are produced in our alimentary system on specific cortical neuronal populations was a real eye opener for me to consider the autonomic nervous system, gut, and liver as part of the brain. For example, the distribution of leptin and orexin receptors is so highly specific that their activation can control brain state in very efficient and specific manners.

I also had to change my views on brain evolution. Hypothalamic hormonal functions are present in all vertebrates, but the six-layered isocortex only appeared in mammals. It is considered that the isocortex evolved on the top of an already fully functional nervous system and the preexisting subcortical modules still represent the main pathways for the action and execution. In human, the cerebral cortex makes up about 80% of the brain's volume and is responsible for many complex phenomena, including perception, thought, language, attention, and memory. What is highly interesting is that as cerebral cortex evolved in a manner that it can now fully utilize the signals provided by these "ancient" and "lower" centers, such as hypothalamus and brainstem and even from the rest of the body, including liver and gut. But new mechanisms of cortico-subcortical interactions evolved and cortex sense these and evolved mechanism to integrate them. Our cortex evolved in synchrony with all these peripheral signals. Our recent work suggests that cortex can detect indicators of homeostatic sleep need and broadcasts signals to the more ancient centers of sleep-wake switching. How this is done is not understood, but now I am very determined to find out. There might be some indirect or direct, excitatory, or inhibitory projections from cortex that influence sleep through direct projections to the lateral preoptic and lateral hypothalamus. Interacting with Tamas and Joy made me reconsider to do more research on the role of cortex and its relationship to subcortical structures.

Contemporary neuroscience is getting very focused and narrow. This can elicit deep understanding of specific topics, but it is vulnerable for derailing research directions.

It takes time to realize that that a particular topic was not as important as it was considered at the time. Research is influenced by temporary trends and fashion. We need interdisciplinary approach to truly understand proportions, importance, and dimensions. One way to understand is step back and remember Churchill: "The further backward you look, the further forward you can see."

Epilogue—Tamas L. Horvath

It has been an incredible experience to have our interactions in the past 2 years with Joy and Zoltán. At the outset, Joy and I were spending a lot of time attempting to come up with a didactic book that provides "book" knowledge to students and aspiring researchers on the brain and on neurosciences. However, from the first meeting, we genuinely stayed away from an intent to pontificate "truths" about neuroscience the way we knew some aspects of it. Then, came Zoltán, who brought in some classical (in its true and positive sense) take on neuroscience, specifically developmental neuroscience. Instead of splitting up parts of the book among the three of us, we decided to freely converse about topics of neuroscience that came to our minds not unfrequently spontaneously. It was an amazing intellectual ride, one that ended up having a major impact on how I wrote my part of the book and how I see now the strengths and shortcomings of contemporary neuroscience. I remain to be the renegade among the three of us questioning fundamental tenets of the state of the art, but I became much more humbled and aware of my enormous limitations in understanding the brain. But I believe that all three of us came away with such impressions. It is a virtual impossibility today to line up a book with logical sets of declarative sentences or even an essay in pretending that we convey "truth" about neuroscience. Instead, I more than ever believe that doubt is the driver of progress in brain research. To be brave to question is a very productive approach to build our understanding and fascination of primitive and complex brain functions.

Epilogue—Joy Hirsch

1 Conversation bootcamp

For the past year and a half, I have meet on Zoom every week with my two friends and scientific colleagues, Tamas Horvath and Zoltán Molnár, for an hour of undirected conversation about science or whatever. The emergent 19 recordings captured our long-time views of our different areas of interest and expertise as well as our real-time academic experiences during the pandemic. We laughed, poked fun at each other, shared personal stories, responded to current events, and found common ground where none was expected. First, and foremost, I emerge from this "conversation bootcamp" with two friends that have shared this odyssey with me. I will miss our meetings when this book is completed. We were bonded together around the common goal of exploring a new view of integrated neuroscience that might be more adept at answering real-world and neuroscience questions. I emerge with the overwhelming realization of how far we have to go. Our conversations uncovered question after question that would not have arisen within our individual silos of expertise and interests. Together we realized that we knew less than we did when we were each alone in our own topics of interest. This was very humbling. The "conversation bootcamp" took us back to scientific "training wheels" where we practiced learning from each other by asking the hard questions and tolerating the paucity of answers. Nonetheless, I think that the questions are the most important product of our conversations and perhaps the largest value for scientific conversations in general.

2 What is a brain?

Who would have ever thought that this could be a fundamental question! It was one of our biggest questions. We have three different views. Tamas conceptualizes the entire body as the brain. His arguments are based on the "neurons everywhere" principle. Zoltán conceptualizes the brain as an evolutionary wonder that embodies 500 million years of optimization with more to come. His thoughts focus on first how to make a brain and then secondly on the principles of how to modify it. I think of the brain of one half of a fundamental social unit and focus on how brains process social information and connect to each other in ways that drive social behaviors and build a social infrastructure. Is there any way that the fusion of these three views might illuminate a novel path forward?

3 What is the value of conversation?

It is self-evident that the interdisciplinary nature of the brain leads us to a paradigm shift that requires consideration of complexities that combine mind, emergent behavior, and the mass of neurons that we refer to as the brain. The fusion of many related disciplines is universally understood as a requirement to understand these complexities. In this way, the coming era in brain science and related sciences connects many of the fundamental disciplines including experimental and theoretical biology, the social sciences, physics, mathematics, humanities, engineering, development, chemistry, medicine, and many, many more. But, how do we do this? What is the mechanism that promotes these fusions?

Novel ideas and solutions require courageous and creative thinking that come from somewhere, but where? Creativity is not a well-behaved discipline and not well-managed by institutional long-term plans or even specific aims. Interestingly, we know very little about how the brain is able to create new relationships and novel concepts. Nonetheless, experience confirms that conversations are often the generators of inspirations and new directions. Here, we explore that process, without understanding it. By presenting examples of our conversations, we invite our readers to "eavesdrop" and hopefully be inspired to elaborate further. Evidence of success would be for others to also connect ideas and questions across the uncharted directions of our conversations.

4 What have I learned?

First, friends are important in science. This is partly because friendships can be a source of motivation, inspiration, and new ideas. This may be an undervalued element of creativity. Second, talking to another scientist who is not in the same field is also important in science, and can also be a source of motivation, inspiration, and new ideas. This "water-cooler" effect, where a mix of people congregate, may also be an undervalued element of creativity. Throughout our many conversations, I was frequently reminded of how much more there is to understand before we can solve the major brain-related problems, and what a small piece of it each of us contributes. And yet, advances in science come in small pieces because details and focus matter. Reconciliation of this apparent paradox between big thinking and focused approaches may be a matter of awareness. The engines of science are fueled by careful tests of specific hypotheses; however, a thoughtful consideration of where and how the details can be connected to larger matters may be an important practice to develop and encourage. The primary advantage for this expansion of related topics comes from the many questions about health and disease that cannot be solved within any single academic silo. Our conversations have over and over again asked seminal questions without answers unless we create operative bridges between us. The knowledge gaps are vast and command creative and uncharted approaches.

Index

Note: Page numbers followed by "*f*" refer to figures.